CONVERGING TECHNOLOGIES FOR IMPROVING HUMAN PERFORMANCE

T0189326

CONVERGING TECHNOLOGIES FOR IMPROVING HUMAN PERFORMANCE

NANOTECHNOLOGY, BIOTECHNOLOGY, INFORMATION TECHNOLOGY AND COGNITIVE SCIENCE

edited by

Mihail C. Roco
The National Science Foundation

and

William Sims Bainbridge
The National Science Foundation

KLUWER ACADEMIC PUBLISHERS
DORDRECHT / BOSTON / LONDON

A C.I.P. Catalogue record for this book is available from the Library of Congress

ISBN 978-90-481-6279-6

Published by Kluwer Academic Publishers,
P.O. Box 17, 3300 AA Dordrecht, The Netherlands.

Sold and distributed in North, Central and South America
by Kluwer Academic Publishers,
101 Philip Drive, Norwell, MA 02061, U.S.A.

In all other countries, sold and distributed
by Kluwer Academic Publishers,
P.O. Box 322, 3300 AH Dordrecht, The Netherlands.

Printed on acid-free paper

NSF/DOC-sponsored report

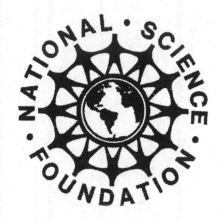

CONVERGING TECHNOLOGIES FOR IMPROVING HUMAN PERFORMANCE

NANOTECHNOLOGY, BIOTECHNOLOGY, INFORMATION TECHNOLOGY AND COGNITIVE SCIENCE (NBIC)

Table of Contents

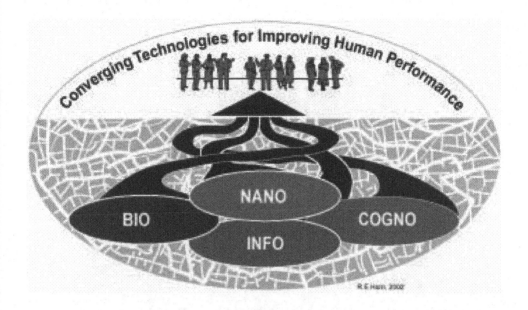

Changing the societal "fabric" towards a new structure
(upper figure by R.E. Horn)

The integration and synergy of the four technologies (nano-bio-info-cogno) originate from the nanoscale, where the building blocks of matter are established. This picture symbolizes the confluence of technologies that now offers the promise of improving human lives in many ways, and the realignment of traditional disciplinary boundaries that will be needed to realize this potential. New and more direct pathways towards human goals are envisioned in working habits, in economic activity, and in the humanities.

NBIC "arrow"

This picture suggests advancement of converging technologies.

EXECUTIVE SUMMARY

M.C. Roco and W.S. Bainbridge

In the early decades of the 21st century, concentrated efforts can unify science based on the unity of nature, thereby advancing the combination of nanotechnology, biotechnology, information technology, and new technologies based in cognitive science. With proper attention to ethical issues and societal needs, converging technologies could achieve a tremendous improvement in human abilities, societal outcomes, the nation's productivity, and the quality of life. This is a broad, cross-cutting, emerging and timely opportunity of interest to individuals, society and humanity in the long term.

The phrase "convergent technologies" refers to the synergistic combination of four major "NBIC" (nano-bio-info-cogno) provinces of science and technology, each of which is currently progressing at a rapid rate: (a) nanoscience and nanotechnology; (b) biotechnology and biomedicine, including genetic engineering; (c) information technology, including advanced computing and communications; (d) cognitive science, including cognitive neuroscience.

Timely and Broad Opportunity. Convergence of diverse technologies is based on *material unity at the nanoscale and on technology integration from that scale.* The building blocks of matter that are fundamental to all sciences originate at the nanoscale. Revolutionary advances at the interfaces between previously separate fields of science and technology are ready to create key *transforming tools* for NBIC technologies. Developments in systems approaches, mathematics, and computation in conjunction with NBIC allow us for the first time to understand the natural world, human society, and scientific research as *closely coupled complex, hierarchical systems*. At this moment in the evolution of technical achievement, *improvement of human performance through integration of technologies* becomes possible.

Examples of payoffs may include improving work efficiency and learning, enhancing individual sensory and cognitive capabilities, revolutionary changes in healthcare, improving both individual and group creativity, highly effective communication techniques including brain-to-brain interaction, perfecting human-machine interfaces including neuromorphic engineering, sustainable and "intelligent" environments including neuro-ergonomics, enhancing human capabilities for defense purposes, reaching sustainable development using NBIC tools, and ameliorating the physical and cognitive decline that is common to the aging mind.

The workshop participants envision important breakthroughs in NBIC-related areas in the next 10 to 20 years. Fundamental research requires about the same interval to yield significant applications. Now is the time to anticipate the research issues and plan an R&D approach that would yield optimal results.

This report addresses key issues: What are the implications of unifying sciences and converging technologies? How will scientific knowledge and current technologies evolve and what emerging developments are envisioned? What visionary ideas can guide research to accomplish broad benefits for humanity? What are the most pressing research and education issues? How can we develop a transforming national strategy to enhance individual capabilities and overall societal outcomes? What should be done to achieve the best results over the next 10 to 20 years?

This report underlines several broad, long-term implications of converging technologies in key areas of human activity, including working, learning, aging, group interaction, and human evolution. If we make the correct decisions and investments today, many of these visions could be addressed within 20 years' time. Moving forward simultaneously along many of these paths could achieve an age of innovation and prosperity that would be a turning point in the evolution of human society. The right of each individual to use new knowledge and technologies in order to achieve personal goals, as well as the right to privacy and choice, are at the core of the envisioned developments.

This report is based on exploratory research already initiated in representative research organizations and on the opinions of leading scientists and engineers using research data.

Strategies for Transformation. It is essential to prepare key organizations and societal activities for the changes made possible by converging technologies. Activities that accelerate convergence to improve human performance must be enhanced, including focused research and development, increased technological synergy from the nanoscale, developing of interfaces among sciences and technologies, and a holistic approach to monitor the resultant societal evolution. The aim is to offer individuals and groups an increased range of attractive choices while preserving such fundamental values as privacy, safety, and moral responsibility. Education and training at all levels should use converging science and technology and prepare people to take advantage of them. We must experiment with innovative ideas to motivate multidisciplinary research and development, while finding ways to address ethical, legal, and moral concerns. In many application areas, such as medical technology and healthcare, it is necessary to accelerate advances that would take advantage of converging technologies.

Towards Unifying Science and Converging Technologies. The evolution of a hierarchical architecture for integrating natural and human sciences across many scales, dimensions, and data modalities will be required. Half a millennium ago, Renaissance leaders were masters of several fields simultaneously. Today, however, specialization has splintered the arts and engineering, and no one can master more than a tiny fragment of human creativity. The sciences have reached a watershed at which they must unify if they are to continue to advance rapidly. Convergence of the sciences can initiate a new renaissance, embodying a holistic view of technology based on transformative tools, the mathematics of complex systems, and unified cause-and-effect understanding of the physical world from the nanoscale to the planetary scale.

Major Themes. Scientific leaders and policy makers across a range of fields prepared written statements for a December 2001 workshop, evaluating the potential impact of NBIC technologies on improving human capabilities at the microscopic, individual, group, and societal levels. During the workshop, participants examined the vast potential in six different areas of relevance:

- *Overall potential of converging technologies.* Representatives of government agencies and the private sector set forth the mission to explore the potential of converging technologies and research needs to improve human performance, as well as the overall potential for revolutionary changes in the economy and society. They identified the synergistic development of nano-, bio-, information- and cognition-based technologies as an outstanding opportunity at the interface and frontier of sciences and engineering in the following decades, and proposed new visions of what is possible to achieve.

- *Expanding human cognition and communication.* Highest priority was given to "The Human Cognome Project," a multidisciplinary effort to understand the structure, functions, and potential enhancement of the human mind. Other priority areas are: personal sensory device interfaces; enriched community through humanized technology; learning how to learn; and enhanced tools for creativity.

- *Improving human health and physical capabilities.* Six priority areas have been identified: nano-bio processors for research and development of treatments, including those resulting from bioinformatics, genomics and proteomics; nanotechnology-based implants and regenerative biosystems as replacements for human organs or for monitoring of physiological well-being; nanoscale machines and comparable unobtrusive tools for medical intervention; multi-modality platforms for increasing sensorial capabilities, particularly for visual and hearing impaired people; brain-to-brain and brain-to-machine interfaces; and virtual environments for training, design, and forms of work unlimited by distance or the physical scale on which it is performed.

- *Enhancing group and societal outcomes.* An NBIC system called "The Communicator" would remove barriers to communication caused by physical disabilities, language differences, geographic distance, and variations in knowledge, thus greatly enhancing the effectiveness of cooperation in schools, corporations, government agencies, and across the world. Other areas of focus are in enhancing group creativity and productivity, cognitive engineering and developments related to networked society. A key priority will be revolutionary new products and services based on the integration of the four technologies from the nanoscale.

- *National security.* Given the radically changing nature of conflict in this new century, seven opportunities to strengthen national defense offered by technological convergence deserve high priority: data linkage and threat anticipation; uninhabited combat vehicles; war fighter education and training; responses to chemical, biological, radiological and explosive threats; war fighter systems; non-drug treatments to enhance human performance; and applications of human-machine interfaces.

- *Unifying science and education.* To meet the coming challenges, scientific education needs radical transformation from elementary school through post-graduate training. Convergence of previously separate scientific disciplines and fields of engineering cannot take place without the emergence of new kinds of people who understand multiple fields in depth and can intelligently work to integrate them. New curricula, new concepts to provide intellectual coherence, and new forms of educational institutions will be necessary.

Beyond the 20-year time span, or outside the current boundaries of high technology, convergence can have significant impacts in such areas as: work efficiency, the human body and mind throughout the life cycle, communication and education, mental health, aeronautics and space flight, food and farming, sustainable and intelligent environments, self-presentation and fashion, and transformation of civilization.

Synopsis of Recommendations

The recommendations of this report are far-reaching and fundamental, urging the transformation of science, engineering and technology at their very roots. The new developments will be revolutionary and must be governed by respect for human welfare and dignity. This report sets goals for societal and educational transformation. Building on the suggestions developed in the five topical groups, and the ideas in the more than 50 individual contributions, the workshop recommended a **national R&D priority area on converging technologies focused on enhancing human performance.** The opportunity is broad, enduring, and of general interest.

a) **Individuals.** Scientists and engineers at every career level should gain skills in at least one NBIC area and in neighboring disciplines, collaborate with colleagues in other fields, and take risks in launching innovative projects that could advance NBIC.

b) **Academe.** Educational institutions at all levels should undertake major curricular and organizational reforms to restructure the teaching and research of science and engineering so that previously separate disciplines can converge around common principles to train the technical labor force for the future.

c) **Private Sector.** Manufacturing, biotechnology, information and medical service corporations will need to develop partnerships of unparalleled scope to exploit the tremendous opportunities from technological convergence, investing in production facilities based on entirely new principles, materials, devices and systems, with increased emphasis on human development.

d) **Government.** The Federal Government should establish a national research and development priority area on converging technologies focused on enhancing human performance. Government organizations at all levels should provide leadership in creating the NBIC infrastructure and coordinating the work of other institutions, and must accelerate convergence by supporting new multidisciplinary scientific efforts while sustaining the traditional disciplines that are essential for success. Ethical, legal, moral, economic, environmental, workforce development, and other societal implications must be addressed from the beginning, involving leading NBIC scientists and engineers, social scientists and a broad coalition of professional and civic organizations. Research on societal implications must be funded, and the risk of potential undesirable secondary effects must be monitored by a government organization in order to anticipate and take corrective action. Tools should be developed to anticipate scenarios for future technology development and applications.

e) **Professional Societies.** The scientific and engineering communities should create new means of interdisciplinary training and communication, reduce the barriers that inhibit individuals from working across disciplines, aggressively highlight opportunities for convergence in their conferences, develop links to a variety of other technical and medical organizations, and address ethical issues related to technological developments.

f) **Other Organizations.** Non-governmental organizations that represent potential user groups should contribute to the design and testing of convergent technologies, in order to maximize the benefits for their diverse constituencies. Private research foundations should invest in NBIC research in those areas that are consistent with their unique missions. The press should increase high-quality coverage of science and technology, on the basis of the new convergent paradigm, to inform citizens so they can participate wisely in debates about ethical issues such as unexpected effects on inequality, policies concerning diversity, and the implications of transforming human capabilities.

A vast opportunity is created by the convergence of sciences and technologies starting with integration from the nanoscale and having immense individual, societal and historical implications for human development. The participants in the meetings who prepared this report recommend *a national research and development priority area on converging technologies focused on enhancing human performance*. This would be a suitable framework for a long-term, coherent strategy in research and education. Science and technology will increasingly dominate the world, as population, resource exploitation, and potential social conflict grow. Therefore, the success of this convergent technologies priority area is essential to the future of humanity.

Overview
Converging Technologies for Improving Human Performance

Nanotechnology, Biotechnology, Information Technology, and Cognitive Science (NBIC)

M.C. Roco and W.S. Bainbridge

1. Background

We stand at the threshold of a new renaissance in science and technology, based on a comprehensive understanding of the structure and behavior of matter from the nanoscale up to the most complex system yet discovered, the human brain. Unification of science based on unity in nature and its holistic investigation will lead to technological convergence and a more efficient societal structure for reaching human goals. In the early decades of the twenty-first century, concentrated effort can bring together nanotechnology, biotechnology, information technology, and new technologies based in cognitive science. With proper attention to ethical issues and societal needs, the result can be a tremendous improvement in human abilities, new industries and products, societal outcomes, and quality of life.

Rapid advances in convergent technologies have the potential to enhance both human performance and the nation's productivity. Examples of payoffs will include improving work efficiency and learning, enhancing individual sensory and cognitive capabilities, fundamentally new manufacturing processes and improved products, revolutionary changes in healthcare, improving both individual and group efficiency, highly effective communication techniques including brain-to-brain interaction, perfecting human-machine interfaces including neuromorphic engineering for industrial and personal use, enhancing human capabilities for defense purposes, reaching sustainable development using NBIC tools, and ameliorating the physical and cognitive decline that is common to the aging mind.

This report addresses several main issues: What are the implications of unifying sciences and converging technologies? How will scientific knowledge and current technologies evolve and what emerging developments are envisioned? What should be done to achieve the best results over the next 10 to 20 years? What visionary ideas can guide research to accomplish broad benefits for humanity? What are the most pressing research and education issues? How can we develop a transforming national strategy to enhance individual capabilities and overall societal outcomes? These issues were discussed on December 3-4, 2001, at the workshop on Converging Technologies to Improve Human Performance, and in contributions submitted after that meeting for this report.

The phrase "convergent technologies" refers to the synergistic combination of four major "NBIC" (nano-bio-info-cogno) provinces of science and technology, each of which is currently progressing at a rapid rate: (a) nanoscience and nanotechnology; (b) biotechnology and biomedicine, including genetic engineering;

(c) information technology, including advanced computing and communications; and (d) cognitive science, including cognitive neuroscience.

This report is based on exploratory research already initiated in representative research organizations and on the opinions of leading scientists and engineers using research data. Contributors to this report have considered possibilities for progress based on full awareness of ethical as well as scientific principles.

Accelerated scientific and social progress can be achieved by combining research methods and results across these provinces in duos, trios, and the full quartet. Figure 1 shows the "NBIC tetrahedron," which symbolizes this convergence. Each field is represented by a vertex, each pair of fields by a line, each set of three fields by a surface, and the entire union of all four fields by the volume of the tetrahedron.

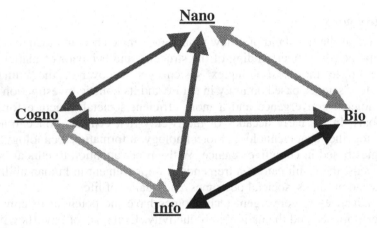

Figure 1.   NBIC tetrahedron.

2. Timely and Broad Opportunity

The sciences have reached a watershed at which they must combine in order to advance most rapidly. The new renaissance must be based on a holistic view of science and technology that envisions new technical possibilities and focuses on people. The unification of science and technology can yield results over the next two decades on the basis of four key principles: material unity at the nanoscale, NBIC transforming tools, hierarchical systems, and improvement of human performance, as described below:

a) Convergence of diverse technologies is based on *material unity at the nanoscale and on technology integration from that scale*. Science can now understand the ways in which atoms combine to form complex molecules, and how these in turn aggregate according to common fundamental principles to form both organic and inorganic structures. Technology can harness natural processes to engineer new materials, biological products, and machines from the nanoscale up to the scale of meters. The same principles will allow us to understand and, when desirable, to control the behavior both of complex microsystems, such as neurons and computer components, and macrosystems, such as human metabolism and transportation vehicles.

b) Revolutionary advances at the interfaces between previously separate fields of science and technology are ready to create key *NBIC transforming tools (nano-, bio, info-, and cognitive-based technologies)*, including scientific instruments, analytical methodologies, and radically new materials systems. The innovative momentum in these interdisciplinary areas must not be lost but harnessed to accelerate unification of the disciplines. Progress can become self-catalyzing if we press forward aggressively; but if we hesitate, the barriers to progress may crystallize and become harder to surmount.

c) Developments in systems approaches, mathematics, and computation in conjunction with work in NBIC areas allow us for the first time to understand the natural world and cognition in terms of *complex, hierarchical systems*. Applied both to particular research problems and to the overall organization of the research enterprise, this complex systems approach provides holistic awareness of opportunities for integration, in order to obtain maximum synergy along the main directions of progress.

d) At this unique moment in the history of technical achievement, *improvement of human performance* becomes possible. Caught in the grip of social, political, and economic conflicts, the world hovers between optimism and pessimism. NBIC convergence can give us the means to deal successfully with these challenges by substantially enhancing human mental, physical, and social abilities. Better understanding of the human body and development of tools for direct human-machine interaction have opened completely new opportunities. Efforts must center on individual and collective human advancement, in terms of an enlightened conception of human benefit that embraces change while preserving fundamental values.

The history of science across the vast sweep of human history undermines any complacency that progress will somehow happen automatically, without the necessity for vigorous action. Most societies at most points in their history were uninterested in science, and they advanced technologically only very slowly, if at all. On rare occasions, such as the pyramid-building age in ancient Egypt or the roughly contemporaneous emergence of intensive agriculture and trade in Babylon, the speed of progress seemed to accelerate, although at a much slower rate than that experienced by Europe and North America over the past five centuries. For modern civilization, the most relevant and instructive precursor remains the classical civilizations of Greece and Rome. Building on the scientific accomplishments of the Babylonians and Egyptians, the Greeks accomplished much in mathematics, astronomy, biology, and other sciences. Their technological achievements probably peaked in the Hellenistic Age as city-states gave way to larger political units, culminating in Roman dominance of the entire Mediterranean area. By the end of the second century, if not long before, scientific and technological progress had slowed with the fall of Rome. Historians debate the degree to which technology advanced during the subsequent Dark Ages and Medieval Period, but clearly, a mighty civilization had fallen into bloody chaos and widespread ignorance.

The Renaissance, coming a thousand years after the decline and fall of the Roman Empire, reestablished science on a stronger basis than before, and technological advancement has continued on an accelerating path since then. The hallmark of the Renaissance was its holistic quality, as all fields of art, engineering, science, and culture shared the same exciting spirit and many of the same intellectual principles. A creative individual, schooled in multiple arts, might be a painter one day, an engineer the next, and a writer the day after that. However, as the centuries passed, the holism of the Renaissance gave way to specialization and intellectual fragmentation. Today, with the scientific work of recent decades showing us at a deeper level the fundamental unity of natural organization, it is time to rekindle the spirit of the Renaissance, returning to the holistic perspective on a higher level, with a new set of principles and theories. This report underlines several broad, long-term implications of converging technologies in key areas of human activity:

- Societal productivity, in terms of well-being as well as economic growth
- Security from natural and human-generated disasters
- Individual and group performance and communication
- Life-long learning, graceful aging, and a healthy life
- Coherent technological developments and their integration with human activities
- Human evolution, including individual and cultural evolution

Fundamental scientific discovery needs at least ten years to be implemented in new technologies, industries, and ways of life. Thus, if we want the great benefits of NBIC convergence within our own lifetimes, now is the right time to begin. The impact of advancing technology on the present quality of life (United Nations Development Program 2001) will be accelerated by NBIC, and new possibilities for human performance will be unleashed.

3. Vision for Enhancing Human Abilities and Societal Performance

Despite moments of insight and even genius, the human mind often seems to fall far below its full potential. The level of human thought varies greatly in awareness, efficiency, creativity, and accuracy. Our physical and sensory capabilities are limited and susceptible to rapid deterioration in accidents or disease and gradual degradation through aging (Stern and Carstensen 2000). All too often we communicate poorly with each other, and groups fail to achieve their desired goals. Our tools are difficult to handle, rather than being natural extensions of our capabilities. In the coming decades, however, converging technologies promise to increase significantly our level of understanding, transform human sensory and physical capabilities, and improve interactions between mind and tool, individual and team. This report addresses key issues concerning how to reach these goals.

Each scientific and engineering field has much to contribute to enhancing human abilities, to solving the pressing problems faced by our society in the twenty-first century, and to expanding human knowledge about our species and the world we inhabit; but combined, their potential contribution is vast. Following are twenty ways the workshop determined that convergent technologies could benefit humanity in a time frame of 10 to 20 years. Each of these scenarios are presented in detail in the body of the report:

- Fast, broadband interfaces directly between the human brain and machines will transform work in factories, control automobiles, ensure military superiority, and enable new sports, art forms and modes of interaction between people.

- Comfortable, wearable sensors and computers will enhance every person's awareness of his or her health condition, environment, chemical pollutants, potential hazards, and information of interest about local businesses, natural resources, and the like.

- Robots and software agents will be far more useful for human beings, because they will operate on principles compatible with human goals, awareness, and personality.

- People from all backgrounds and of all ranges of ability will learn valuable new knowledge and skills more reliably and quickly, whether in school, on the job, or at home.

- Individuals and teams will be able to communicate and cooperate profitably across traditional barriers of culture, language, distance, and professional specialization, thus greatly increasing the effectiveness of groups, organizations, and multinational partnerships.

- The human body will be more durable, healthier, more energetic, easier to repair, and more resistant to many kinds of stress, biological threats, and aging processes.

- Machines and structures of all kinds, from homes to aircraft, will be constructed of materials that have exactly the desired properties, including the ability to adapt to changing situations, high energy efficiency, and environmental friendliness.

- A combination of technologies and treatments will compensate for many physical and mental disabilities and will eradicate altogether some handicaps that have plagued the lives of millions of people.

- National security will be greatly strengthened by lightweight, information-rich war fighting systems, capable uninhabited combat vehicles, adaptable smart materials, invulnerable data networks, superior intelligence-gathering systems, and effective measures against biological, chemical, radiological, and nuclear attacks.

- Anywhere in the world, an individual will have instantaneous access to needed information, whether practical or scientific in nature, in a form tailored for most effective use by the particular individual.

- Engineers, artists, architects, and designers will experience tremendously expanded creative abilities, both with a variety of new tools and through improved understanding of the wellsprings of human creativity.

- The ability to control the genetics of humans, animals, and agricultural plants will greatly benefit human welfare; widespread consensus about ethical, legal, and moral issues will be built in the process.

- The vast promise of outer space will finally be realized by means of efficient launch vehicles, robotic construction of extraterrestrial bases, and profitable exploitation of the resources of the Moon, Mars, or near-Earth approaching asteroids.

- New organizational structures and management principles based on fast, reliable communication of needed information will vastly increase the effectiveness of administrators in business, education, and government.

- Average persons, as well as policymakers, will have a vastly improved awareness of the cognitive, social, and biological forces operating their lives, enabling far better adjustment, creativity, and daily decision making.

- Factories of tomorrow will be organized around converging technologies and increased human-machine capabilities as "intelligent environments" that achieve the maximum benefits of both mass production and custom design.

- Agriculture and the food industry will greatly increase yields and reduce spoilage through networks of cheap, smart sensors that constantly monitor the condition and needs of plants, animals, and farm products.

- Transportation will be safe, cheap, and fast, due to ubiquitous realtime information systems, extremely high-efficiency vehicle designs, and the use of synthetic materials and machines fabricated from the nanoscale for optimum performance.

- The work of scientists will be revolutionized by importing approaches pioneered in other sciences, for example, genetic research employing principles from natural language processing and cultural research employing principles from genetics.

- Formal education will be transformed by a unified but diverse curriculum based on a comprehensive, hierarchical intellectual paradigm for understanding the architecture of the physical world from the nanoscale through the cosmic scale.

If we make the correct decisions and investments today, any of these visions could be achieved within 20 years' time. Moving forward simultaneously along many of these paths could achieve a golden age that would be a turning point for human productivity and quality of life. Technological convergence could become the framework for human convergence (Ostrum et al. 2002). The twenty-first century could end in world peace, universal prosperity, and evolution to a higher level of compassion and accomplishment. It is hard to find the right metaphor to see a century into the future, but it may be that humanity would become like a single, distributed and interconnected "brain" based in new core pathways of society. This will be an enhancement to the productivity and independence of individuals, giving them greater opportunities to achieve personal goals.

Table 1 shows a simplified framework for classifying improving human performance areas as they relate to an individual (see also Spohrer 2002, in this volume).

Table 1. Main improvement areas relative to an individual

Relative position	Improvement area
External (outside the body), environmental	• New products: materials, devices and systems, agriculture and food • New agents: societal changes, organizations, robots, chat-bots, animals • New mediators: stationary tools and artifacts • New places: real, virtual, mixed
External, collective	• Enhanced group interaction and creativity • Unifying science education and learning
External, personal	• New mediators: mobile/wearable tools and artifacts
Internal (inside the body), temporary	• New ingestible medicines, food
Internal, permanent	• New organs: new sensors and effectors, implantables • New skills: converging technologies, new uses of old sensors and effectors • New genes: new genetics, cells

4. Strategies for Transformation

Science and engineering as well as societal activities are expected to change, regardless of whether there are policies to guide or promote such changes. To influence and accelerate changes in the most beneficial directions, it is not enough to wait patiently while scientists and engineers do their traditional work. Rather, the full advantages of NBIC developments may be achieved by making special efforts to break down barriers between fields and to develop the new intellectual and

Figure 2. Vision of the world as a distributed, interconnected "brain" with various architectural levels that can empower individuals with access to collective knowledge while safeguarding privacy.

physical resources that are needed. The workshop identified the following general strategies for achieving convergence:

a) We should prepare key organizations and social activities for the envisioned changes made possible by converging technologies. This requires establishing long-term goals for major organizations and modeling them to be most effective in the new setting.

b) Activities must be enhanced that accelerate convergence of technologies for improving human performance, including focused research, development, and design; increasing synergy from the nanoscale; developing interfaces among sciences and technologies; and taking a holistic approach to monitor the resultant societal evolution. The aim is to offer individuals and groups an increased range of attractive choices while preserving fundamental values such as privacy, safety, and moral responsibility. A research and development program for exploring the long-term potential is needed.

c) Education and training at all levels should use converging technologies as well as prepare people to take advantage of them. Interdisciplinary education programs, especially in graduate school, can create a new generation of scientists and engineers who are comfortable working across fields and collaborating with colleagues from a variety of specialties. Essential to this effort is the integration of research and education that combines theoretical training with experience gained in the laboratory, industry, and world of application. A sterling example is NSF's competition called Integrative Graduate Education and Research Training (IGERT). A number of comparable graduate education projects need to be launched at the intersections of crucial fields to build a scientific community that will achieve the convergence of technologies that can greatly improve human capabilities.

d) Experimentation with innovative ideas is needed to focus and motivate needed multidisciplinary developments. For example, there could be a high-visibility annual event, comparable to the sports Olympics, between information technology interface systems that would compete in terms of speed, accuracy, and other measurements of enhanced human performance. Professional societies could set performance targets and establish criteria for measuring progress toward them.

e) Concentrated multidisciplinary research thrusts could achieve crucially important results. Among the most promising of such proposed endeavors are the Human Cognome Project to understand the nature of the human mind, the development of a "Communicator" system to optimize human teams and organizations, and the drive to enhance human physiology and physical performance. Such efforts probably require the establishment of networks of research centers dedicated to each goal, funded by coalitions of government agencies and operated by consortia of universities and corporations.

f) Flourishing communities of NBIC scientists and engineers will need a variety of multiuser, multiuse research and information facilities. Among these will be data infrastructure archives, that employ advanced digital technology to serve a wide range of clients, including government agencies, industrial designers, and university laboratories. Other indispensable facilities would include regional nanoscience centers, shared brain scan resources, and engineering simulation supercomputers. Science is only as good as its instrumentation, and information is an essential tool of engineering, so cutting-edge infrastructure must be created in each area where we desire rapid progress.

g) Integration of the sciences will require establishment of a shared culture that spans across existing fields. Interdisciplinary journals, periodic new conferences, and formal partnerships between professional organizations must be established. A new technical language will need to be developed for communicating the unprecedented scientific and engineering challenges based in the mathematics of complex systems, the physics of structures at the nanoscale, and the hierarchical logic of intelligence.

h) We must find ways to address ethical, legal, and moral concerns, throughout the process of research, development, and deployment of convergent technologies. This will require new mechanisms to ensure representation of the public interest in all major NBIC projects, to incorporate ethical and social-scientific education in the training of scientists and engineers, and to ensure that policy makers are thoroughly aware of the scientific and engineering implications of the issues they face. Examples are the moral and ethical issues involved in applying new brain-related scientific findings (*Brain Work* 2002). Should we make our own ethical decisions or "are there things we'd rather not know" (Kennedy 2002)? To live in harmony with nature, we must understand natural processes and be prepared to protect or harness them as required for human welfare. Technological convergence may be the best hope for the preservation of the natural environment, because it integrates humanity with nature across the widest range of endeavors, based on systematic knowledge for wise stewardship of the planet.

i) It is necessary to accelerate developments in medical technology and healthcare in order to obtain maximum benefit from converging technologies, including molecular medicine and nano-engineered medication delivery systems, assistive devices to alleviate mental and emotional disabilities, rapid sensing and preventive measures to block the spread of infectious and environmental diseases, continuous detection and correction of abnormal individual health indications, and integration of genetic therapy and genome-aware treatment into daily medical practice. To accomplish this, research laboratories, pharmaceutical companies, hospitals and health maintenance organizations, and medical schools will need to expand greatly their institutional partnerships and technical scope.

General Comments

There should be specific partnerships among high-technology agencies and university researchers in such areas as space flight, where a good foundation for cutting edge technological convergence already exists. But in a range of other areas, it will be necessary to build scientific communities and research projects nearly from scratch. It could be important to launch a small number of well-financed and well-designed demonstration projects to promote technological convergence in a variety of currently low-technology areas.

The U.S. economy has benefited greatly from the rapid development of advanced technology, both through increased international competitiveness and through growth in new industries. Convergent technologies could transform some low-technology fields into high-technology fields, thereby increasing the fraction of the U.S. economy that is both growing and world-preeminent.

This beneficial transformation will not take place without fundamental research in fields where such research has tended to be rare or without the intensity of imagination and entrepreneurship that can create new products, services, and entire new industries. We must begin with a far-sighted vision that a renaissance in science and technology can be achieved through the convergence of nanotechnology, biotechnology, information technology, and cognitive science.

5. Towards Unifying Science and Converging Technology

Although recent progress in the four NBIC realms has been remarkable, further rapid progress in many areas will not happen automatically. Indeed, science and engineering have encountered several barriers, and others are likely to appear as we press forward. In other areas, progress has been hard-won, and anything that could accelerate discovery would be exceedingly valuable. For example, cognitive neuroscience has made great strides recently unlocking the secrets of the human brain, with such computer-assisted techniques as functional magnetic resonance imaging (fMRI). However, current methods already use the maximum magnetic field strength that is considered safe for human beings. The smallest structures in the brain that can routinely be imaged with this technique are about a cubic millimeter in size, but this volume can contain tens of thousands of neurons, so it really does not let scientists see many of the most important structures that are closer to the cellular level. To increase the resolution further will require a new approach, whether novel computer techniques to extract more information from fMRI data or a wholly different method to study the structure and function of regions of the brain, perhaps based on a marriage of biology and nanotechnology.

Another example is in the area of information science, where progress has depended largely upon the constant improvement in the speed and cost-effectiveness of integrated circuits. However, current methods are nearing their physical limits, and it is widely believed that progress will cease in a few years unless new approaches are found. Nanotechnology offers realistic hope that it will be possible to continue the improvement in hardware for a decade or even two decades longer than current methods will permit. Opinion varies on how rapidly software capabilities are improving at the present time, but clearly, software efficiency has not improved at anything like the rate of hardware, so any breakthrough that increases the rate of software progress would be especially welcome. One very promising direction to look for innovations is biocomputing, a host of software methods that employ metaphors from such branches of biology as genetics. Another is cognitive science, which can help computer scientists develop software inspired by growing understanding of the neural architectures and algorithms actually employed by the human brain.

Many other cases could be cited in which discoveries or inventions in one area will permit progress in others. Without advances in information technology, we cannot take full advantage of biotechnology in areas such as decoding the human genome, modeling the dynamic structure of protein molecules, and understanding how genetically engineered crops will interact with the natural environment. Information technology and microbiology can provide tools for assembling nanoscale structures and incorporating them effectively in microscale devices. Convergence of nonorganic nanoscience and biology will require breakthroughs in the ways we conceptualize and teach the fundamental processes of chemistry in complex systems, which could be greatly facilitated by cognitive science research on scientific thinking itself.

Thus, in order to attain the maximum benefit from scientific progress, the goal can be nothing less than a fundamental transformation of science and engineering. Although the lists of potential medium-term benefits have naturally stressed applications, much of the unification must take place on the level of fundamental science. From empirical research, theoretical analysis, and computer modeling we will have to develop overarching scientific principles that unite fields and make it possible for scientists to understand complex phenomena. One of the reasons sciences have not merged in the past is that their subject matter is so complex and challenging to the human intellect. We must find ways to rearrange and connect scientific findings so that scientists from a wider range of fields can comprehend and apply them within their own work. It will therefore be necessary to support fundamental scientific research in each field that can become the foundation of a bridge to other fields, as well as support fundamental research at the intersections of fields.

Fundamental research will also be essential in engineering, including computer engineering, because engineers must be ready in the future to take on entirely new tasks from those they have traditionally handled. The traditional tool kit of engineering methods will be of limited utility in some of the most important areas of technological convergence, so new tools will have to be created. This has already begun to happen in nanotechnology, but much work remains to be done developing engineering solutions to the problems raised by biology, information, and the human mind.

It is possible to identify a number of areas for fundamental scientific research that will have especially great significance over the coming twenty years for technological convergence to improve human performance. Among these, the following four areas illustrate how progress in one of the NBIC fields can be energized by input from others:

- *Entirely new categories of materials, devices, and systems for use in manufacturing, construction, transportation, medicine, emerging technologies, and scientific research.* Nanotechnology is obviously preeminent here, but information technology plays a crucial role in both research and design of the structure and properties of materials and in the design of complex molecular and microscale structures. It has been pointed out that industries of the future will use engineered biological processes to manufacture valuable new materials, but it is also true that fundamental knowledge about the molecular-level processes essential to the growth and metabolism of living cells may be applied, through analogy, to development of new inorganic materials. Fundamental materials science research in mathematics, physics, chemistry, and biology will be essential.

- *The living cell, which is the most complex known form of matter with a system of components and processes operating at the nanoscale.* The basic properties and functions are established at the first level of organization of biosystems, that is, at the nanoscale. Recent work at the intersection of biotechnology and microelectronics, notably the so-called gene-on-a-chip approach, suggests that a union of nanotechnology, biotechnology, and computer science may be able to create "bio-nano processors" for programming complex biological pathways that will mimic cellular processes on a chip. Other research methodologies may come from the ongoing work to understand how genes are expressed in the living body as physical structures and chemical activities. Virtual reality and augmented reality computer technology will allow scientists to visualize the cell from inside, as it were, and to see exactly what they are doing as they manipulate individual protein molecules and cellular nanostructures.

- *Fundamental principles of advanced sensory, computational, and communications systems, especially the integration of diverse components into the ubiquitous and global network.* Breakthroughs in nanotechnology will be necessary to sustain the rapid improvement of computer hardware over the next 20 years. From biology will come important insights about the behavior of complex dynamic systems and specific methods of sensing organic and chemical agents in the environment. Cognitive science will provide insights into ways to present information to human beings so they can use it most effectively. A particularly challenging set of problems confronting computer and information science and engineering at the present time is how to achieve reliability and security in a ubiquitous network that collects and offers diverse kinds of information in multiple modalities, everywhere and instantly at any moment.

- *The structure, function, and occasional dysfunction of intelligent systems, most importantly, the human mind.* Biotechnology, nanotechnology, and computer simulations can offer powerful new techniques for studying the dynamic behavior of the brain, from the receptors and other structures far smaller than a single neuron, up through individual neurons, functionally specific modules composed of many neurons, the major components of the brain, and then the entire brain as a complex but unified system. Cognition cannot be understood without attention also to the interaction of the individual with the environment, including the ambient culture. Information technology will be crucial in processing data about the brain, notably the difficult challenge of understanding the mature human brain as a product of genetics and development. But it will also be essential to experiment with artificial intelligent systems, such as neural networks, genetic algorithms, autonomous agents, logic-based learning programs, and sophisticated information storage and retrieval systems.

The complementarity of the four NBIC areas is suggested by the statement of workshop participant W.A. Wallace:

> If the *Cognitive Scientists* can think it
> the *Nano* people can build it
> the *Bio* people can implement it, and
> the *IT* people can monitor and control it

Each of the four research challenges described above focuses on one of the NBIC areas (nanotechnology, biotechnology, information technology, and cognitive science) and shows how progress can be catalyzed by convergence with the other areas. They are not merely convenient didactic examples, but represent fascinating questions, the answers to which would enable significant improvements in human performance. However, convergence will be possible only if we overcome substantial intellectual barriers.

Especially demanding will be the development of a hierarchical architecture for integrating sciences across many scales, dimensions, and data modalities. For a century or more, educated people have understood that knowledge can be organized in a hierarchy of sciences, from physics as a base, up through chemistry and biology, to psychology and economics. But only now is it really possible to see in detail how each level of phenomena both rests upon and informs the one below. Some partisans for independence of biology, psychology, and the social sciences have argued against "reductionism," asserting that their fields had discovered autonomous truths that should not be reduced to the laws of other sciences. But such a discipline-centric outlook is self-defeating, because as this report makes clear, through recognizing their connections with each other, all the sciences can progress more effectively. A trend towards unifying knowledge by combining natural sciences, social sciences, and humanities using cause-and-effect explanation has already begun (NYAS 2002), and it should be reflected in the coherence of science and engineering trends (Roco 2002, in this report) and in the integration of R&D funding programs.

The architecture of the sciences will be built through understanding of the architecture of nature. At the nanoscale, atoms and simple molecules connect into complex structures like DNA, the subsystems of the living cell, or the next generation of microelectronic components. At the microscale, cells such as the neurons and glia of the human brain interact to produce the transcendent phenomena of memory, emotion, and thought itself. At the scale of the human body, the myriad processes of chemistry, physiology, and cognition unite to form life, action, and individuals capable of creating and benefiting from technology.

Half a millennium ago, Renaissance artist-engineers like Leonardo da Vinci, Filippo Brunelleschi, and Benvenuto Cellini were masters of several fields simultaneously. Today, however, specialization has splintered the arts and engineering, and no one can master more than a tiny fragment of human creativity. We envision that convergence of the sciences can initiate a new renaissance, embodying a holistic view of technology based on transformative tools, the mathematics of complex systems, and unified understanding of the physical world from the nanoscale to the planetary scale.

6. Major Themes

A planning meeting was held May 11, 2001, at the National Science Foundation to develop the agenda for the December workshop and to identify key participants from academia, industry, and government. Scientific leaders and policymakers across a range of fields were asked to prepare formal speeches for plenary sessions, and all participants were invited to contribute written statements evaluating the potential impact of NBIC technologies on improving human capabilities at the microscopic, individual, group, and societal levels.

Participants in the December 2001 workshop on Convergent Technologies to Improve Human Performance submitted more than fifty written contributions, each of which is like a single piece in a jigsaw puzzle. Together, they depict the future unification of nanotechnology, biotechnology, information technology, and cognitive science, with the amazing benefits these promise. Roughly half of these written contributions, which we call *statements*, describe the current situation and suggest strategies for building upon it. The other half describe *visions* of what could be accomplished in 10 or 20 years. During the workshop, participants examined the vast potential of NBIC in five different areas of relevance, as well as the overall potential for changing the economy, society, and research needs:

a) Overall Potential of Converging Technologies. In plenary sessions of the workshop, representatives of government agencies and the private sector set forth the mission to explore the potential of converging technologies to improve human performance. They identified the synergistic development of nano-, bio-, information- and cognition-based technologies as the outstanding opportunity at the interface and frontier of sciences in the following decades. They proclaimed that it is essential to courageously identify new technologies that have great potential, to develop transforming visions for them, and to launch new partnerships between government agencies, industry, and educational institutions to achieve this potential. Government has an important role in setting long-term research priorities, respecting the ethical and social aspects of potential uses of technology, and ensuring economic conditions that facilitate the rapid invention and deployment of beneficial technologies. Technological superiority is the fundamental basis of the economic prosperity and national security of the United States, and continued progress in NBIC technologies is an essential component for government agencies to accomplish their designated missions. Science and engineering must offer society new visions of what it is possible to achieve through interdisciplinary research projects designed to promote technological convergence.

b) Expanding Human Cognition and Communication. This group of workshop participants examined needs and opportunities in the areas of human cognitive and perceptual functions, communication between individuals and machines programmed with human-like characteristics, and the ways that convergent technologies could enhance our understanding and effective use of human mental abilities. The group identified five areas where accelerated efforts to achieve technological convergence would be especially worthwhile. Highest priority was given to what Robert Horn called The Human Cognome Project, a proposed multidisciplinary effort to understand the structure, functions, and potential enhancement of the human mind. The four other priority areas were personal sensory device interfaces, enriched community through humanized technology, learning how to learn, and enhanced tools for creativity.

c) Improving Human Health and Physical Capabilities. This group of workshop participants also focused primarily on the individual, but on his or her physical rather than mental abilities. Essential to progress in this area is comprehensive scientific understanding of the fundamental chemical and biological processes of life. Control of metabolism in cells, tissue, organs, and organisms is sought. Direct conversion of bio-molecular signals and useful neural codes to man-made motors will open opportunities to direct brain control of devices via neuromorphic engineering. Six technological capabilities for improvement of human health and physical performance received high priority: bio-nano machines for development of treatments, including those resulting from bioinformatics, genomics and proteomics; nanotechnology-based implants as replacements for human organs (Lavine et al. 2002) or for monitoring of physiological well-being; nanoscale robots and comparable unobtrusive tools for medical intervention; extending brain-to-brain and brain-to-machine interfaces using connections to the human neural system; multi-modality platforms for vision- and hearing-impaired people; and virtual environments for training, design, and forms of work unlimited by distance or the physical scale on which it is performed.

d) Enhancing Group and Societal Outcomes. This group of workshop participants examined the implications of technological convergence for human social behavior, social cognition, interpersonal relations, group processes, the use of language, learning in formal and informal settings, and the psychophysiological correlates of social behavior. A wide range of likely benefits to communities and the nation as a whole has been identified, and a specific vision has been proposed of how these benefits could be achieved through a focused research effort to develop a system this group called The Communicator. This NBIC technology would remove barriers to communication caused by disabilities, language differences, geographic distance, and variations in knowledge, thus greatly enhancing the effectiveness of cooperation in schools, in corporations, in government agencies, and across the world. Converging technologies will lead to revolutionary new industries, products and services based on the synergism and integration of biology, information, and cognitive sciences from the nanoscale.

e) National Security. This group of workshop participants examined the radically changing nature of conflict in this new century and the opportunities to strengthen national defense offered by technological convergence. It identified seven highly diverse goals: data linkage and threat anticipation; uninhabited combat vehicles; war fighter education and training; responses to chemical, biological, radiological, and explosive threats; war fighter systems; non-drug treatments to enhance human performance; exoskeletons for physical performance augmentation; preventing brain changes caused by sleep deprivation; and applications of brain-machine interfaces. These highly varied goals could be achieved through specific convergences of NBIC technologies.

f) Unifying Science and Education. The final group examined the opportunities for unifying science and the current limitations of scientific education, which

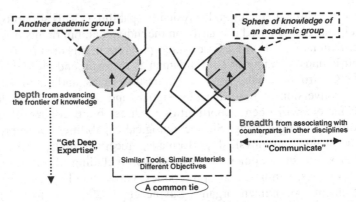

Figure 3.   Combining depth with breath in NBIC education and research of various groups.

is poorly designed to meet the coming challenges. The group documented the need for radical transformation in science education from elementary school through postgraduate training. Part of the answer will come from the convergence of NBIC technologies themselves, which will offer valuable new tools and modalities for education. But convergence of previously separate scientific disciplines and fields of engineering cannot take place without the emergence of new kinds of personnel who understand multiple fields in depth and can intelligently work to integrate them (Figure 3; see Tolles 2002, in this volume). New curricula, new concepts to provide intellectual coherence, and new types of educational institutions will be necessary.

Thus, based on the contributions of individual participants and the work of the six subgroups, the workshop identified the major areas where improved human performance is needed, and identified both short-term and longer-term opportunities to apply convergent technologies to these needs. Table 2 summarizes the key visionary projects discussed in this report. Progress was made in developing a transforming management plan for what should be done to integrate the sciences and engineering in accordance with the convergent technologies vision, including advice to government policymakers. In addition, the workshop recognized specific needs to develop meaningful partnerships and coherent interdisciplinary activities.

7. Future Prospects

Nanotechnology, biotechnology, and information technology are moving closer together, following an accelerated path of unparalleled breakthroughs. Their focus on human dimensions is still emerging but promises to dominate the next decades. Despite efforts of workshop organizers, given the breadth of the topic, it was impossible to recruit leading experts in all the areas where the convergence of NBIC technologies is likely to have a significant impact in 10 to 20 years. In addition, work has really not begun in some of the key application areas, and new areas are likely to emerge that have not yet attracted the attention of many scientists and engineers. Thus, the section below presents the following *admittedly speculative* additional ideas on how technological convergence may transform human abilities two decades and more in the future. Many of the ideas that follow emerged during the workshop, and others were suggested in discussions with participants afterward.

Work Efficiency

Improvement of human physical and mental performance, at both the individual and group level, can increase productivity greatly. Several concepts are in development that could enhance working environments (cf. IBM 2002). To remain competitive, American industry must continue to find ways to improve quality and efficiency (Mowery 1999; Jorgenson and Wessner 2002). Nanotechnology promises to become an efficient length scale for manufacturing (NSTC 2002) because rearranging matter at the nanoscale via weak molecular interactions would require less energy and material. The recent trend toward intensive electronic monitoring and just-in-time inventories has reduced waste, but tightening the efficiency of manufacturing and distribution supply chains could prove to be a one-time-only improvement in profitability that could not be duplicated in the future (National Research Council 2000).

However, application of new generations of convergent technology has the

Table 2. Key visionary ideas and projects discussed in this report

Theme	Key visionary ideas/projects
A. Overall Potential of Converging Technologies	NBIC strategy for technological and economical competitiveness
	New patterns for S&T, economy, and society
	Enhancing individual and group abilities, productivity, and learning
	Sustainable and "intelligent" environments
	Changing human activities towards the "innovation age"
B. Expanding Human Cognition and Communication	Human cognome project and cognitive evolution
	Brain-to-brain interactions and group communication
	Spatial cognition and visual language using converging technologies
	Enhanced tools for learning and creativity
	Predictive science of societal behavior
C. Improving Human Health and Physical Capabilities	Healthcare, body replacements, and physiological self-regulation
	Brain-machine interfaces and neuromorphing engineering
	Improving sensorial capacities and expanding functions
	Improving quality of life of disabled people
	Aging with dignity and life extension
D. Enhancing Group and Societal Outcomes	The Communicator: enhancing group interaction and creativity
	Cognitive engineering and enhancing productivity
	Revolutionary products, including "aircraft of the future"
	Networked society, with bio-inspired culture
E. National Security	Enhancing physical and mental capacity of a soldier
	Enhancing readiness and threat anticipation tools
	Globally linked detection devices
	Uninhabited combat vehicles
F. Unifying Science and Education	Unifying science from the nanoscale and integrative principles
	Cognitive, civic, and ethical changes in a networked society
	Breadth, depth, "trading zones," and reshaping education at all levels
	Changing the human culture

potential to provide better value to customers at lower cost to producers, offering the possibility of further profitability improvements. For example, even more intensive use of information technology in conjunction with nanotechnology, biotechnology, and cognitive sciences could reduce waste and pollution costs and permit very rapid reconfiguration of manufacturing processes and product lines (National Research Council 1998). Business and industry are already beginning to restructure themselves on a global scale as network-based organizations following fundamentally new management principles.

Biology in conjunction with nanoscale design and IT control has the potential to contribute both abstract models and specific physical processes to the development of customer-centric production that blends the principles of custom-design craftsmanship (which maximizes customer satisfaction) with the principles of assembly-line mass production (which minimizes production costs). In the gestation of higher animals, a single fertilized egg cell differentiates rapidly into specialized cells that grow into very different organs of the body, controlled in a complex manner by the messenger chemicals produced by the cells themselves. Whether based in nanotechnology, information technology, biotechnology, or cognitive-based technology, new adaptive production systems could be developed that automatically adjust design features in a way analogous to the growing embryo, without the need to halt production or retool. Convergence of these four technologies could also develop many bio-inspired processes for "growing" key components of industrial products, rather than wastefully machining them out of larger materials or laboriously assembling them from smaller parts (cf. National Research Council 1999).

The Human Body and Mind Throughout the Life Cycle

Improving perceptual capabilities, biohybrid systems, exoskeletons, and metabolic enhancement can be considered for human performance augmentation. Medical implants for sensory replacement, including multiple sensory modalities for visually and hearing-impaired persons, and direct brain-machine interfaces are real possibilities. Controlled metabolism in cells, specific tissues, organs, or the entire body is possible. One application would be increased endurance and resistance to sleep deprivation; another is a method of optimizing oxygenization of blood when metabolism is compromised in a critical medical situation. Others would be realtime genetic testing so that individually tailored drugs can be provided to patients, and an artificial pancreas that would monitor and adjust the release of hormones in the human body.

Increasing intellectual capabilities requires understanding the brain and simulating its processes. Knowledge about the structure, function, and occasional dysfunction of the human mind will provide new ways to increase cognitive capabilities (Steve et al. 2002; National Research Council 1988). Reverse engineering of the human brain may be accomplished in the next two decades that would allow for better understanding of its functions. An artificial brain (Cauller and Penz 2002) could be a tool for discovery, especially if computers could closely simulate the actual brain. It would be revolutionary to see if aspects of human consciousness could be transferred to machines (Kurzweil 1999) in order to better interact with and serve humans.

Sustaining human physical and mental abilities throughout the life span would be facilitated by progress in neuroscience (Stern and Carstensen 2000) and cellular biology at the nanoscale. An active and dignified life could be possible far into a person's second century, due to the convergence of technologies (cf. Saxl 2002). Gene therapy to cure early aging syndromes may become common, giving vastly improved longevity and quality of life to millions of people (Bonadio 2002; Heller 2002; Connolly 2002).

Communication and Education

New communication paradigms (brain-to-brain, brain-machine-brain, group) could be realized in 10 to 20 years. Neuromorphic engineering may allow the transmission of thoughts and biosensor output from the human body to devices for signal processing. Wearable computers with power similar to that of the human brain will act as personal assistants or brokers, providing valuable information of every kind in forms optimized for the specific user. Visual communication could complement verbal communication, sometimes replacing spoken language when speed is a priority or enhancing speech when needed to exploit maximum mental capabilities (Horn 2002; Hewlett Packard 2002).

People will be able to acquire a radically different instinctive understanding of the world as a hierarchy of complex systems rooted in the nanoscale. Advances in cognitive science will enable nanoscience education, by identifying the best ways for students to conceptualize nanostructures and processes at increasingly advanced stages in their learning (National Institute of Mental Health 2002). Education at all levels will exploit augmented reality, in which multimedia information displays are seamlessly integrated into the physical world. Strategies for hierarchical, architectural, and global analysis and design of complex systems will help integrate the curriculum of schools and inform management decisions across a diverse range of fields.

Mental Health

In many respects, perhaps the most difficult challenge we face in improving human performance is understanding and remediating mental illness (Anderson 1997). For fully the past two centuries, psychiatry has alternated between periods of optimism and pessimism, as well as between competing psychological, social, physiological, chemical, and genetic theories of mental illness. We can hope that these disputes will be resolved through physiological and psychological understanding of mental processes, and that scientific convergence will achieve lasting cures through a combination of biological and cognitive treatments, all assisted by information and nanoscale technologies.

Nanotechnology will provide means to deliver medications to the exact location within the brain where they are needed, thus minimizing negative side effects elsewhere in the nervous system. The convergence of cognitive science with nano-, bio-, and information technologies should permit systematic evaluation of the bewildering range of current psychiatric theories and therapies, and allow clinicians to improve the best treatments. It is also possible that convergent communications and robotics technologies may produce an entirely new category of prosthetic or assistive devices that can compensate for cognitive or emotional deficiencies.

Aeronautics and Space Flight

NBIC synergies could greatly expand capabilities for piloted adaptive aircraft, unmanned aircraft, and human space flight. Nanostructured materials and advanced electronics have the promise of reducing the weight of spacecraft by three quarters in the next 10 to 20 years. Specific subsystems for human space flight may also be revolutionized by the same combination of technologies, for example durable but light and self-repairing spacesuits, high-performance electronics with low demands for electric power, and low-cost but high-value large orbiting structures. If the problems of orbital launch costs and efficient subsystems can be solved, then human society can effectively exploit Earth orbital space, the Moon, asteroids, and the planet Mars. Several participants in the workshop noted the potential for intelligent machines of the future to take on progressively more human characteristics, so we can well imagine that the first pioneers that take "humanity" far into space will be descendents of Pathfinder and the Voyagers that will be endowed with intelligence and communication capabilities reflecting human behavior.

Food and Farming

Farmers have long appreciated the advantages of science and technology; the convergence of nanotechnology, biotechnology, and information technology could significantly improve their effectiveness. For example, nanoscale genetics may help preserve and control food production. Inexpensive nano-enabled biosensors could monitor the health and nutrition of cattle, transmitting the data into the farmer's personal computer that advises him about the care the animals need. In the same way, sensors distributed across farmland could advise the farmer about the need for water and fertilizer, thus avoiding wastage and achieving the most profitable acreage crop yield (National Research Council 1997). Bio-nano convergence can provide new ways of actually applying the treatment to the crops, increasing the efficiency of fertilizers and pesticides.

Use of nano-enabled biosensors would monitor freshness to help grocers avoid selling stale goods and to avoid the wastage of discarding perfectly good packaged food that has merely reached an arbitrary shelf life date. The consumer should have access to the same information, both before and after purchase. Many consumers are dissatisfied with the limited information about ingredients on many packaged foods, and the total lack of information about foods served in restaurants. Convergent technologies could provide portable instruments, for example packaged into a pen-like device or perhaps a ring, that could instantly tell the consumer how much sodium, fats, or allergenic substances a food contains.

Sustainable and Intelligent Environments

Sustainable resources of food, water, energy, and materials are achievable through converging technologies. Exact manufacturing, exact integration in biosystems, and IT control will help stabilize the supply of resources. Value will stem from information, including that embodied in the complex structure of manufactured items made from the nanoscale out of common chemical elements, rather than rare metals or nonrenewable energy supplies. Sensing the environment and biosystems of the world will become essential in global environmental monitoring and remediation. New sources for a distributed energy system are envisioned, as well as new solutions such as highly efficient photosynthetic proteins, membranes, and devices.

Interactive and "intelligent" environments for human activities are envisioned, responding to advancements in areas such as neuro-ergonomics and the needs of persons with disabilities.

External surfaces of buildings could automatically change shape and color to adjust to different conditions of temperature, lighting, wind, and precipitation. Once the science, manufacturing processes, and economic markets have developed sufficiently, adaptive materials need not be especially expensive, especially when their increased performance and energy efficiency are factored in. For example, nanotechnology materials and IT-assisted design could produce new, durable house paints that change color, reflecting heat on hot days and absorbing heat on cold days. Indoors, ordinary walls could be vast computer displays, capable of enhancing the residents' aesthetic experience by displaying changing virtual artworks and wallpapers. Adaptive materials could obtain their energy from temperature differentials between different surfaces (thermocouples) or naturally occurring vibrations (piezoelectric), rather than requiring electrical input. The ability to engineer inexpensive materials on the nanoscale will be crucial, and information technology can help design the materials as well as be designed into some of the adaptive systems. There also will be a role for cognitive science, because architects need to take account of human needs and the often unexpected ways that human beings respond to particular design features.

Self-Presentation and Fashion

Government-supported academic researchers frequently ignore many economically important industries, in part because those industries traditionally have not involved advanced technology but also perhaps because they were not perceived as "serious" fields. Among these are clothing fashions, jewelry, and cosmetics. Stereotypes aside, these are multibillion dollar industries that could benefit from the new opportunities afforded by convergent technologies. In social life, physical attractiveness is very important. Anything that enhances a person's beauty or dignity improves that individual's performance in relations with other people.

Convergence of nanotechnology and biotechnology with cognitive science could produce new kinds of cosmetics that change with the user's moods, enhancing the person's emotional expressiveness. Components of wearable computers could be packaged in scintillating jewelry, automatically communicating thoughts and feelings between people who are metaphorically and electronically "on the same wave length." Biotechnology could produce new materials that would be combined in manufacturing with nanotechnology-based information technology to produce clothing that automatically adjusts to changing temperatures and weather conditions. Perhaps the colors and apparent textures of this "smart clothing" would adjust also to the wearer's activities and social environment.

Transformation of Civilization

The profound changes of the next two decades may be nothing compared to the utter transformation that may take place in the remainder of the 21st century. Processes both of decentralization and integration would render society ever more complex, resulting in a new, dynamic social architecture. There would be entirely new patterns in manufacturing, the economy, education, and military conflict.

People may possess entirely new capabilities for relations with each other, with machines, and with the institutions of civilization. In some areas of human life, old customs and ethics will persist, but it is difficult to predict which realms of action and experience these will be. Perhaps wholly new ethical principles will govern in areas of radical technological advance, such as the acceptance of brain implants, the role of robots in human society, and the ambiguity of death in an era of increasing experimentation with cloning. Human identity and dignity must be preserved. In the same way in which machines were built to surpass human physical powers in the industrial revolution, computers can surpass human memory and computational speed for intended actions. The ultimate control will remain with humans and human society. With proper attention to safeguards, ethical issues, and societal needs, quality of life could increase significantly.

New professions for humans and new roles for machines may arise to mediate between all this complexity and the individual person. Art, music, and literature may reach new levels of subtlety and sophistication, enhancing the mental qualities of life and the innate human appreciation for beauty.

A networked society of billions of human beings could be as complex compared to an individual human being as a human being is to a single nerve cell. From local groups of linked enhanced individuals to a global collective intelligence, key new capabilities would arise from relationships created with NBIC technologies. Such a system would have distributed information and control and new patterns of manufacturing, economic activity, and education. It could be structured to enhance individuals' creativity and independence. Far from unnatural, such a collective social system may be compared to a larger form of a biological organism. Biological organisms themselves make use of many structures such as bones and circulatory system. The networked society enabled through NBIC convergence could explore new pathways in societal structures, in an increasingly complex system (Bar-Yam 1997).

It may be possible to develop a predictive science of society and to apply advanced corrective actions, based on the convergence ideas of NBIC. Human culture and human physiology may undergo rapid evolution, intertwining like the twin strands of DNA, hopefully guided by analytic science as well as traditional wisdom. As Table 3 suggests, the pace of change is accelerating, and scientific convergence may be a watershed in history to rank with the invention of agriculture and the Industrial Revolution.

8. Recommendations

The recommendations of this report are far-reaching and fundamental, urging the transformation of science at its very roots. But the recommendations also seek to preserve the wonderful accomplishments of science and sustain the momentum of discovery that has been energized by generations of scientists. Only by evolving can science continue to thrive and make the vast contributions to society that it is capable of in the coming decades. There are outstanding opportunities that were not available in the past. The new developments will be revolutionary and must be governed by respect for human welfare and dignity.

Specific Areas for Research and Education Investment

The research and education needs are both deep and broad. In order to connect disciplines at their interfaces, understand and assemble matter from its building blocks, while focusing on a broad systems perspective and improving human performance, research and education must have deep scientific roots and superior communication among the fields of human endeavor.

Table 3. History of some very significant augmentations
to human performance:
Improving our ability to collectively improve ourselves (see also Spohrer 2002)

Generations	Several Key Advancements (human kind, tools and technology, communication)
-m	Cell, body and brain development
- 100,000	Old Stone Age (Paleolithic), Homo Erectus, speech
-10,000	Homo Sapiens, making tools
-500	Mesolithic, creating art
-400	Neolithic, agricultural products, writing, libraries
-40	Universities
-24	Printing
-16	Renaissance in S&T, accurate clocks
-10	Industrial revolution
-5	Telephone
-4	Radio
-3	TV
-2	Computers
-1	Microbiology, Internet
0	Reaching at the building blocks of matter (nanoscience) Biotechnology products Global connection via Internet; GPS/sensors for navigation
½	Unifying science and converging technologies from the nanoscale Nanotechnology products Improving human performance advancements Global education and information infrastructure
1	Converging technology products for improving human physical and mental performance (new products and services, brain connectivity, sensory abilities, etc.) Societal and business reorganization
n	Evolution transcending human cell, body, and brain?

The following general integrative approaches have been identified as essential to NBIC:

- Development of NBIC tools for investigation and transformational engineering at four levels: nano/microscopic, individual, group, and society
- Integration of fundamental concepts of NBIC across all scales, beginning with the nanoscale
- Investigation of converging technologies that is systems- and holistic-based

- Focus of future technological developments on implications for improving human performance

These principles concern the research methods, theoretical analyses, systemic perspective, and human benefit dimensions of scientific and technological integration. Sharing research techniques and engineering tools is one way that scientists in traditionally different fields can integrate their work. Another is utilization of similar ideas, mathematical models, and explanatory language. Expected to be a major challenge in approaching complex systems is the hierarchical architecture in which various components are integrated and used. Consideration of the human implications of converging technologies will include examination of potential unexpected consequences of NBIC developments, including ethical and legal aspects.

Recommendations to Individuals and Organizations

This report has educational and transformational goals. Building on the suggestions developed in the five topical groups and on the ideas in the more than 50 individual contributions, workshop participants recommended a **national R&D priority area on converging technologies focused on enhancing human performance.** The main transforming measures are outlined in section 4 of this summary. The opportunity now is broad, enduring, and of general interest. The report contributors addressed the roles that individuals, academe, the private sector, the U.S. Government, professional societies, and other organizations should play in this converging technology priority area:

a) **Individuals.** Scientists and engineers at every career level should gain skills in at least one NBIC area and in neighboring disciplines, collaborate with colleagues in other fields, and take risks in launching innovative projects that could advance technology convergence for enhancing human performance.

b) **Academe.** Educational institutions at all levels should undertake major curricular and organizational reforms to restructure the teaching of science and engineering so that previously separate disciplines can converge around common principles to train the technical labor force for the future. The basic concepts of nanoscience, biology, information, and cognitive sciences should be introduced at the beginning of undergraduate education; technical and humanistic degrees should have common courses and activities related to NBIC and the human dimensions of science and technology. Investigations of converging technologies should focus on the holistic aspects and synergism. The hierarchical architecture in which various components are integrated and used is expected to be a major challenge.

c) **Private Sector.** Manufacturing, biotechnology, and information service corporations will need to develop partnerships of unparalleled scope to exploit the tremendous opportunities from technological convergence, engaging in joint ventures with each other, establishing research linkages with universities, and investing in production facilities based on entirely new principles and materials, devices, and systems.

d) **Government.** A national research and development priority area should be established to focus on converging technologies that enhance human performance. Organizations should provide leadership to coordinate the work of other institutions and must accelerate convergence by supporting new multidisciplinary scientific efforts while sustaining the traditional disciplines that are essential for success. Special effort will be required to identify future technological developments; explore their implications for human performance; study unexpected consequences of NBIC developments; and consider ethical, legal, and policy issues. Governments must provide support for education and training of future NBIC workers and to prepare society for the major systemic changes envisioned for a generation from now. Policymakers must envision development scenarios to creatively stimulate the convergence. Ethical, legal, moral, economic, environmental, workforce development, and other societal implications must be addressed from the beginning, involving leading NBIC scientists and engineers, social scientists and a broad coalition of professional and civic organizations. Research on societal implications must be funded, and the risk of potential undesirable secondary effect must be monitored by a government organization in order to anticipate and take corrective actions. Tools should be developed to anticipate scenarios for future technology development and applications. The transforming measures outlined in section 4 above suggest the dimensions of the Federal Government role.

e) **Professional Societies.** The scientific community should create new means of interdisciplinary training and communication, reduce the barriers that inhibit individuals from working across disciplines, aggressively highlight opportunities for convergence in their conferences, develop links to a variety of other technical organizations, and address ethical issues related to technological developments. Through mechanisms like conferences and publications, professional societies can seed NBIC ideas in learning organizations, funding agencies, and the society at large.

f) **Other Organizations.** Nongovernmental organizations that represent potential user groups should contribute to the design and testing of convergent technologies and recommend NBIC priorities, in order to maximize the benefits for their diverse constituencies. Private research foundations should invest in NBIC research in those areas that are consistent with their particular missions. The public media should increase high-quality coverage of science and technology, on the basis of the new convergent paradigm, to inform citizens so they can participate wisely in debates about ethical issues such as the unexpected effects on social equality, policies concerning diversity, and the implications of transforming human nature.

A vast opportunity is created by the convergence of sciences and technologies starting with integration from the nanoscale, having immense individual, societal, and historical implications for human development. Therefore, the contributors to this report recommend *a national research and development priority area on converging technologies focused on enhancing human performance.* Advancing knowledge and transforming tools will move our activities from simple repetitions to creative, innovative acts and transfer the focus from machines to human development. Converging technologies are at the confluence of key disciplines and areas of application, and the role of government is important because no other participant can cover the breadth and level of required collective effort. Without special efforts for coordination and integration, the path of science might not lead to the fundamental unification envisioned here. Technology will increasingly dominate the world, as population, resource exploitation, and potential social conflict grow. Therefore, the success of this convergent technologies priority area is essential to the future of humanity.

References

Anderson, N.C. 1997. Linking mind and brain in the study of mental illnesses. *Science* 275:1586-1593.

Bar-Yam, Y. 1997. *Dynamics of complex systems*. Cambridge, Perseus Press.

Bonadio, J. 2002. Gene therapy: Reinventing the wheel or useful adjunct to existing paradigms? In Chapter C of this report.

Brain work: The neuroscience newsletter. 2002. Neuroethics: Mapping the field. Editor's note, in Vol. 12 No. 3, May-June 2002, p. 1.

Caplow, T., L. Hicks, and B.J. Wattenberg. 2001. *The first measured century: An illustrated guide to trends in America, 1900-2000*. Washington, D.C.: American Enterprise Institute Press.

Cauller, L. and A. Penz. 2002. Artificial brains and natural intelligence. In Chapter C of this report.

Connolly, P. 2002. Nanobiotechnology and life extension. In Chapter C of this volume.

Horn, R. 2002. *Visual language*. Stanford U. http://www.stanford.edu/~rhorn/ index.html; http://macrovu.com/

Heller, M. 2002. The nano-bio connection and its implication for human performance. In Chapter C of this report.

Hewlett Packard. 2002. *Cool town*. http://www.exploratorium.edu/ pr/alliance.html.

IBM. 2002. User system ergonomics research. http://www.almaden.ibm.com/ cs/user.html.

Jorgenson, D.W., and C.W. Wessner (eds.). 2002. *Measuring and sustaining the new economy*. Washington, D.C.: National Academy Press.

Kennedy, D. 2002. Are there things we'd rather not know? *Brain work: The neuroscience newsletter,* Vol. 12 No. 3, May-June 2002, p. 6.

Kurzweil, R. 1999. *The age of spiritual machines*. New York: Viking.

Lavine, M., L. Roberts, and O. Smith. 2002. Bodybuilding: The bionic human. *Science* 295:995-1033.

Mowery, D.C. (ed.). 1999. *U.S. industry in 2000: Studies in competitive performance*. Washington, D.C.: National Academy Press.

National Institute of Mental Health. 2002. *Learning and the brain*. http://www.edupr.com/brain4.html; http://www.edupr.com/bsched.html.

National Research Council. 1988. *Enhancing human performance*. Washington, D.C.: National Academy Press.

National Research Council (Committee on Assessing Crop Yields). 1997. *Precision agriculture in the 21st century*. Washington, D.C.: National Academy Press.

National Research Council (Committee on Visionary Manufacturing Challenges). 1998. *Visionary manufacturing challenges for 2020*. Washington, D.C.: National Academy Press.

National Research Council (Committee on Biobased Industrial Products). 1999. *Biobased industrial products: Priorities for research and commercialization*. Washington, D.C.: National Academy Press.

National Research Council (Committee on Supply Chain Integration). 2000. *Surviving supply chain integration*. Washington, D.C.: National Academy Press.

National Science and Technology Council (NSTC, Subcommittee on Nanoscale Science, Engineering and Technology). 2002. *National Nanotechnology Initiative: The initiative and its implementation plan (detailed technical report associated with the supplemental report to the President's FY 2003 budget).* White House: Washington, DC.

New York Academy of Sciences (NYAS). 2002. *Unifying knowledge: The convergence of natural and human science.* New York: NYAS Annals.

Ostrom, E., T. Dietz, P.C. Stern, S. Stonich, and E.U. Weber. 2002. *The drama of the commons.* Washington D.C.: National Academy Press.

Roco, M.C. and W.S. Bainbridge (eds). 2001. *Societal implications of nanoscience and nanotechnology.* Boston: Kluwer Academic Publ. Also available at http://www.wtec.org/loyola/nano/NSET.Societal.Implications.

Roco, M.C. 2002. Coherence and divergence of megatrends in science and technology. In Chapter A of this volume.

Saxl, O. 2002. Summary of conference on nanobiotechnology, life extension and the treatment of congenital and degenerative disease, institute of nanotechnology, U.K.

Spohrer, J. 2002. NBICs (nano-bio-info-cogno-socio) convergence to improve human performance: opportunities and challenges. In Chapter B of this report.

Stern, Paul C., and Laura L. Carstensen (eds.). 2000. *The aging mind: Opportunities in cognitive research.* Washington, D.C.: National Academy Press. Also available at http://www.nap.edu/catalog/9783.html.

Steve, S., R. Orpwood, M. Malot, X, and E. Zrenner. 2002. Mimicking the brain. *Physics World* 15(2):27-31.

United Nations Development Program. 2001. Making new technologies work for human development. Part of the *2001 Human Development Report*, UK: Oxford University Press. Also available at http://www.undp.org/hdr2001/.

GENERAL STATEMENTS AND VISIONARY PROJECTS

The following six sets of contributions (chapters A to F) present key statements and visions from academe, private sector and government illustrating what technological convergence could achieve in the next 10 to 20 years. In each set, statements are grouped at the beginning that consider the current situation in the particular area and project ways we could build on it to achieve rapid progress. The later contributions in the set present visions of what might be achieved toward the end of the two-decade period. In the first of these six sets, government leaders and representatives of the private sector provide the motivation for this effort to understand the promise of converging technologies. The second and third sets of contributions identify significant ways in which the mental and physical abilities of individual humans could be improved. The third and fourth sets examine prospects on the group and societal level, one considering ways in which the internal performance of the society could benefit and the other focusing on the defense of the society against external threats to its security. The sixth and final set of essays considers the transformation of science and engineering themselves, largely through advances in education.

A. MOTIVATION AND OUTLOOK

THEME A SUMMARY

Panel: P. Bond, J. Canton, M. Dastoor, N. Gingrich, M. Hirschbein, C.H. Huettner, P. Kuekes, J. Watson, M.C. Roco, S. Venneri, R.S. Williams

In a sense, this section of the report gives the authors their assignment, which is to identify the technological benefits of convergence that could be of greatest value to human performance and to consider how to achieve them. Five of the statements were contributed by representatives of government agencies: The Office of Science and Technology Policy, The Department of Commerce, The National Aeronautics and Space Administration, the National Institutes of Health, and the National Science Foundation. The remaining three were contributed from private sector organizations: The American Enterprise Institute, Hewlett Packard, and the Institute for Global Futures. But these eight papers are far more than mission statements because they also provide an essential outlook on the current technological situation and the tremendous potential of convergence.

1. It is essential to identify new technologies that have great potential to improve human performance, especially those that are unlikely to be developed as a natural consequence of the day-to-day activities of single governmental, industrial, or educational institutions. Revolutionary technological change tends to occur outside conventional organizations, whether through social movements that promulgate new goals, through conceptual innovations that overturn old paradigms of how a goal can be

achieved, or through cross-fertilization of methods and visions across the boundaries between established fields (Bainbridge 1976). Formal mechanisms to promote major breakthroughs can be extremely effective, notably the development of partnerships between government agencies to energize communication and on occasion to launch multiagency scientific initiatives.

2. Government has an important role in setting long-term priorities and in making sure a national environment exists in which beneficial innovations will be developed. There must be a free and rational debate about the ethical and social aspects of potential uses of technology, and government must provide an arena for these debates that is most conducive to results that benefit humans. At the same time, government must ensure economic conditions that facilitate the rapid invention and deployment of beneficial technologies, thereby encouraging entrepreneurs and venture capitalists to promote innovation. Of course, government cannot accomplish all this alone. In particular, scientists and engineers must learn how to communicate vividly but correctly the scientific facts and engineering options that must be understood by policymakers and the general public, if the right decisions are to be made.

3. While American science and technology benefit the entire world, it is vital to recognize that technological superiority is the fundamental basis of the economic prosperity and national security of the United States. We are in an Age of Transitions, when we must move forward if we are not to fall behind, and we must be ready to chart a course forward through constantly shifting seas and winds. Organizations of all kinds, including government itself, must become agile, reinventing themselves frequently while having the wisdom to know which values are fundamental and must be preserved. The division of labor among institutions and sciences will change, often in unexpected ways. For many years, scholars, social scientists, and consultants have been developing knowledge about how to manage change (Boulding 1964; Drucker 1969; Deming 1982; Womack and Jones 1996), but vigorous, fundamental research will be needed throughout the coming decades on the interaction between organizations, technology, and human benefit.

4. Government agencies need progress in NBIC in order to accomplish their designated missions. For example, both spacecraft and military aircraft must combine high performance with low weight, so both NASA and the Department of Defense require advances in materials from nanotechnology and in computing from information technology. Furthermore, in medicine and healthcare, for example, national need will require that scientists and engineers tackle relatively pedestrian problems, whose solutions will benefit people but not push forward the frontiers of science. But practical challenges often drive the discovery of new knowledge and the imagination of new ideas. At the same time, government agencies can gain enhanced missions from NBIC breakthroughs. One very attractive possibility would be a multiagency initiative to improve human performance.

5. Science must offer society new visions of what it is possible to achieve. The society depends upon scientists for authoritative knowledge and professional judgment to maintain and gradually improve the well-being of citizens, but

scientists must also become visionaries who can imagine possibilities beyond anything currently experienced in the world. In science, the intrinsic human need for intellectual advancement finds its most powerful expression. At times, scientists should take great intellectual risks, exploring unusual and even unreasonable ideas, because the scientific method for testing theories empirically can ultimately distinguish the good ideas from the bad. Across all of the sciences, individual scientists and teams should be supported in their quest for knowledge. Then interdisciplinary efforts can harvest discoveries across the boundaries of many fields, and engineers will harness them to accomplish technological progress.

The following eight statements develop these and other ideas more fully, thereby providing the motivation for the many chapters that follow. They also provide a perspective on the future by identifying a number of megatrends that appear to be dominant at this point in human history and by suggesting ways that scientists and policymakers should respond to these trends. Their advice will help Americans make history, rather than being subjects of it, strengthening our ability to shape our future. The statements include a message from the White House Office of Science and Technology Policy (OSTP) concerning the importance of this activity to the nation, a message from the Department of Commerce on its potential impact on the economy and U.S. competitiveness, a vision for converging technologies in the future, examples of activities already underway at NASA and NIH, industry and business perspectives on the need for a visionary effort, and an overview of the trend toward convergence of the megatrends in science and engineering.

References

Bainbridge, W.S. 1976. *The spaceflight revolution*. New York: Wiley-Interscience.

Boulding, K.E. 1964. *The meaning of the twentieth century: The great transition*. New York: Harper and Row.

Deming, W.E. 1982. *Quality, productivity, and competitive position*. Cambridge, MA: MIT Center for Advanced Engineering Study.

Drucker, P.F. 1969. *The age of discontinuity: Guideline to our changing society*. New York: Harper and Row.

Roco, M.C., R.S. Williams, and P.Alivisatos, eds. 2000. *Nanotechnology research directions*. Dordrecht, Netherlands: Kluwer Academic Publishers.

Roco, M.C., and W.S. Bainbridge, eds. 2001. *Societal implications of nanoscience and nanotechnology*. Dordrecht, Netherlands: Kluwer Academic Publishers.

Siegel, R.W., E. Hu, and M.C. Roco, eds. 1999. *Nanostructure science and technology*. Dordrecht, Netherlands: Kluwer Academic Publishers.

Womack, J.P., and D. Jones. 1996. *Lean thinking*. New York: Simon and Schuster.

NATIONAL STRATEGY TOWARDS CONVERGING SCIENCE AND TECHNOLOGY

Charles H. Huettner, OSTP, White House

Good morning. I want to express to you on behalf of Dr. John Marburger, who is the President's science advisor and the Director of the Office of Science and Technology Policy (OSTP), his regrets for not being able to be with you, particularly because this workshop is a very important first step towards the future in which different sciences come together.

The role of the OSTP is to identify cross-cutting, high-risk technologies that don't reside in a particular department or agency and to sponsor them, thus helping them move across government agencies. Nanotechnology is a clear example of the kinds of technologies that have great potential and yet need government-wide review and focus.

Obviously, nanotechnology is just one of a number of emerging technologies. We are living in a very exciting time. Just think of what has happened with information technology over the last 10 years. It has allowed us to have the Internet, a global economy, and all of the things that we know about and have been living through. In just this past year, the field of biology has experienced tremendous advances with the human genome project. New this year in the budget is the national nanotechnology initiative, and similar kinds of progress and accomplishment are anticipated there.

Could these technologies and others merge to become something more important than any one individually? The answer to that question obviously is that they must. Convergence means more than simply coordination of projects and groups talking to one another along the way. It is imperative to integrate what is happening, rise above it, and get a bigger picture than what is apparent in each individual section.

There is an institute at Harvard called the Junior Fellows, formed many, many years ago by a forward thinker at Harvard and endowed with a beautiful building with a wonderful wine cellar. Senior Fellows, who were the Nobel Laureates of the university, and Junior Fellows, who were a select group of people picked from different disciplines, came together there for dinner from time to time. Sitting together at one Junior Fellows dinner I attended several years ago were musicians, astrophysicists, and astronomers discussing how certain musical chords sound good and others don't, and how those sorts of harmonics actually could help to explain the solar system, the evolution of galaxies, and so forth. Essentially, this is what the two-day NBIC workshop is doing, bringing together thinkers from different disciplines to find common ground and stimulate new thinking. When professionals as diverse as musicians and astrophysicists can discover mutually resonant concepts, think about what we can do with the kinds of technologies that we have today. That is why this NBIC workshop is so important.

You are the national technology leaders, or you are connected with them. You are the beginnings of an important group coming together. Nuclear and aerospace technology, psychology, computer science, chemistry, venture capital, medicine, bioengineering, social sciences — you're all here, and you represent not only the government, but also industry and academia. I thought it was tremendously creative, the way that the working sessions were broken down around people's needs because, in the end, that's why science is here. Science is here to serve people. So, it is very important for the breakout groups to look at human cognition and communications and human physical performance by focusing on how to solve human needs.

Take this opportunity to begin the cross-fertilization and understanding of each other's disciplines. The language of each technology is different. The key ideas that define them are different. The hopes and visions are different. The needs to accomplish those are different. But the network that we can form and the learning that we can have as a result of today's efforts can somehow bridge those gaps and begin the understanding.

I applaud you for being here today. I challenge you to learn and think beyond your discipline to help to establish the inner technology visions, connections, and mechanisms that will solve the human problems of our world. This is the beginning of the future, and we at OSTP are both anxious to help and anxious to learn from you.

CONVERGING TECHNOLOGIES AND COMPETITIVENESS

The Honorable Phillip J. Bond, Undersecretary for Technology, Department of Commerce

Good morning, and thank you all. It is a pleasure to be here as a co-host, and I want to give you all greetings on behalf of Secretary of Commerce Don Evans, whom I am thrilled and privileged to serve with in the Bush administration. Thank you, Mike Roco and Joe Bordogna for bringing us all together. Charlie Huettner, please give my best wishes to Jack Marburger. Dr. Marburger and I were confirmation cousins, going through our Senate hearings and then floor consideration together.

It is a rare thing to see people inside the Washington Beltway coming together to actually think long-term in a town that is usually driven by the daily headlines. I believe it was George Will who observed that most people inside the Beltway survive on the intellectual capital they accumulated before they came inside the Beltway. I certainly hope that's not true in my case. I do want to encourage you and join you. Let us lift our eyes, look at the future, and really seize the opportunity for some of the policy implications.

I stand before you today not as a scientist, but as an advocate. My background as the head of Hewlett-Packard's office here in Washington, before that with an IT association, and then on the Hill, and before that with Dick Cheney at the Pentagon, implies that I am supposed to know something about moving the gears of government toward positive policy outcomes. With that in mind, I now have the privilege of overseeing the National Institute of Standards and Technology (NIST), the Office of Technology Policy, and the National Technical Information Service that I am sure many of you periodically go to for information, as well as the National Medal of Technology.

I am sure that many of you saw the news this morning that one of our past National Medal of Technology winners has unveiled what was previously code-named Ginger, which I now understand is the Segway Human Transporter — Dean Kamen's new project. So, next time we can all ride our two-wheelers to the meeting. At any rate, I want to pledge to you to really try to provide the kind of support needed over the long term on the policy front.

Historical perspective is useful for a meeting such as this, and for me this is best gained in very personal terms. My grandparents, Ralph and Helen Baird, just passed away. He died earlier this year at 101 and she two years ago at 99. They taught me about the importance of science and technology to the human condition. Before they passed on, they sat down and made a videotape reviewing the things they had seen in their life.

In that arena, what was particularly relevant is the fact that Ralph had been a science teacher. Both of my grandparents saw men learn to fly and to mass-produce horseless carriages. They told great stories about living in Kansas and getting on the community phone, ringing their neighbors and saying, "Quick, run down to the road. One's coming. Run down to see one of these gizmos rolling by." They saw the generation and massive transmission of electricity, the harnessing of the power of the atom, the space-travel to our moon, the looking back in time to the origins of our universe, the development of instantaneous global communications, and most recently, the deciphering of the human genome and cloning of very complex organisms. Each of these is extraordinary in its technical complexity but also profound in terms of its economic and social significance.

This is one of the challenges we have for you in the discussions. To borrow from Churchill, as everybody seems to do, this is "the end of the beginning." As we head into the 21st Century, we are going to have not only accelerating change, but accelerating moral and ethical challenges. Again here, I take a very personal view of this. My daughters Jackie and Jesse are 10 and 7. So when I look at the future and think about the ethical possibilities and possibilities of robo-sapiens, as *Wired* magazine talks about, I think in terms of what my daughters will face and how we as a society can reap the harvest of technology and remove the chaff of unethical uses of that technology. We have a real balancing act moving forward. The future of all of us — and my daughters' futures — are on the line.

Other speakers have mentioned the exciting fields that you're going to be looking at today and how they converge. I will leave most of the description of that to others, including the always provocative and mesmerizing Newt Gingrich and my friend Stan Williams from HP, and to your breakout discussions. However, as a political appointee, let me do what I do best, and that is to observe the obvious.

Obviously, powerful technologies are developing. Each is powerful individually, but the real power is synergy and integration, all done at the nanoscale. There's plenty of room at the bottom. Intel recently announced it expects to produce a terahertz chip about six or seven years out — 25 times the number of transistors as the top-of-the-line Pentium 4. Within the next few years we're going to be looking at computers that are really personal brokers or information assistants. These devices will be so small that we'll wear them and integrate them. They will serve as information brokers. Again, when I think about my daughters, if current trends hold, one of those information brokers will be looking at science and horses and the other will be looking at hairstyles — but to each their own. Seriously, that day is coming fast, based on breakthroughs in producing computer chips with extremely small components. If we do policy right, with each breakthrough will come technology transfer, commercialization, economic growth, and opportunity that will pay for the next round of research.

In all of this, at least as a policy person, I try to separate hype from hope. But the more I thought about that, the more I determined that in this political town, maybe the separation isn't all that important, because hype and hope end up fueling the social passion that forms our politics. It gets budgets passed. It makes things possible for all of you. Without some passion in the public square, we will not achieve many of our goals. Those goals are mind-boggling — what we used to think of as miraculous — the deaf to hear, the blind to see, every child to be fed. And that's just for starters.

Always, each advance in technology carries a two-edged sword. As a policy person I need your help. One hundred years ago, the automobile was not immediately embraced; it was rejected as a controversial new innovation. Eventually it was accepted, then we had a love affair with it, and now it's perhaps a platonic relationship. Our journey with these other technologies is going to have similar bumps in the road. And so, as you set out today, I think you should include these three important considerations in your mission:

- to achieve the human potential of everybody
- to avoid offending the human condition
- to develop a strategy that will accelerate benefits

Earlier, we talked about the network effect of bringing you all together, and these new technologies are going to enhance group performance in dramatic ways, too. We really must look at some of the ethical challenges that are right around the corner or even upon us today. Our strategy must establish priorities that foster scientific and technical collaboration, and ensure that our nation develops the necessary disciplines and workforce. We need a balanced but dynamic approach that protects intellectual property, provides for open markets, allows commercialization, and recognizes that American leadership is very much at stake.

Look all around the globe at the work that's going on at the nanoscale. American leadership is at stake, but we need a global framework for moving forward. The federal government, of course, has an important role: ensuring a business environment that enables these technologies to flourish, to work on that global aspect through the institutions of government, to continue to provide federal support for R&D. I am proud that President Bush recommended a record investment in R&D. I know there are concerns about the balance of the research portfolio. We need your help on that. President Bush specifically requested a record increase in the nano budget, over $604 million, almost double what it was two years ago.

The federal government has a clear fiscal role to play but also should use the bully pulpit to inspire young kids like one daughter of mine who does love science right now, so that they will go ahead and pursue careers like yours to reach the breakthroughs, so we will have more people like 39-year-old Eric Cornell at NIST, one of our recent winners of a Nobel Prize for Physics.

I think we can achieve our highest aspirations by working together as we are today — and we've got some of the best minds gathered around this table. But my message is distilled to this: If we set the right policies and we find the right balance, we can reap the rewards and avoid the really atrocious unethical possibilities. At every step — whether it's funding, advocacy, policy formation, public education, or commercialization — we're going to need you scientists and engineers to be intimately involved. I look forward to being a part of this promising effort. Thank you.

VISION FOR THE CONVERGING TECHNOLOGIES

Newt Gingrich

My theme is to argue that you want to be unreasonable in your planning.

I was struck with this at the session Mike Roco invited me to about six months ago, where somebody made a very impassioned plea against promising too much too quickly and not exaggerating. In 1945, Vannevar Bush wrote what was a quite unreasonable article for his day, about the future of computational power. Einstein's letter to Franklin Delano Roosevelt in September of 1939 was an extremely unreasonable letter. Edward Teller told me recently that he got in a big argument with Niels Bohr about whether or not it was possible to create an atomic bomb. Bohr asserted emphatically, it would take all of the electrical production capacity of an entire country. Teller said they didn't meet again until 1944 when they were at Los Alamos and Bohr yelled down the corridor "You see, I was right." By Danish standards, the Manhattan Project was using all the power of an entire country.

Vannevar Bush's classic article is a profound, general statement of what ultimately became the ARPANET, Internet, and today's personal computation system. At the time it was written, it was clearly not doable. And so, I want to start with the notion that at the visionary level, those who understand the potential have a real obligation to reach beyond any innate modesty or conservatism and to paint fairly boldly the plausible achievement.

Now, in this case you're talking about converging technology for improving human performance. Perhaps you should actually put up on a wall somewhere all of the achievable things in each zone, in the next 20 years, each of the stovepipes if you will. And then back up and see how you can move these against each other. What does the combination make true?

Think about the nanoscale in terms of a whole range of implications for doing all sorts of things, because if you can in fact get self-assembly and intelligent organization at that level, you really change all sorts of capabilities in ways that do in fact boggle the imagination, because they are that remarkable. If you bring that together with the biological revolution, the next 20 years of computation, and what we should be learning about human cognition, the capability can be quite stunning. For example, there's no reason to believe we can't ultimately design a new American way of learning and a new American way of thinking about things.

You see some of that in athletics, comparing all the various things we now do for athletes compared to 40 years ago. There is a remarkable difference, from nutrition to training to understanding of how to optimize the human body, that just wasn't physically possible 40 years ago. We didn't have the knowledge or the experience. I would encourage you first of all to put up the possibilities, multiply them against each other, and then describe what that would mean for humans, because it really is quite astounding.

I was an army brat in an era when we lived in France. In order to call back to the United States you went to a local post office to call the Paris operator to ask how many hours it would be before there would be an opening on the Atlantic cable. When my daughter was an *au pair*, I picked up my phone at home to call her cell phone in a place just south of Paris. Imagine a person who, having gotten cash out of an ATM, drives to a self-serve gas station, pays with a credit card, drives to work on the expressway listening to a CD while talking on a digital cell phone, and then says, "Well, what does science do for me?"

This brings me to my second point about being unreasonable. *When you lay out the potential positive improvements for the nation, for the individual, for the society, you then have to communicate that in relatively vivid language.*

People like Isaac Asimov, Arthur C. Clarke, and Carl Sagan did an amazing amount to convince humans that science and technology were important. Vannevar Bush understood it at the beginning of the Second World War. But if those who know refuse to explain in understandable language, then they should quit griping about the ignorance of those who don't know. Science can't have it both ways. You can't say, "This is the most important secular venture of mankind; it takes an enormous amount of energy to master it, and by the way, I won't tell you about it in a language you can understand." Scientists have an obligation as citizens to go out and explain what they need and what their work will mean.

I am 58 and I am already thinking about Alzheimer's disease and cancer. The fact that George Harrison has died and was my age makes mortality much more vivid. So, I have a vested interest in accelerating the rate of discovery and the application of that discovery. The largest single voting block is baby boomers, and they would all understand that argument. They may not understand plasma physics or the highest level of the human genome project. But they can surely understand the alternative between having Alzheimer's and not having it.

If you don't want Alzheimer's, you had better invest a lot more, not just in the National Institutes of Health (NIH) but also at the National Science Foundation (NSF) and a variety of other places, because the underlying core intellectual disciplines that make NIH possible all occur outside NIH. And most of the technology that NIH uses occurs outside of NIH. The argument has to be made by someone. If the scientific community refuses to make it, then you shouldn't be shocked that it's not made.

Let me suggest at a practical level what I think your assignments are once you've established a general vision. If you bring the four NBIC elements together into a converging pattern, you want to identify the missing gaps. What are the pieces that are missing? They may be enabling technologies, enabling networking, or joint projects.

Here again, I cite the great work done at the (Defense) Advanced Research Projects Agency ([D]ARPA). Scientists there consciously figured out the pieces that were missing to make computation easy to use and then began funding a series of centers of excellence that literally invented the modern world. You would not have gotten modern computing without ARPA, at least for another 30 years. Part of what they did that was so powerful was start with a general vision, figure out the pieces that were blocking the vision, and get them funded.

The predecessor to the Internet, ARPANET, wouldn't have occurred without two things: one was ARPA itself which had the funding, and the second was a vision that we should not be decapitated by a nuclear strike. People tend to forget that the capacity to surf on the Web in order to buy things is a direct function of our fear of nuclear war.

It helps to have the *vision* of very large breakthrough systems and some pretty long-term source of consistent funding. I've argued for the last three years that if we are going to talk about global warming, we ought to have several billion dollars set aside for the kind of climatology capabilities that will be comparable to the international geophysical year, and it would really give us the knowledge to move things a long way beyond our current relative guesses. If you look at the difference between the public policy implications of the Kyoto agreement, in the $40 trillion range, and the amount of money you could plausibly invest if you had an opportunity-based atmospheric and climatological research program, the differences are just stunning. For far less than one percent of the cost we would in theory pay to meet Kyoto, you would have a database and a knowledge base on climatology that would be stunning.

That's outside current budgeting, because current budgeting is an incremental-increase pork barrel; it is not an intellectual exercise. I would argue that's a profound mistake. So, it's very important for you to figure out what are the large projects as a consequence of which we would be in a different league of capabilities. I would suggest, too, that both the international geophysical year and its stunning impact on the basic understanding of geology may be the most decisive change in paradigms in 20th century, at least in terms that everybody agreed it was right. I would also suggest to you the example of ARPANET, which ultimately enabled people to invent the World Wide Web. For today's purpose, take the NBIC convergence and work back to identify the large-scale projects that must be underway in order to create parallel kinds of capabilities.

I want to make further points about being unreasonable. Scientists really have an obligation to communicate in vivid, simple language the possibilities so that the President, the Vice-President and the various people who make budget decisions are forced to reject that future if they settle for lower budgets. It's really important that people understand what's at stake. It is my experience that consistently, politicians underestimate the potential of the future.

If we in fact had the right level of investment in aeronautics, we would not currently be competing with Airbus. We would be in two different worlds. Considering all the opportunities to dramatically change things out of nanoscale technology combined with large-scale computing, there's no doubt in my mind if we were willing to make a capital investment, we would create a next-generation aviation industry that would be stunningly different. It would be, literally, beyond competition by anything else on the planet. Our military advantage in Afghanistan compared with the 1979 Soviet capabilities isn't courage, knowledge of military history, or dramatically better organizational skills, but a direct function of science and technology. We need to say that, again and again.

I'll close with two thoughts. First, my minimum goal is to triple the NSF budget and then have comparable scalable increases. One of the major mistakes I made as Speaker of the House is that I committed to doubling NIH without including other parts of science. In retrospect, it was an enormous mistake. We should have proportionally carried the other scientific systems, many of which are smaller, to a substantial increase. I'm probably going to do penance for the next decade by arguing that we catch up. Second, in the media there is some talk that the Administration may offer cuts in science spending in order to get through this current budget. Let me just say this publicly as often as I can. That would be madness.

If we want this economy to grow, we have to be the leading scientific country in the world. If we want to be physically safe for the next 30 years, we have to be the leading scientific country in the world. If we want to be healthy as we age, we have to be the leading scientific country in the world. It would be literally madness to offer anything except an increase in science funding. And if anybody here is in the Administration, feel free to carry that back. I will say this publicly anywhere I can, and I will debate anyone in the Administration on this.

Congress finds billions for pork and much less for knowledge. That has to be said over and over. It's not that we don't have the money. You watch the pork spending between now and the time Congress leaves. They'll find plenty of appropriations money, if there is enough political pressure. Scientists and engineers have to learn to be at least as aggressive as corn farmers. A society that can make a profound case for ethanol can finance virtually anything, and I think we have to learn that this is reality.

Now, a lot of scientists feel above strongly advocating government funding for their work. Fine, then you won't get funded. Or you'll get funded because somebody else was a citizen. However, I don't accept the notion that scientists are above civic status, and that scientists don't have a citizen's duty to tell the truth as they understand it and argue passionately for the things they believe in.

I have this level of passion because I believe what you're doing is so profoundly real. It's real in the sense that there are people alive today that would have died of illnesses over the last week if it weren't for the last half-century of science. There are capabilities today that could allow us to create a fuel cell system in Afghanistan, as opposed to figuring out how to build a large central electric distribution system for a mountainous country with small villages. With satellite technology, we could literally create a cell phone capability for most of the country instantaneously as opposed to going back to copper.

I just visited in Romania ten days ago and saw a project that goes online December 2002 to provide 156 K mobile capability, and the Romanians think they'll be at the third generation of cellular phones at a 1.2 million capability by January of 2003. In effect, I think Romania may be the first country in the world that has a 100% footprint for the 1.2 meg cellphone business.

We ought to talk, not about re-creating 1973 Afghanistan, but about how to create a new, better, modern Afghanistan where the children have access to all kinds of information, knowledge, and capabilities. My guess is it will not be a function of money. You watch the amount of money we and the world community throw away in the next six years in Afghanistan, and the relatively modest progress it buys. Take the same number of dollars, and put them into a real connectivity, a real access to the best medicine, a real access to logical organization, and you will have a dramatically healthier country in a way that would improve the life of virtually every Afghan.

Real progress requires making the connection between science and human needs. Vannevar Bush's great effort in the Second World War was to take knowledge and match it up with the military requirements in a way that gave us radical advantages; the submarine war is a particularly good example. The key was bringing science into the public arena at the state of possibility. Most of the technological advances that were delivered in 1944 did not exist in 1940. They were invented in real-time in places like MIT and brought to bear in some cases within a week or two of being invented.

I think we need that sense of urgency, and we need the sense of scale, because that's what Americans do well. We do very big things well, and we do things that are very urgent well. If they are not big enough and we bureaucratize them, we can often extend the length of time and money it takes by orders of magnitude. Thus, to be unreasonable in our planning can actually be quite realistic. We have entered a period I call The Age of Transitions, when science can achieve vast, positive improvements for the individual and the society, if we communicate the vision effectively.

The Age of Transitions: Converging Technologies

Overview

1. We are already experiencing the dramatic changes brought on by computers, communications, and the Internet. The combination of science and technology with entrepreneurs and venture capitalists has created a momentum of change which is extraordinary. Yet these changes will be overshadowed in the next 20 years by the emergence of an even bigger set of changes based on a combination of biology, information, and nanoscience (the science of objects at a billionth of a meter, from one to four hundred atoms in size). This new and as yet unappreciated wave of change will combine with the already remarkable pattern of change brought on by computers, communication, and the Internet to create a continuing series of new breakthroughs, resulting in new goods and services. We will be constantly in transition as each new idea is succeeded by an even better one. This will be an Age of Transitions, and it will last for at least a half-century.

2. In the Age of Transitions, the ways we acquire goods and services are rapidly evolving in the private sector and in our personal lives. Government and bureaucracy are changing at a dramatically slower rate, and the gaps between the potential goods and services, productivity, efficiencies, and conveniences being created and the traditional behaviors of government and bureaucracies are getting wider.

3. The language of politics and government is increasingly isolated from the language of everyday life. Political elites increasingly speak a language that is a separate dialect from the words people use to describe their daily lives and their daily concerns. The result in part is that the American people increasingly tune out politics.

4. Eventually a political movement will develop a program of change for government that will provide greater goods and services at lower and lower costs. When that movement can explain its new solutions in the language of everyday life, it will gain a decisive majority as people opt for better lives through better solutions by bringing government into conformity with the entrepreneurial systems they are experiencing in the private sector.

5. Understanding the Age of Transitions is a very complex process and requires thought and planning. It involves applying principles to create better solutions for delivery of government goods and services and developing and communicating a program in the language of everyday life, so that people

hear it and believe it despite the clutter and distractions of the traditional language of politics and government.

Introduction

We are living through two tremendous patterns of scientific-technological change: an overlapping of a computer-communications revolution and a nanotechnology-biology-information revolution. Each alone would be powerful; combined, the two patterns guarantee that we will be in constant transition as one breakthrough or innovation follows another.

Those who study, understand, and invest in these patterns will live dramatically better than those who ignore them. Nations that focus their systems of learning, healthcare, economic growth, and national security on these changes will have healthier, more knowledgeable people in more productive jobs creating greater wealth and prosperity and living in greater safety through more modern, more powerful intelligence and defense capabilities.

Those countries that ignore these patterns of change will fall further behind and find themselves weaker, poorer, and more vulnerable than their wiser, more change-oriented neighbors.

The United States will have to continue to invest in new science and to adapt its systems of health, learning, and national security to these patterns of change if we want to continue to lead the world in prosperity, quality of life, and military-intelligence capabilities.

At a minimum, we need to double the federal research budget at all levels, reform science and math learning decisively, and to modernize our systems of health, learning, and government administration.

Periods of transition are periods of dramatic cost crashes. We should be able to use the new patterns of change to produce greater health and greater learning at lower cost. Government administration can be more effective at lower cost. Our national security will experience similar crashes in cost.

This combination of better outcomes at lower cost will not be produced by liberal or conservative ideology. It will be produced by the systematic study of the new patterns and the use of new innovations and new technologies.

Simply Be a More Powerful Industrial Era

Computing is a key element in this revolution. The numbers are stunning. According to Professor James Meindl, the chairman of the Georgia Tech Microelectronics Department, the first computer built with a transistor was Tradic in 1955, and it had only 800 transistors. The Pentium II chip has 7,500,000 transistors. In the next year or so, an experimental chip will be built with one billion transistors. Within 15 to 20 years, there will be a chip with one trillion transistors. However that scale of change is graphed, it is enormous, and its implications are huge. It is estimated that we are only one-fifth of the way into developing the computer revolution.

Yet focusing only on computer power understates the scale of change. Communications capabilities are going to continue to expand dramatically, and that may have as big an impact as computing power. Today, most homes get Internet access at 28,000 to 56,000 bits per second. Within a few years, a combination of new technologies for compressing information (allowing you to get more done in a

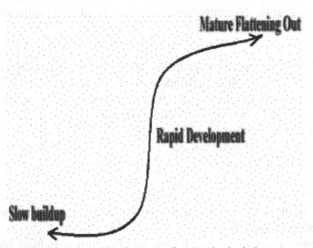

Figure A.1.   The S-curve of technological change.

given capacity) with bigger capacity (fiberoptic and cable) and entirely new approaches (such as satellite direct broadcast for the Internet) may move household access up to at least six million bits per second and some believe we may reach the 110 million bits needed for uncompressed motion pictures. Combined with the development of high definition television and virtual systems, an amazing range of opportunities will open up. This may be expanded even further by the continuing development of the cell phone into a universal utility with voice, Internet, credit card, and television applications all in one portable hand-held phone.

The S-curve of Technological Change

The communications-computer revolution and the earlier Industrial Revolution are both examples of the concept of an "S"-curve. The S-curve depicts the evolution of technological change. Science and technology begin to accelerate slowly, and then as knowledge and experience accumulates, they grow much more rapidly. Finally, once the field has matured, the rate of change levels off. The resulting pattern looks like an S. An overall S-curve is made up of thousands of smaller breakthroughs that create many small S-curves of technological growth.

The Two S-Curves of the Age of Transitions

We are starting to live through two patterns of change. The first is the enormous computer and communications revolution described above. The second, only now beginning to rise, is the combination of the nanotechnology-biology-information revolution. These two S curves will overlap. It is the overlapping period that we are just beginning to enter, and it is that period that I believe will be an Age of Transitions.

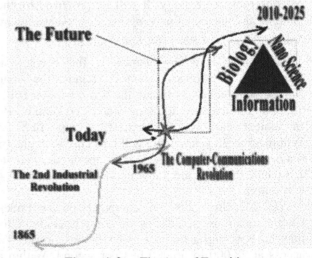

Figure A.2.   The Age of Transitions.

The Nano World, Biology, and Information as the Next Wave of Change

Focusing on computers and communications is only the first step toward understanding the Age of Transitions. While we are still in the early stages of the computer-communications pattern of change, we are already beginning to see a new, even more powerful pattern of change that will be built on a synergistic interaction among three different areas: the nano world, biology, and information.

The nano world may be the most powerful new area of understanding. "Nano" is the space measuring between one atom and about 400 atoms. It is the space in which quantum behavior begins to replace the Newtonian physics you and I are used to. The word "nano" means one-billionth and is usually used in reference to a nanosecond (one billionth of a second) or a nanometer (one billionth of a meter). In this world of atoms and molecules, new tools and new techniques are enabling scientists to create entirely new approaches to manufacturing and healthcare. Nanotechnology "grows" materials by adding the right atoms and molecules. Ubiquitous nanotechnology use is probably 20 years away, but it may be at least as powerful as space or computing in its implications for new tools and new capabilities.

The nano world also includes a series of materials technology breakthroughs that will continue to change the way we build things, how much they weigh, and how much stress and punishment they can take. For example, it may be possible to grow carbon storage tubes so small that hydrogen could be safely stored without refrigeration, thus enabling the creation of a hydrogen fuel cell technology, with dramatic implications for the economy and the environment. These new materials may make possible a one-hour flight from New York to Tokyo, an ultra-lightweight car, and a host of other possibilities. Imagine a carbon tube 100 times as strong as steel and only one-sixth as heavy. It has already been grown in the NASA Ames Laboratory. This approach to manufacturing will save energy, conserve our raw materials, eliminate waste products, and produce a dramatically healthier environment. The implications for the advancement of environmentalism and the irrelevancy of oil prices alone are impressive.

The nano world makes possible the ability to grow molecular "helpers" (not really "tools" because they may be organic and be grown rather than built). We may be able to develop anti-cancer molecules that penetrate cells without damaging them and hunt cancer at its earliest development. Imagine drinking with your normal orange juice 3,000,000 molecular rotor rooters to clean out your arteries without an operation.

The nano world opens up our understanding of biology, and biology teaches us about the nano world because virtually all biological activities take place at a molecular level. Thus, our growing capabilities in nano tools will dramatically expand our understanding of biology. Our growing knowledge about molecular biology will expand our understanding of the nano world.

Beyond the implications of biology for the nano world, in the next decade, the Human Genome Project will teach us more about humans than our total knowledge to this point. The development of new technologies (largely a function of physics and mathematics) will increase our understanding of the human brain in ways previously unimaginable. From Alzheimer's to Parkinson's to schizophrenia, there will be virtually no aspect of our understanding of the human brain and nervous system that cannot be transformed in the next two decades.

We are on the verge of creating intelligent synthetic environments that will revolutionize how medical institutions both educate and plan. It will be possible to practice a complicated, dangerous operation many times in a synthetic world with feel, smell, appearance, and sound, that are all precisely the same as the real operation. The flight and combat simulators of today are incredibly better than the sand tables and paper targets of forty years ago. An intelligent, synthetic environment will be an even bigger breakthrough from our current capabilities. It will be possible to design a building or an organization in the synthetic world before deciding whether to actually build it. The opportunities for education will be unending.

Finally, the information revolution (computers and communications) will give us vastly better capabilities to deal with the nano world and with biology.

It is the synergistic effect of these three systems (the nano world, biology, and information) that will lead to an explosion of new knowledge and new capabilities and create intersecting S-curves. We will simultaneously experience the computer/communications revolution and the nano/biology/information revolution. These two curves will create an Age of Transitions.

This rest of this paper attempts to outline the scale of change being brought about by the Age of Transitions, the principles that underlie those changes, and how to apply those principles in a strategic process that could lead to a governing majority.

Politics and Government in the Age of Transitions

In the foreseeable future, we will be inundated with new inventions, new discoveries, new startups, and new entrepreneurs. These will create new goods and services. The e-customer will become the e-patient and the e-voter. As expectations change, the process of politics and government will change. People's lives will be more complex and inevitably overwhelming. Keeping up with the changes that affect them and their loved ones exhausts most people. They focus most of their time and energy on the tasks of everyday life. In the future, when they achieve success in their daily tasks, people will turn to the new goods and services, the new job and investment opportunities, and the new ideas inherent in the entrepreneurial creativity of the Age of Transitions. No individual and no country will fully understand all of the changes as they occur or be able to adapt to them flawlessly during this time. On the other hand, there will be a large premium placed on individuals, companies, and countries that are able to learn and adjust more rapidly.

The political party or movement that can combine these three zones into one national dialogue will have an enormous advantage, both in offering better goods and services, and in attracting the support of most Americans.

The new products and services created by the Age of Transitions are creating vast opportunities for improving everyday life. The government has an opportunity to use these new principles to develop far more effective and appropriate government services. Politicians have the chance to explain these opportunities in a language most citizens can understand and to offer a better future, with greater quality of life, by absorbing the Age of Transitions into government and politics.

The average citizen needs to have political leadership that understands the scale of change we are undergoing, that has the ability to offer some effective guidance about how to reorganize daily life, and that simultaneously has the ability to reorganize the government that affects so much of our daily life. Inevitably, the Age of Transitions will overwhelm and exhaust people. Only after they have dealt with their own lives do they turn to the world of politics and government.

When we look at politics, we are discouraged and in some cases repulsed by the conflict-oriented political environment; the nitpicking, cynical nature of the commentaries; and the micromanaged, overly detailed style of political-insider coverage. The more Americans focus on the common sense and the cooperative effort required for their own lives, and the more they focus on the excitement and the wealth-creating and opportunity-creating nature of the entrepreneurial world, the more they reject politics and government as an area of useful interest.

Not only do politics and government seem more destructive and conflict oriented, but the language of politics seems increasingly archaic and the ideas increasingly trivial or irrelevant. People who live their lives with the speed, accuracy, and convenience of automatic teller machines (ATMs) giving them cash at any time in any city, cell phones that work easily virtually everywhere, the ease of shopping on the Web and staying in touch through email, will find the bureaucratic, interest-group-focused, and arcane nature of political dialogue and government policy to be painfully outmoded. Politicians' efforts to popularize the obsolete are seen as increasingly irrelevant and are therefore ignored.

This phenomenon helps explain the January 2000 poll in which 81 percent of Americans said that they had not read about the presidential campaign in the last 24 hours, 89 percent said that they had not thought about a presidential candidate in the same period, and 74 percent said that they did not have a candidate for president (up 10 percent from the previous November).

The average voter's sense of distance from politics is felt even more strongly by the entrepreneurial and scientific groups that are inventing the future. They find the difference between their intensely concentrated, creative, and positive focus of energy and the negative, bickering nature of politics especially alienating, so they focus on their own creativity and generally stay aloof from politics unless a specific interest is threatened or a specific issue arouses their interest.

Projects that focus on voter participation miss the nature of a deliberate avoidance by voters of politics. In some ways, this is a reversion to an American norm prior to the Great Depression and World War II. For most of American history, people focused their energies on their own lives and their immediate

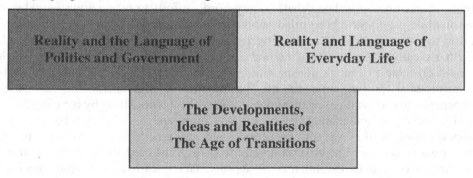

Figure A.3.   Zones of social reality in the Age of Transitions.

communities. The national government (and often even the state government) seemed distant and irrelevant. This was the world of very limited government desired by Jefferson and described by Tocqueville in *Democracy in America*. With the exception of the Civil War, this was the operating model from 1776 until 1930. Then the Depression led to the rise of Big Government, the World War II led to even bigger government, and the Cold War sustained a focus on Washington. When there was a real danger of nuclear war and the continuing crisis threatened the survival of freedom, it was natural for the president to be the central figure in America and for attention to focus on Washington. With the collapse of the Soviet Union, there has been a gradual shift of power and attention from Washington and back to the state and local communities. There has been a steady decline in popular attention paid to national politics.

When Republicans designed a positive campaign of big ideas in the 1994 Contract With America, some nine million additional voters turned out (the largest off-year, one-party increase in history). When Jesse Ventura offered a real alternative (at least in style) in 1998, younger voters turned out in record numbers. The voter as a customer tells the political-governmental system something profound by his or her indifference. The political leadership is simply failing to produce a large enough set of solutions in a lay language worth the time, attention, and focus of increasingly busy American citizens.

After a year of traveling around 23 states in America and spending time with entrepreneurs, scientists, and venture capitalists, I am increasingly convinced that the American voters are right.

Let us imagine a world of 1870 in which the private sector had completed the transcontinental railroad and the telegraph but the political-governmental elites had decided that they would operate by the rules of the Pony Express and the stagecoach. In private life and business life, you could telegraph from Washington to San Francisco in a minute and ship a cargo by rail in seven days. However, in political-governmental life, you had to send written messages by pony express that took two weeks and cargo by stagecoach that took two months. The growing gap between the two capabilities would have driven you to despair about politics and government as being destructive, anachronistic systems.

Similarly, imagine that in 1900 a Washington Conference on Transportation Improvement had been created, but the political-governmental elite had ruled that the only topic would be the future of the horseshoe and busied themselves with a brass versus iron horseshoe debate. Henry Ford's efforts to create a mass-produced automobile would have been ruled impractical and irrelevant. The Wright brothers' effort to create an airplane would have been derided as an absurd fantasy. After all, neither clearly stood on either the brass or the iron side of the debate. Yet which would do more to change transportation over the next two decades: the political-governmental power structure of Washington or the unknown visionaries experimenting without government grants and without recognition by the elites?

Consider just one example of the extraordinary and growing gap between the opportunities of the Age of Transitions and the reactionary nature of current government systems. The next time you use your ATM card, consider that you are sending a code over the Internet to approve taking cash out of your checking account. It can be done on a 24 hours a day, seven days a week anywhere in the country. Compare that speed, efficiency, security, and accuracy with the paper-dominated, fraud- and waste-ridden Healthcare Financing Administration (HCFA) with its 133,000 pages of regulations (more pages than the tax code). As a symbol of a hopelessly archaic model of bureaucracy there are few better examples than HCFA.

This growing gap between the realities and language of private life and the Age of Transitions, on the one hand, and the increasingly obsolete language and timid proposals of the political governmental system, on the other hand, convinces more and more voters to ignore politics and focus on their own lives and on surviving the transitions.

This is precisely the pattern described by Norman Nie and colleagues in *The Changing American Voter* (1979). They described a pool of latent voters who in the 1920s found nothing in the political dialogue to interest them. These citizens simply stayed out of the process as long as it stayed out of their lives. The Depression did not mobilize them. They sat out the 1932 election. Only when the New Deal policies of Franklin Delano Roosevelt penetrated their lives did they become involved. In 1936, Alf Landon, the Republican nominee, actually received a million more votes than Herbert Hoover had gotten in 1932. However, FDR received seven million more votes than he had gotten in his first election. It was this massive increase in participation that made the polls inaccurate and created the Democratic majority, which in many ways survived until the 1994 election. The Republican victory of 1994 drew nine million additional voters over its 1990 results by using bold promises in a positive campaign to engage people who had been turned off by politics.

A similar opportunity awaits the first political party and political leader to make sense of the possibilities being created by the Age of Transitions and to develop both a language and a set of bold proposals that make sense to average Americans in the context of their own lives and experiences.

This paper should be seen as the beginning of a process rather than as a set of answers. Political-governmental leaders need to integrate the changes of the Age of Transitions with the opportunities these changes create to improve people's lives, develop the changes in government necessary to accelerate those improvements, and explain the Age of Transitions era — and the policies it requires — in the language of everyday life, so that people will understand why it is worth their while to be involved in politics and subsequently improve their own lives. Getting this done will take a lot of people experimenting and attempting to meet the challenge for a number of years. That is how the Jeffersonians, the Jacksonians, the early Republicans, the Progressives, the New Dealers, and the Reagan conservatives succeeded. Each, over time, created a new understanding of America at an historic moment. We aren't any smarter, and we won't get it done any faster; however, the time to start is now, and the way to start is to clearly understand the scale of the opportunity and the principles that make it work.

Characteristics of an Age of Transitions

Thirty-six years after Boulding's first explanation of the coming change, and thirty-one years after Drucker explained how to think about discontinuity, some key characteristics have emerged. This section outlines 18 characteristics and gives examples of ways political and governmental leaders can help develop the appropriate policies for the Age of Transitions. It should first be noted that there is an overarching general rule: assume there are more changes coming.

It is clear that more scientists, engineers, and entrepreneurs are active today than in all of previous human history. Venture capitalists are developing powerful models for investing in and growing startup companies. In the process, they are acquiring more and more capital as the markets shift away from the smokestack industries and toward new models. It is also clear that there is a growing world market in which more entrepreneurs of more nationalities are competing for more customers than ever in human history.

All this growing momentum of change simply means that no understanding, no reform, no principle will be guaranteed to last for very long. Just as we get good at one thing or come to understand one principle, it will be challenged by an emerging new idea or achievement from a direction we haven't even considered.

Within that humbling sense that the change is so large that we will never really know in our lifetime the full analysis of this process, here are 18 powerful characteristics for developing government policy and politics in the Age of Transitions:

1. **Costs will crash.** A major pattern will be a continuing, and in many cases steep, decline in cost. An ATM is dramatically cheaper than a bank teller. A direct-dial phone call is much less expensive than an operator-assisted call. My brother used Priceline.com and received four airline tickets for his family for the price of one regular ticket. We have not even begun to realize how much costs will decline, even in the fields of health and healthcare, education and learning, defense procurement, and government administration. We also have not yet learned to think in terms of purchasing power instead of salary. Yet the pattern is likely to be a huge change in both purchasing power and behavior for both citizens and government. Those who are aggressive and alert will find remarkable savings by moving to the optimum cost crashes faster than anyone else. As a result, they will dramatically expand their purchasing power.

2. **Systems will be customer-centered and personalized.** Customers of Amazon.com and other systems already can look up precisely the books or movies that interest them, and after a while, the company is able to project their interests and alert them to similar products in their inventory. Consumers can consider personal Social Security Plus accounts who already have personal Roth IRAs and 401Ks. Individuals can consider purchasing personal learning and personal health systems just as they purchase electronic airline tickets on the Internet. Anything that is not personalized and responsive to changing individual needs will rapidly be replaced by something that is.

3. **24-7 will be the world of the future.** Customer access 24 hours a day, 7 days a week, will become the standard of the future. ATMs symbolize this emerging customer convenience standard, providing cash to card-holders any day, round the clock. Yet today's schools combine an agricultural-era nine- or ten-month school year (including the summer off for harvesting) with an industrial era 50-minute class, with a "foreman" at the front of the room facing a class of "workers" in a factory-style school day, in a Monday-

to-Friday work week. Learning in the future will be embedded in the computer and on the Internet and will be available on demand with a great deal of customization for each learner. Similarly, government offices will have to shift to meeting their constituents' needs at their convenience rather than demanding that the constituents make themselves available at the bureaucrat's convenience. These are big changes, and they are unavoidable given the emerging technologies and the e-customer culture that is evolving.

4. **Convenience will be a high value.** As customers get used to one-click shopping (note the shopping cart approach on Amazon), they will demand similar convenience from government. People will increasingly order products and services to be delivered to their homes at their convenience. They will initially pay a premium for this convenience, but over time they will conclude that it is a basic requirement of any business that they deal with, and costs will go down. After a while, e-customers will begin to carry these attitudes into their relationship with bureaucracy, and as e-voters they will favor politicians who work to make their lives easier (i.e., more convenient).

5. **Convergence of technologies will increase convenience, expand capabilities, and lower costs.** The various computation and communication technologies will rapidly converge with cell phones, computers, land-lines, mobile systems, satellite capabilities, and cable, all converging into a unified system of capabilities that will dramatically expand both capabilities and convenience.

6. **Processes will be expert system-empowered.** When you look up an airline reservation on the Internet, you are dealing with an expert system. In virtually all Internet shopping you are actually interacting with such a system. The great increase in capability for dealing with individual sales and individual tastes is a function of the growing capacity of expert systems. These capabilities will revolutionize health, learning, and government once they are used as frequently as they currently are in the commercial world. If it can be codified and standardized, it should be done by an expert system rather than a person: that is a simple rule to apply to every government activity.

7. **Middlemen will disappear.** This is one of the most powerful rules of the Age of Transitions. In the commercial world, where competition and profit margins force change, it is clear that customers are served more and more from very flat hierarchies, with very few people in the middle. In the protected guilds (medicine, teaching, law, and any group that can use its political power to slow change) and in government structures, there are still very large numbers of middlemen. This will be one of the most profitable areas for political-governmental leaders to explore. In the Age of Transitions, the customer should be foremost, and every unnecessary layer should be eliminated to create a more agile, more rapidly changing, more customer-centered, and less expensive system.

8. **Changes can come from anywhere.** The record of the last thirty years has been of a growing shift toward new ideas coming from new places. Anyone can have a good idea, and the key is to focus on the power of the idea rather

than the pedigree of the inventor. This directly challenges some of the peer review assumptions of the scientific community, much of the screening for consultants used by government, much of the credentialing done by education and medicine, and much of the contractor certification done by government. This principle requires us to look very widely for the newest idea, the newest product, and the newest service, and it requires testing by trial and error more than by credentialing or traditional assumptions.

9. **Resources will shift from opportunity to opportunity.** One of the most powerful engines driving the American economy has been the rise of an entrepreneurial venture capitalism that moves investments to new opportunities and grows those opportunities better than any other economy in the world. There is as yet no comparable government capacity to shift resources to new start-ups and to empower governmental entrepreneurs. There are countless efforts to reform and modernize bureaucracies, but that is exactly the wrong strategy. Venture capitalists very seldom put new money into old corporate bureaucracies. Even many established corporations are learning to create their own startups because they have to house new ideas and new people in new structures if they are really to get the big breakthroughs. We need a doctrine for a venture capitalist-entrepreneurial model of government that includes learning, health, and defense.

10. **The rapid introduction of better, less expensive products will lead to continual replacement.** Goods and services will take on a temporary nature as their replacements literally push them out of the door. The process of new, more capable, and less expensive goods and services, and in some cases, revolutionary replacements that change everything (as Xerox did to the mimeograph, and as the fax machine, e-mail, and PC have done) will lead to a sense of conditional existence and temporary leasing that will change our sense of ownership.

11. **The focus will be on success.** Entrepreneurs and venture capitalists have a surprisingly high tolerance for intelligent failure. They praise those who take risks, even if they fail, over those who avoid risks, even if they avoid failure. To innovate and change at the rate the Age of Transitions requires, government and politicians have to shift their attitudes dramatically. (It would help if the political news media joined them in this.) Today it is far more dangerous for a bureaucrat to take a risk than it is to do nothing. Today the system rewards people (with retirement and noncontroversy) for serving their time in government. There are virtually no rewards for taking the risks and sometimes failing, sometimes succeeding. Yet in all the areas of science, technology, and entrepreneurship, the great breakthroughs often involve a series of failures. (Consider Edison's thousands of failed experiments in inventing the electric light and how they would have appeared in a congressional hearing or a news media expose.) Setting a tone that supports trying and rewards success while tolerating intelligent failure would do a great deal to set the stage for a modernized government.

12. **Venture capitalists and entrepreneurs will focus on opportunities.** This is similar to focusing on success but refers to the zone in which energy and resources are invested. It is the nature of politics and government to focus on

problems (schools that fail, hospitals that are too expensive, people who live in poverty) when the real breakthroughs come from focusing on opportunities (new models of learning that work, new approaches to health and healthcare that lower the cost of hospitals, ways to get people to work so that they are no longer in poverty). Venture capitalists are very good at shifting their attention away from problem zones toward opportunity zones. Politicians and the political news media tend to do the opposite. Yet the great opportunities for change and progress are in the opportunities rather than the problems.

13. **Real breakthroughs will create new products and new expectations.** Before Disney World existed, it would have been hard to imagine how many millions would travel to Orlando. Before the Super Bowl became a cultural event, it was hard to imagine how much of the country would stop for an entire evening. Before faxes, we did not need them, and before e-mail, no one knew how helpful it would be. One of the key differences between the public and private sector is this speed of accepting new products and creating new expectations. The public sector tends to insist on using the new to prop up the old. For two generations we have tried to get the computer into the classroom with minimal results. That's because it is backward. The key is to get the classroom into the computer and the computer to the child's home, so that learning becomes personal and 24/7. Doctors still resist the information technologies that will revolutionize health and healthcare and that will lower administrative costs and decrease unnecessary deaths and illnesses dramatically. In the private sector, competition and the customer force change. In government and government-protected guilds, the innovations are distorted to prop up the old, and the public (that is the customer) suffers from more expensive and less effective goods and services.

14. **Speed matters: new things will need to get done quickly.** There is a phrase in the Internet industry, "launch and learn," which captures the entrepreneurial sense of getting things done quickly and learning while doing so. One Silicon Valley entrepreneur noted he had moved back from the East because he could get things done in the same number of days in California as the number of months it would have taken where he had been. Moving quickly produces more mistakes, but it also produces a real learning that only occurs by trying things out. The sheer volume of activity and the speed of correcting mistakes as fast as they are discovered allows a "launch and learn" system to grow dramatically faster than a "study and launch" system. This explains one of the major differences between the venture capitalist-entrepreneurial world and the world of traditional corporate bureaucracies. Since governments tend to study and study without ever launching anything truly new, it is clear how the gap widens between the public and private sectors in an Age of Transitions. Today it takes longer for a presidential appointee to be cleared by the White House and approved by the Senate than it takes to launch a startup company in Silicon Valley.

15. **Start small but dream big.** Venture capital and entrepreneurship are about baby businesses rather than small businesses. Venture capitalists know that in a period of dramatic change, it is the occasional home run rather than a

large number of singles that really make the difference. The result is that venture capitalists examine every investment with a focus on its upside. If it does not have a big enough growth potential, it is not worth the time and energy to make the investment. Government tends to make large, risk-averse investments in relatively small, controllable changes. This is almost the exact opposite of the venture capital-entrepreneurial model. The question to ask is, "If this succeeds, how big will the difference be, and if the difference isn't very substantial, we need to keep looking for a more powerful proposal."

16. **Business-to-business is the first big profit opportunity.** While most of the attention in the Internet market is paid to sales to the final customer, the fact is that that market is still relatively small and relatively unprofitable. However, there is no question that Internet-based systems such as Siebel and Intelisys are creating business-to-business opportunities that will dramatically lower the cost of doing business. Every government, at every level, should be rationalizing its purchasing system and moving on to the net to eliminate all paper purchasing. The savings in this area alone could be in the 20 percent to 30 percent range for most governments. The opportunities for a paperless system in health and healthcare could lead to a crash in costs rather than a worry about rising costs.

17. **Applying quality and lean thinking can save enormous amounts.** Whether it is the earlier model of quality espoused by Edwards Deming, or the more recent concept of lean thinking advocated by James Womack and Daniel Jones, it is clear that there is an existing model for systematically thinking through production and value to create more profitable, less expensive approaches. The companies that have really followed this model have had remarkable success in producing better products at lower expense, yet it is almost never used by people who want to rethink government.

18. **Partnering will be essential.** No company or government can possibly understand all the changes in an Age of Transitions. Furthermore, new ideas will emerge with great speed. It is more profitable to partner than to try to build in-house expertise. It allows everyone to focus on what they do best while working as a team on a common goal. This system is prohibited throughout most of government, and yet it is the dominant organizing system of the current era of startups. As government bureaucracies fall further and further behind the most dynamic of the startups (in part because civil service salaries cannot compete with stock options for the best talent), it will become more and more important to develop new mechanisms for government-private partnering.

These initial principles give a flavor of how big the change will be and of the kind of questions a political-governmental leader should ask in designing a program for the Age of Transitions. These principles and questions can be refined, expanded, and improved, but they at least let leaders start the process of identifying how different the emerging system will be from the bureaucratic-industrial system that is at the heart of contemporary government.

The Principles of Political-Governmental Success in an Age of Transitions

In the Age of Transitions, the sheer volume of new products, new information, and new opportunities will keep people so limited in spare time that any real breakthrough in government and politics will have to meet several key criteria:

1. **Personal.** It has to involve a change that occurs in individual people's lives in order for them to think it is worth their while, because it will affect them directly. Only a major crisis such as a steep recession or a major war will bring people back to the language of politics. In the absence of such a national crisis, political leaders will not be able to attract people into the zone of government and politics unless they use the new technologies and new opportunities of the Age of Transitions to offer solutions that will dramatically improve people's lives.

2. **Big Ideas.** The changes offered have to be large enough to be worth the time and effort of participation. People have to convinced that their lives or their families' lives will be affected significantly by the proposals, or they will simply nod pleasantly at the little ideas but will do nothing to get them implemented.

3. **Common Language.** New solutions have to be explained in the language of everyday life because people will simply refuse to listen to the traditional language of political and governmental elites. People have become so tired of the bickering, the conflict, and the reactionary obsolete patterns of traditional politics that they turn off the minute they hear them. New solutions require new words, and the words have to grow out of the daily lives of people rather than out of the glossary of intellectual elites or the slogans of political consultants.

4. **Practical.** The successful politics of the Age of Transitions will almost certainly be pragmatic and practical rather than ideological and theoretical. People are going to be so busy and so harried that their first question is going to be "will it work?" They will favor conservative ideas that they think will work, and they will favor big government ideas that they think will work. Their first test will be, "Will my family and I be better off?" Their second test will be, "Can they really deliver and make this work?" Only when a solution passes these two tests will it be supported by a majority of people. Note that both questions are pragmatic; neither is theoretical or ideological.

5. **Positive.** The successful politicians of the Age of Transitions will devote 80 percent of their time to the development and communication of large positive solutions in the language of everyday life and the gathering of grassroots coalitions and activists to support their ideas. They will never spend more than 20 percent of their effort on describing the negative characteristics of their opponents. When they do describe the destructive side of their opponents, it will be almost entirely in terms of the costs to Americans of the reactionary forces blocking the new solutions and the better programs (study FDR's 1936 and 1940 campaigns for models of this lifestyle definition of the two sides: the helpful and the harmful. FDR was tough on offense, but more importantly, he cast the opposition in terms of how they hurt the lives of ordinary people.)

6. **Electronic.** The successful large, personal, positive, practical movement of the Age of Transitions will be organized on the Internet and will be interactive. Citizens will have a stake in the movement and an ability to offer ideas and participate creatively in ways no one has ever managed before. The participatory explosion of the 1992 Perot campaign, in which tens of thousands of volunteers organized themselves, and the Internet-based activism of the closing weeks of the 1998 Ventura campaign are forerunners of an interactive, Internet-based movement in the Age of Transitions. None has yet occurred on a sustainable basis, for two reasons:

   a) First, no one has come up with a believable solution big enough to justify the outpouring of energy beyond brief, personality-focused campaign spasms lasting weeks or a few months.

   b) Second, no one has mastered the challenge of building a citizen-focused genuinely interactive system that allows people to get information when they want it, offer ideas in an effective feedback loop, and organize themselves to be effective in a reasonably efficient and convenient manner. When the size of the solution and the sophistication of the system come together, we will have a new model of politics and government that will be as defining as the thirty-second commercial and the phone bank have been.

The Political Challenge for the Coming Decade in America

For change to be successful, it is essential that we sincerely and aggressively communicate in ways that are inclusive, not exclusive. Our political system cannot sustain effectiveness without being inclusive. There are two principle reasons this strategy must be pursued:

1. A majority in the Age of Transitions will be inclusive. The American people have reached a decisive conclusion that they want a unified nation with no discrimination, no bias, and no exclusions based on race, religion, sex, or disability. A party or movement that is seen as exclusionary will be a permanent minority. The majority political party in the Age of Transitions will have solutions that improve the lives of the vast majority of Americans and will make special efforts to recruit activists from minority groups, to communicate in minority media, and to work with existing institutions in minority communities. For Republicans, this will mean a major effort to attract and work with every American of every background. Only a visibly, aggressively inclusive Republican Party will be capable of being a majority in the Age of Transitions.

2. The ultimate arbiter of majority status in the next generation will be the Hispanic community. The numbers are simple and indisputable. If Hispanics become Republican, the Republican Party is the majority party for the foreseeable future; if Hispanics become Democrat, the Republican Party is the minority party for at least a generation. On issues and values, Hispanics are very open to the Republican Party. On historic affinity and networking among professional politicians and activist groups, Democrats have an edge among Hispanics. There should be no higher priority for American politicians than reaching out to and incorporating Hispanics at every level in

every state. George W. Bush, when he was governor of Texas, and Governor Jeb Bush have proven that Republicans can be effectively inclusive and create a working partnership with Hispanics. Every elected official and every candidate should follow their example.

Conclusion

These are examples of the kind of large changes that are going to be made available and even practical by the Age of Transitions. The movement or political party that first understands the potential of the Age of Transitions, develops an understanding of the operating principles of that Age, applies them to creating better solutions, and then communicates those solutions in the language of everyday life will have a great advantage in seeking to become a stable, governing majority.

This paper outlines the beginning of a process as big as the Progressive Era or the rise of Jacksonian Democracy, the Republicans, the New Deal, or the conservative movement of Goldwater and Reagan. This paper outlines the beginning of a journey, not its conclusion. It will take a lot of people learning, experimenting, and exploring over the next decade to truly create the inevitable breakthrough.

References

Boulding, K.E. 1964. *The meaning of the twentieth century: The great transition.* New York: Harper and Row.

Deming, W.E. 1982. *Quality, productivity, and competitive position.* Cambridge, Massachusetts: MIT Center for Advanced Engineering Study.

Drucker, P.F. 1969. *The age of discontinuity: Guideline to our changing society.* New York: Harper and Row.

Kohn, L.T., J.M. Corrigan, and M.S. Donaldson (Committee on Healthcare in America, Institute of Medicine). 1999. *To err is human: Building a safer health system.* Washington, D.C.: National Academy Press.

Nie, N., S. Verba, and J.R. Petrovik. 1979. *The changing American voter.* Cambridge, MA: Harvard University Press.

Tocqueville, A. de. 1848. *Democracy in America.* New York: Pratt, Woodford.

Womack, J.P., and D. Jones. 1996. *Lean thinking.* New York: Simon and Schuster.

ZONE OF CONVERGENCE BETWEEN BIO/INFO/NANO TECHNOLOGIES: NASA'S NANOTECHNOLOGY INITIATIVE

S. Venneri, M. Hirschbein, M. Dastoor, National Aeronautics and Space Administration

NASA's mission encompasses space and Earth science, fundamental biological and physical research (BPR), human exploration and development of space (HEDS), and a responsibility for providing advanced technologies for aeronautics and space systems. In space science, agency missions are providing deeper insight into the evolution of the solar system and its relationship to Earth; structure and evolution of the universe at large; and both the origins and extent of life throughout the cosmos. In Earth science, a fundamental focus is to provide, through observations and models, the role of the physical, chemical, and biological processes in long-term climate change as well as push the prediction capability of short-term weather. In addition, NASA's challenge is to understand the biosphere and its evolution and future health in the face of change wrought by humankind.

The goal of NASA for BPR is to conduct research to enable safe and productive human habitation of space as well as to use the space environment as a laboratory to test the fundamental principals of biology, physics, and chemistry. For HEDS, a long-term presence in low Earth orbit is being accomplished with the space station. In the longer term, humans will venture beyond low earth orbit, probably first to explore Mars, following a path blazed by robotic systems.

A critical element of science missions and HEDS is safe and affordable access to space and dramatically reduced transit times for in-space transportation systems. In pursuance of this mission, NASA needs tools and technologies that must push the present state of the art. NASA spacecraft must function safely and reliably, on their own, far from Earth, in the extremely harsh space environment in terms of radiation and temperature variance coupled with the absence of gravity. This places demands on NASA technologies that are highly unique to the Agency. NASA's aeronautics goals are focused on developing technology to support new generations of aircraft that are safer, quieter, more fuel efficient, environmentally cleaner, and more economical than today's aircraft; as well as on technology to enable new approaches to air systems management that can greatly expand the capacity of our air space and make it even safer than it is today.

Virtually all of NASA's vision for the future of space exploration — and new generations of aircraft — is dependent upon mass, power requirements, and the size and intelligence of components that make up air and space vehicles, spacecraft, and rovers. Dramatic increases in the strength-to-weight ratio of structural materials offers the potential to reduce launch and flight costs to acceptable levels. Such structural materials can also lead to increases in payload and range for aircraft, which can translate into U.S. dominance of the world marketplace. Packing densities and power consumption are absolutely critical to realizing the sophisticated on-board computing capability required for such stressing applications as autonomous exploration of Europa for evidence of simple life forms or their precursors. The integration of sensing, computing, and wireless transmission will enable true health management of reusable launch vehicles and aircraft of the future.

To do this, NASA aircraft and space systems will have to be much more capable than they are today. They will have to have the characteristics of autonomy to "think for themselves": they will need self-reliance to identify, diagnose, and correct internal problems and failures; self-repair to overcome damage; adaptability to function and explore in new and unknown environments; and extreme efficiency to operate with very limited resources. These are typically characteristics of robust biological systems, and they will also be the characteristics of future aerospace systems. Acquisition of such intelligence, adaptability, and computing power go beyond the present capabilities of microelectronic devices.

The current state-of-the-art microelectronics is rapidly approaching its limit in terms of feature size (0.1 microns). Future enhancements will need novel alternatives to microelectronics fabrication and design as we know them today. Nanotechnology will afford a new class of electronics. In addition to possessing the benefits inherent in smaller feature size, nanotechnology will harness the full power of quantum effects that are operable only at nanoscale distances. Hence, not only should we expect a performance enhancement at the quantitative level, due to the higher packing density of nanoscale components, but also the emergence of qualitatively new functionalities associated with harnessing the full power of quantum effects. The hybridization of nanolithography and bioassembly could serve as the basis of an engineering revolution in the fabrication of complex systems.

We are already seeing the potential of nanotechnology through the extensive research into the production and use of carbon nanotubes, nano-phase materials, and molecular electronics. For example, on the basis of computer simulations and available experimental data, some specific forms of carbon nanotubes appear to possess extraordinary properties: Young's modulus over one Tera Pascal (five times that of steel) and tensile strength approaching 100 Giga Pascal (over 100 times the strength of steel). Recent NASA studies indicate that polymer composite materials made from carbon nanotubes could reduce the weight of launch vehicle — as well as aircraft — by half. Similarly, nanometer-scale carbon wires have 10,000 times better current carrying capacity than copper, which makes them particularly useful for performing functions in molecular electronic circuitry that are now performed by semiconductor devices in electronic circuits. Electronic devices constructed from molecules (nanometer-scale wires) will be hundreds of times smaller than their semiconductor-based counterparts.

However, the full potential of nanotechnology for the systems NASA needs is in its association with biology. Nanotechnology will enable us to take the notion of "small but powerful" to its extreme limits, but biology will provide many of the paradigms and processes for doing so. Biology has inherent characteristics that enable us to build the systems we need: selectivity and sensitivity at a scale of a few atoms; ability of single units to massively reproduce with near-zero error rates; capability of self-assembly into highly complex systems; ability to adapt form and function to changing conditions; ability to detect damage and self repair; and ability to communicate among themselves. Biologically inspired sensors will be sensitive to a single photon. Data storage based on DNA will be a trillion times more dense than current media, and supercomputers modeled after the brain will use as little as a billionth of the power of existing designs. Biological concepts and nanotechnology will enable us to create both the "brains and the body" of future systems with the characteristics that we require. Together, nanotechnology, biology, and information technology form a powerful and intimate scientific and technological triad.

Such technologies will enable us to send humans into space for extended durations with greater degrees of safety. While the vehicle they travel in will have much greater capability and display the same self-protective characteristics of spacecraft, nanotechnology will enable new types of human health monitoring systems and healthcare delivery systems. Nanoscale, bio-compatible sensors can be distributed throughout the body to provide detailed information of the health of astronauts at the cellular level. The sensors will have the ability to be queried by external monitoring systems or be self-stimulated to send a signal, most likely through a chemical messenger. NASA is currently working with the National Cancer Institute (NCI) to conduct research along these specific lines.

Currently, NASA's program is split primarily between the Office of Aerospace Technology (OAT) with a focus on nanotechnology and the newly formed Office of Biological and Physical Research (OBPR) with a focus on basic research in nanoscience related to biomedical applications. Furthermore, the OAT Program integrates nanotechnology development in three areas:

1. materials and structures
2. nanoelectronics and computing
3. sensors and spacecraft components

A summary of the content of these programs follows.

Materials and Structures

A major emphasis for NASA over the next five years will be the production scale-up of carbon nanotubes; the development of carbon nanotube-reinforced polymer matrix composites for structural applications; and the development of analysis, design, and test methods to incorporate these materials into new vehicle concepts and validate their performance and life. NASA also will explore the use of other materials, such as boron nitride, for high-temperature applications and will research the use of crystalline nanotubes to ultimately exploit the full potential of these materials. In the long term, the ability to create biologically inspired materials and structures provides a unique opportunity to produce new classes of self-assembling material systems without the need to machine or process materials. Some unique characteristics anticipated from biomimetics (that is, "mimicking" biology) include multifunctional material systems, hierarchical organization, adaptability, self healing/self-repair, and durability. Thus, by exploiting the characteristics of biological systems, mechanical properties of new materials can be tailored to meet complex, rigorous design requirements and revolutionize aerospace and spacecraft systems.

Nanoelectronics and Computing

Biologically inspired neural nets have been developed in laboratory demonstrations that allow computers to rapidly account for loss of aircraft control elements, understand the resulting aerodynamics, and then teach the pilot or autopilot how to avoid the loss of the vehicle and crew by an innovative use of the remaining aerodynamic control. Such approaches, coupled with the advances in computing power anticipated from nanoelectronics, will revolutionize the way aerospacecraft deal with condition-based maintenance, aborts, and recovery from serious in-flight anomalies. While aircraft do not require electronic devices that can tolerate the space radiation environment, spacecraft exploration for the Space Science and HEDS Enterprises, e.g., vehicles exploring Mars, the outer planets, and their moons, will require such capabilities. NASA mission planners view such capability as enabling them to conduct *in-situ* science (without real-time Earth operators), where huge amounts of data must be processed, converted to useful information, and then sent as knowledge to Earth without the need for large bandwidth communication systems. A longer-term vision incorporates the added complexity of morphing devices, circuits, and systems whose characteristics and functionalities may be modified in flight. NASA will support work at the underlying device level, in which new device configurations with new functionalities may be created through intra-device switching.

Sensors and Spacecraft Components

NASA's challenge to detect ultra-weak signals from sources at astronomical distances make every photon or particle a precious commodity that must be fully analyzed to retrieve all of the information it carries. Nanostructured sensing elements, in which each absorbed quantum generates low-energy excitations that record and amplify the full range of information, provide an approach to achieve this goal. NASA will also develop field and inertial sensors with many orders of magnitude enhancement in the sensitivity by harnessing quantum effects of photons, electrons, and atoms. A gravity gradiometer based on interference of atom beams is currently under development by NASA with the potential space-based mapping of the interior of the Earth or other astronomical bodies. Miniaturization of entire spacecraft will entail reduction in the size and power required for all system functionalities, not just sensors. Low-power, integrable nano devices are needed for inertial sensing, power generation and management, telemetry and communication, navigation and control, propulsion, and *in situ* mobility, and so forth. Integrated nano-electro-mechanical systems (NEMS) will be the basis for future avionics control systems incorporating transducers, electromagnetic sources, active and passive electronic devices, electromagnetic radiators, electron emitters, and actuators.

Basic Nanoscience

Foremost among the technological challenges of long-duration space flight are the dangers to human health and physiology presented by the space environment. Acute clinical care is essential to the survival of astronauts, who must face potentially life-threatening injuries and illnesses in the isolation of space. Currently, we can provide clinical care and life support for a limited time, but our only existing option in the treatment of serious illness or injury is expeditious stabilization and evacuation to Earth. Effective tertiary clinical care in space will require advanced, accurate diagnostics coupled with autonomous intervention and, when necessary, invasive surgery. This must be accomplished within a complex man-machine interface, in a weightless environment of highly limited available space and resources, and in the context of physiology altered by microgravity and chronic radiation exposure. Biomolecular approaches promise to enable lightweight, convenient, highly focused therapies guided with the assistance of artificial intelligence enhanced by biomolecular computing. Nanoscopic, minimally invasive technology for the early diagnosis and monitoring of disease and targeted inter-vention will save lives in space and on Earth. Prompt implementation of specifically targeted treatment will insure optimum use and conservation of therapeutic resources, making the necessity for invasive interventions less likely and minimizing possible therapeutic complications.

BIOMEDICINE EYES 2020

John Watson, National Institutes of Health

I will present ideas from my experience with targeted, goal-oriented research programs and traditional investigator-initiated research projects. I strongly endorse both approaches. For NBIC to reach its potential, national science and engineering priorities should be set to complement investigator-initiated research projects. We should consider including in our NBIC thinking "human performance and health" (not just performance alone) to provide the most for our future quality of life.

How many of us know someone who has undergone angioplasty? A vision for ten and twenty years is under consideration: tomorrow's needs, tomorrow's patients, and tomorrow's diverse society. Well, what about today's needs, today's patients, and today's diverse society? It is riskier to talk about targeting a research goal to solve today's problems than to focus on promising basic research for solving as yet undefined problems.

We do not know what causes atherosclerosis. Surgically bypassing atherosclerotic plaques was shown to have clinical benefit. Using a small balloon to push the plaques into a coronary artery wall, thus opening the lumen, was met with lots of skepticism. If we had waited until we knew all the atherosclerosis basic science, millions of patients would not have benefited from angioplasty.

Picking up on Newt Gingrich's comments about providing some constructive unreasonableness to the conversation, let me suggest expanding our thinking to science and engineering, not science alone. Also, one can compliment our executive branch and Congress for setting national priorities. For discussion today, I will use the example of Congress establishing as a national priority use of mechanical systems to treat heart failure.

If NBIC is to blend into the fifth harmonic envisioned by Newt Gingrich, some national priorities are needed to complement unplanned, revolutionary discoveries. For instance, urinary incontinence a major health problem for today's patients. If the nation had a science and engineering capacity focused on urinary incontinence, this very personal problem would be virtually eliminated. As Mr. Gingrich stated, basic research can be associated with a specific goal..

Table A.1 is a list of the greatest engineering achievements of the past century. The primary selection criterion in constructing this list was worldwide impact on quality of life. Electrification was the number one selection, because the field was fully engineered to improve efficiency, to lower cost, and to provide benefit for virtually everyone. You will notice that healthcare technologies is number 16. NBIC technologies could focus on this field in this century and help move it into the top 10, to the enormous benefit of human performance, health, and overall quality of life.

Table A.1. Greatest Engineering Achievements of the Twentieth Century

1. Electrification	11. Highways
2. Automobile	12. Spacecraft
3. Airplane	13. Internet
4. Water Supply	14. Imaging
5. Electronics	15. Household Appliances
6. Radio and TV	16. Health Technologies
7. Agricultural Mechanization	17. Petroleum Technologies
8. Computers	18. Laser and Fiber Optics
9. Telephones	19. Nuclear Technologies
10. Air Conditioning & Refrigeration	20. High-performance Materials

Setting priorities involves national needs, process, and goals. The Congressional legislative process is quite effective for targeting priorities. The human genome is an example of a work in progress. Today I would like to focus on the field of prevention and repair of coronary heart disease (CHD), where the clinical benefits timeline for today's patients is a little clearer. Successfully addressing priorities such as these usually requires a few decades of sustained public (tax payer) support.

Following hearings in the 1960s, Congress identified advanced heart failure as a growing public health concern needing new diagnostic and treatment strategies. It called for NIH to establish the Artificial Heart Program. Following a decade of system component research, the National Heart, Lung, and Blood Institute (NHLBI) initiated the left ventricular assist device (LVAD) program in 1977. Research and development was targeted towards an implantable system with demonstrated two-year reliability that improved patients' heart function and maintained or improved their quality of life. A series of research phases based on interim progress reviews was planned over a 15-year timeline.

A few years earlier, the NHLBI established less invasive imaging of coronary artery disease as a top priority. A similar program was established that produced less invasive, high-resolution ultrasound, MRI, and CAT scanning for evaluating cardiac function and assessing obstructive coronary artery disease. While this was not an intended outcome, these imaging systems virtually eliminated the need for exploratory surgery. The purpose of long timelines for national programs is not to exclude individual or group-initiated research, and both can have tremendous benefit when properly nurtured.

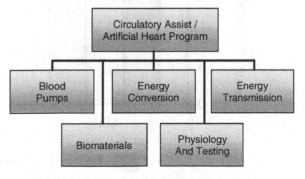

Figure A.4.   NHLBI program organization.

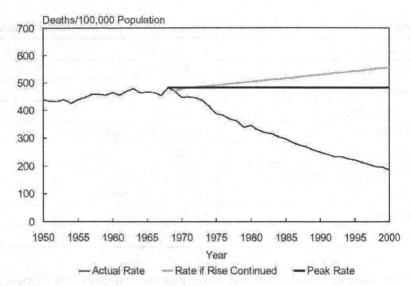

Figure A.5. Coronary heart disease statistics from 1950—1998, age-adjusted to the 2000 standard. CHD accounted for 460,000 deaths in 1998. It would have accounted for 1,144,000 if the rate had remained at its 1963 peak. Comparability ratio applied to rates for 1968-1978.

Heart failure remains a public health issue. At any given time, about 4.7 million Americans have a diagnosed condition of this kind, and 250,000 die each year. The death rates and total deaths from cardiovascular disease have declined for several decades (Fig. A.5). However, during this same time frame, death rates from congestive heart failure (CHF) increased for men and women of all races (Fig. A.6). The most recent interim look at this field estimates that 50,000 to 100,000 patients per year could benefit from left ventricular assist (90 percent of the patients) and

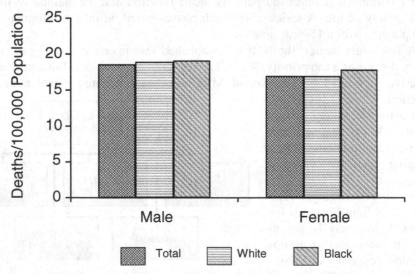

Figure A.6. Age-adjusted death rates for congestive heart failure by race and sex, U.S. 1997. Death rates for CHF are relatively similar in blacks and in whites, but are slightly higher in males than in females.

- **BVS 5000**

- **4,250 patients**

- **33% Discharged**

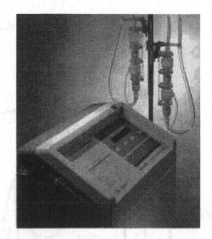

Figure A.7. Postcardiotomy heart dysfunction.

total artificial heart systems (10 percent of the patients), as reported by the Institute of Medicine in *The Artificial Heart* (1991).

The first clinical systems were designed to support, for days or weeks, the blood circulation of patients with dysfunctional hearts following cardiac surgery. This short-term support would enable the hearts of some patients to recover and establish normal function. More than 4,000 patients treated by a product of this program resulted in 33% being discharged to their homes (Fig. A.7). Prior to this experience, only five to ten percent of these patients were discharged.

Clinicians learned that assist devices could "bridge" patients to cardiac transplant. For advanced heart failure and circulatory collapse, implantable ventricular assist devices restore the patient's circulation, allowing patients to leave the intensive care unit and regain strength before undergoing cardiac transplantation. Many patients received support for over one year, some for two or three years, with one patient supported for over four years. Table A.2 lists a tabulation of some 6,000 patients and the assist device used to discharge them to their homes (50 to 70 percent with cardiac transplants). The question remains, will these systems meet the overall program objective of providing destination therapy for heart failure patients?

Table A.2. Bridge-to-Cardiac Transplant

Device	Number of Patients
Heartmate	3000
Novacor	1290
Thoratec	1650
Cardiowest	206
Discharged	50-70%

To answer this question, the Randomized Evaluation of Mechanical Assistance

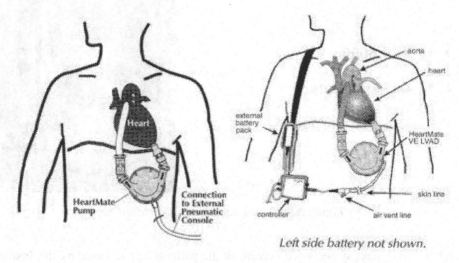

Figure A.8.   HeartMate IP and VE.

for the Treatment of Congestive Heart Failure (REMATCH) clinical trial was
conducted. The Heartmate left ventricular assist (LVAD) system was used (Fig.
A.8). This trial was a true cooperative agreement based on mutual trust among
academia, the private sector, and the government. This was a single blind trial, with
the company and the principle investigator blinded to the aggregate results of the
trial as it was underway. The NHLBI established a Data and Safety Monitoring
Board (DSMB) to confidentially review the progress of the trial and evaluate every
adverse event. At each meeting, the DSMB recommended to NHLBI if the trial

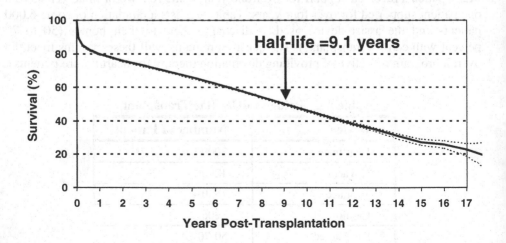

Figure A.9.   Heart transplant survival.

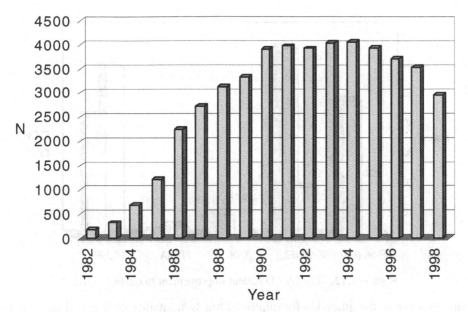

Figure A.10. ISHLT Registry of annualized heart transplant volume.

should continue and what was needed to improve recruitment and the quality of the data. The NHLBI made the final decisions about the conduct of the trial.

It should be noted here that the burden of heart failure on healthcare is increasing. Heart transplants provide remarkable survival and quality of life, but only for some patients, because the limited donor pool provides hearts for only about 2000 patients a year. Figure A.9 is based on a registry of some 52,000 heart transplant patients. The mean survival is nine years, with some patients surviving 15 years or more. These

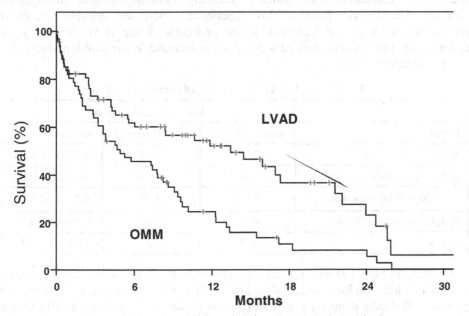

Figure A.11. The LVAD patient improvements in survival.

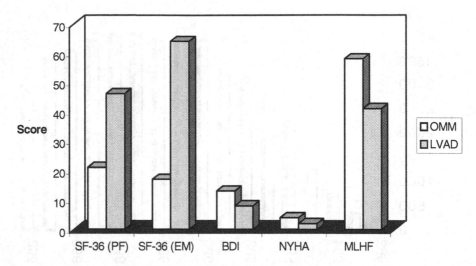

Figure A.12. The LVAD patient improvement in quality of life.

patients serve as the guideline for improved function, quality of life, and survival for alternative therapies (Fig. A.9).

The REMATCH primary end-point was set at a 33 percent improvement in survival for LVAD patients who are not eligible for cardiac transplantation over two years. The goal was for patients to experience improved function without a decrease in quality of life compared to the randomized control group. Cost and cost-effectiveness will also be analyzed as the data becomes available.

The LVAD patients demonstrated a 48 percent improvement in survival (Fig. A.11), significant functional gains, and suggestions of improved quality of life (Fig. A.12), compared with patients receiving optimal medical management (OMM). The LVAD patients also experienced increased adverse events of infections, bleeding, and technical device problems (Table A.3). At two years, updated data (not shown) showed a 200 percent increase in survival but also a high number of device failures.

Table A.3. LVAD Patients' Adverse Events

	Rate per patient-year		
Event	OMM (n=60)	LVAD (n=67)	Ratio (95% CI)
All	2.75	6.45	2.35 (1.86-2.95)
Bleeding (Nonneurological)	0.06	0.56	9.47 (2.3-38.9)
Neurological Dysfunction	0.09	0.39	4.35 (1.31-14.5)
Peripheral Embolic Event	0.06	0.14	2.29 (0.48-10.8)
Sepsis	0.3	0.6	2.03 (0.99-4.13)

Overall, REMATCH patients have a higher mortality than is measured for AIDS or breast, colon, and lung cancer. Based on REMATCH results, LVAD systems will prevent 270 deaths annually per 1000 patients treated — four times as effective as beta blockers and ace inhibitors, with a quality of life similar to ambulatory heart failure patients (Table A.4). All of the evidence suggests that these factors could improve, with fewer adverse events, following further research and clinical experience.

The potential of LVAD systems is highlighted in the following two examples of patients from the REMATCH trial. The first example is a 35-year-old women. Following her implant, she has married and is enjoying her husband, home, and dogs. The second patient is a 67-year-old man who collapsed on the golf course. He now claims he is playing better golf than ever against those "40-year-old flat bellies."

This program would not have occurred without priority-setting by Congress. The clinical need is still substantial. Without sustained public support, the needed research and development capacity would not have materialized. NBIC holds even greater promise but will not achieve its potential without setting some national long-term research objectives.

Table A.4. REMATCH Results for LVAD Systems

LVAD Mortality Impact	Quality of Life	Adverse Events
LVAD Rx would avert 270 deaths annually per 1000 patients treated	Improved compared to ESHF, yet normalcy not restored	LVAD morbidity still considerable
Nearly 4 times the observed impact of beta-blockers and ACEI (70 deaths prevented per 1000 patients)	Physical function scores similar to hemodialysis and ambulatory heart failure	Infections and mechanical failure obvious targets for device and management improvement
Magnitude of effect commensurate with complexity of intervention	Emotional role scores better than clinical depression and similar to ambulatory heart failure	Rate of neurological events encouraging

BALANCING OPPORTUNITIES AND INVESTMENTS FOR NBIC

R. Stanley Williams and Philip J. Kuekes, Hewlett Packard Labs

Over the course of the last several millennia, human beings have learned that major tasks can be performed much more efficiently by dividing up the workload and sharing it among individuals and groups with specialized skills. Larger and more complex tasks require societies with more capable tools and communications skills. As we view the beginning of the 21st century, the tasks we want to perform have become so complex and the tools we have created so sophisticated, that we are challenged to even describe them coherently. It is time to take a holistic view of how we relate to our technologies and develop strategic approaches to integrating them in a fashion that makes them more adaptable and responsive to human desires and capabilities.

In 2001, we are seeing the simultaneous beginnings of three great technological and industrial revolutions that will spring from advances in fundamental research during the past two decades:

Information Science — the understanding of the physical basis of information and the application of this understanding to most efficiently gather, store, transmit, and process information.

Nanoscale Science — the understanding and control of matter on the nanometer length scale to enable qualitatively new materials, devices, and systems.

Molecular Biology — the understanding of the chemical basis of life and the ability to utilize that chemistry.

The knowledge base in each of these areas has the capacity to increase exponentially for several decades into the future, assuming that the research enterprise is maintained. Each field, by itself, offers tremendous opportunities and potential dangers for society, but the fact that there are three simultaneous technology revolutions is literally unprecedented in human history.

The greatest prospects and challenges will occur in the overlap areas that combine two or all three of the new technologies. The great difficulties are that (1) each area by itself is so large and intricate that no single human being can be an expert in all of it, and (2) that each area has developed a language and culture that is distinct and nearly incomprehensible to those working in the other areas. Thus, we find that the most significant problems are often not those related to any particular technology but are based on the basic inadequacies of human understanding and communication. This all-important human factor requires that we better understand and apply cognition. *Cognitive science* will become an increasingly important field for research and utilization in order to more effectively employ the technologies springing from information, nanoscience, and molecular biology. In turn, these technologies will enable major advances in the study and applications of cognition by allowing the construction and emulation of physical models of brain function.

A concrete example can help to illustrate the potential of these overlapping technologies. Since 1960, the efficiency of computing has increased approximately two orders of magnitude every decade. However, this fact has rarely been factored into solving a grand challenge by trading off computation for other types of work as an effort proceeded. This is largely because humans are used to maintaining a particular division of labor for at least a human generation. When paradigms change at a rate that is faster, humans have a difficult time adjusting to the situation. Thus, instead of a smooth adoption of technological improvements, there are often revolutionary changes in problem-solving techniques. When the human genome project began, the shotgun approach for gene sequencing was not employed, because the speed of computing was too slow and the cost was too high to make it a viable technique at that time. After a decade of steady progress utilizing, primarily, chemical analysis, advances in computation made it possible to sequence the genome in under two years utilizing a very different procedure. Thus, the optimum division of labor between chemical analysis and computation changed dramatically during the solution of the problem. In principle, that change could have been exploited to sequence the genome even faster and less expensively if the division of labor had been phased in over the duration of the effort.

As long as technologies progress at an exponential pace for a substantial period of time, those improvements should be factored into the solution of any grand challenge. This will mean that the division of labor will constantly change as the technologies evolve in order to solve problems in the most economical and timely fashion. For computation, the exponential trend of improvement will certainly continue for another ten years, and, depending on the pace of discovery in the nano- and information-sciences, it could continue for another four to five decades. Similar advances will occur in the areas of the storage, transmission, and display of information, as well as in the collection and processing of proteomic and other biological information. The route to the fastest solution to nearly any grand challenge may lie in a periodic (perhaps biannual) multivariate re-optimization of how to allocate the labor of a task among technologies that are changing exponentially during execution of the challenge.

These thrusts in 21st century science are being recognized by those in academia. Some university deans are calling them the "big O's": *nano, bio,* and *info.* These are seen as the truly hot areas where many university faculty in the sciences and engineering want to work. In looking further into the future, we believe that *cogno* should join the list of the big O's.

One way in which academe responds to new opportunities is by creating new disciplines at the intersections between the established divisions. Materials science was created early in the last century at the boundary between chemistry and structural engineering and has evolved as a separate and highly rigorous discipline. Computer science was created in the middle of the last century at the boundary of electrical engineering and mathematics. Now we are beginning to see new transdisciplinary groups coming together, such as chemists and computer scientists, to address new problems and opportunities. One of the problems we face at the turn of this century is that as device components in integrated circuits continue to shrink, they are becoming more difficult to control, and the factories required to build them are becoming extraordinarily expensive. The opportunity is that chemists can inexpensively manufacture components, i.e., molecules, very precisely at the nanometer scale and do so at an extremely low cost per component. Therefore, the new discipline of molecular electronics is arising out of the interactions between computer scientists and chemists. However, developing this new field requires the rigor of both disciplines, the ability to communicate successfully between them, and the proper negotiation process that allows them to optimally share the workload of building new computers. Chemists can make relatively simple structures out of molecules, but they necessarily contain some defects, whereas computer scientists require extremely complex networks that operate perfectly. Economic necessity brings these two very different fields together in what is essentially a negotiation process to find the globally optimal solution of building a working computer from nanometer scale objects at a competitive cost.

There are other very interesting examples of different sciences just beginning to leverage each other. In the bio-info arena, Eric Winfree at the California Institute of Technology is using DNA for self-assembly of complex structures by designing base-pair sequences to construct nano-scaffolding. There is also the whole area of the interaction between biology and information science known as bioinformatics. With the discovery and recording of the human genome and other genomes, we essentially have the machine language of life in front of us. In a sense, this is the instruction set of a big computer program that we do not otherwise understand: we have only the binary code, not the source code. There is a huge amount of work to reverse-engineer this binary code, and we are going to have to rely on computing power to understand what these programs are doing.

Another arena of extreme importance is the bio-nano intersection, since at the fundamental level these both deal with the same size scale. There will be tremendous opportunities to design and build measurement devices that can reach to the scale of molecules and give us a lot more knowledge about biology than we have now. But the reverse is also true. We are going to learn new ways to manipulate matter at the nanoscale from our huge investment in biology. The current goal of molecular electronics is to combine simple physical chemistry with computer design. But biomolecules have incredible functionality based on four billion years of R&D on very interesting nano-structures. The world is going to make a huge investment over the next few years in the biosciences, and we will be able to leverage much of that knowledge in engineering new nanoscale systems.

Work on the relationship between cognition and information goes back the Turing test (i.e., a test that determines if a computer can fool a human being into thinking it is a person during a short conversation) — ideas Turing had even before computers existed. As more powerful computers have become cheaper, we now have cars that talk to us. How will the next generation of people respond when all kinds of devices start talking to them semi-intelligently, and how will society start reacting to the "minds" of such devices? As well as the coming impact of info on cogno, we have already seen the impact of cogno on info. Marvin Minsky, in his *Society of Mind*, looked at the cognitive world and what we know about the brain and used that to work out a new model of computation.

With nanotechnology literally trillions of circuit elements will be interconnected. There is a set of ideas coming out of the cognitive science community involving connectionist computing, which only starts to make sense when you have such a huge number of elements working together. Because of nanotechnology, we will be able to start experimentally investigating these connectionist computing ideas. The other connection of nanotechnology with the cognitive sciences is that we will actually be able to have nonintrusive, noninvasive brain probes of conscious humans. We will be able to understand tremendously more about what is going on physically in the brains of conscious minds. This will be possible because of measuring at the nanoscale, and because quantum measurement capability will provide exquisitely accurate measurements of very subtle events. Over the next couple of decades, our empirical, brain-based understanding in the cognitive sciences is going to increase dramatically because of nanotechnology. The hardest challenge will be the bio-cogno connection. Ultimately, this will allow us to connect biology to what David Chalmers recognizes as the hard problem — the problem of the actual nature of consciousness.

The topic of discussion at this workshop is literally "How do we change the world?" What new can be accomplished by combining nanoscience, bioscience, information science, and cognitive science? Will that allow us to qualitatively change the way we think and do things in the 21st century? In the course of discussions leading up to this workshop, some of us identified nano, bio, and information sciences as being the key technologies that are already turning into 21st century industrial revolutions. Where do the cognitive sciences fit in? One of the major problems that we have in dealing with technology is that we do not know how we know. There is so much we do not understand about the nature of knowledge and, more importantly, about the nature of communication. Behind innovative technologies and industrial revolutions there is another dimension of human effort. In order to harness the new scientific results, integrate them, and turn them into beneficial technologies, we need to strengthen the cognitive sciences and begin the task of integrating the four big O's.

THE IMPACT OF CONVERGENT TECHNOLOGIES AND THE FUTURE OF BUSINESS AND THE ECONOMY

James Canton, Institute for Global Futures

The convergence of nanotechnology, biotechnology, information technology, and cognitive science, which together are referred to here as "convergent technologies," will play a dominant role in shaping the future economy, society, and industrial infrastructure. According to the Commerce Department, over one third of GDP is contributed by information technology. This data would suggest that with new technology being introduced daily, the share of GDP driven by technology will increase. Emerging technologies, especially convergent technologies discussed here, are the engines of the future economy. The objective of enhancing human performance is vital to the well-being of individuals and to the future economic prosperity of the nation. The convergent technologies model has yet to be fully mapped. The convergence of nano-, bio-, and information technologies and cognitive science is in the embryonic stages of our understanding. We need to examine the factors driving convergent technologies and the possible impacts on

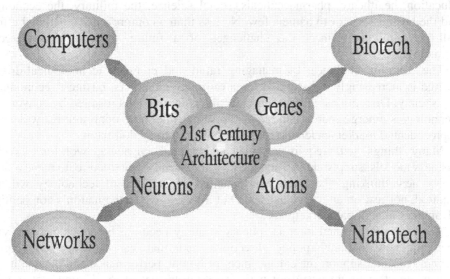

Figure A.13. 21st century architecture.

business and the economy. There is a need to better prepare the nation, coordinate efforts, and work collaboratively towards a national initiative to focus our efforts. How we manage the realtime impact of radical innovation on the social and economic infrastructure of the United States will determine the future wealth, prosperity, and quality of life of the nation. This is no less important than the capacity of the United States to play a global leadership role via the leveraging of next-generation innovations like convergent technologies.

Inventing the Future

Already, massive socio-economic new directions have appeared due to emerging technologies. Examples include the Internet's impact on business, genomics' impact on healthcare, and the wireless impact on personal communications. Some convergence is happening organically, as the evolution of interdisciplinary science, a systems-approach, and the necessity of sharing tools and knowledge is bringing separate disciplines together. The tyranny of reductionism, too long the unwritten law of modern science, is changing, incorporating a more holistic convergent model. We need to take this effort to a new level of fast innovation, inter-science coordination, and action.

The enhancement of human performance via the deployment of convergent technologies requires new work to focus on the synergy of interdependent arenas of science. The benefits to the nation and its citizens may be great in offering lifestyle choices for individuals and incentives for business that do not exist today. New lifestyles, workstyles, and economic business models may be born of this work. The benefits, the payoff we envision, should be the betterment of people and the sustainability of our economy.

It may be possible to influence the ways convergent technologies will change economics and society, on a national scale, by providing leadership and support for a nationwide, collaborative development effort. A national initiative to enhance human performance will be needed. This effort should have many stakeholders in education, healthcare, pharmaceuticals, social science, the military, the economy, and the business sector to name a few. No less than a comprehensive national effort will be required to meet the challenges of a future shaped by convergent technologies.

The daunting challenge of managing rapid and complex technological-driven change is increasingly a disruptive force on today's markets, business, economics, and society. Disruptions will cut more deeply as innovations fostered by convergent technologies emerge more quickly. At the same time, new opportunities will offer unprecedented market leadership for those prepared to exploit them.

Many things will require change: educational curricula, workforce skills, business models, supply chains, and the post-industrial infrastructure, to name a few. Savvy new thinking about the real potential of convergent technology will be required, not just on an individual scale but also relative to the nation's competitive advantages in a global marketplace.

A comprehensive and interdisciplinary strategy needs to be developed that will open up new national policy directions and that can leverage convergent technologies and support the enhancement of human performance and the quality of human life. The future wealth of nations, certainly that of the United States, may well be based on the national readiness we set in motion today to facilitate the adaptation of our society to the challenges and opportunities of convergent technologies.

Managing Fast Change: The Power Tools of the Next Economy

The exponential progress in technology undeniably influences every aspect of business, the economy, and society. Accelerated change is the daily reality we face — and events are speeding up. These convergent technologies are exponentially increasing in months, not years or decades. Consider that Internet traffic doubles every six months; wireless capacity doubles every nine months; optical capacity doubles every 12 months; storage doubles every 15 months; and chip performance (per Moore's Law) doubles every 18 months.

Will we as a nation be ready to adapt to this pace of change? Will we as a nation be ready to be a global leader in a world where radical technological, social, and economic change occurs overnight, not over a century as in the past? There are vast social policy questions and challenges we have yet to ponder, yet to debate, and yet to understand.

Trying to manage fast and complex change is always a messy business for organizations and people, and even more so for nations. Large systemic change most often happens around a crisis like war or the identification of a potential threat or opportunity. Planned change can backfire. So can policy that attempts to predict the future rather than allow the market economy and free enterprise to rule. Yet there is a role for raising awareness and better directing science policy and private sector coordination that must reflect the changing times.

One would argue that the need to bridge the gap between policy and the fast-changing global market economy may be critically important to the nation's future prosperity and global leadership. A more directed technology policy that is both in sync with the marketplace and capable of rapid responsive change — enabling all sectors of society — would be the preferred direction for the future.

There have been instances where a planned change process was beneficial for the nation such with government management of telecommunications giant ATT as a regulated monopoly. Some innovations are too valuable not to promote in the public's interest. Certainly, supply has driven demand often, such as with the telegraph, train routes, and the telephone. Even the Internet, though never considered by its inventors as the power tool it is today, was built ahead of demand. Enlightened public policymakers understood the immense value of these technologies to shape the economic opportunity of a nation. There are some today who argue with merit for turning the next generation of the Internet, broadband, into a utility so that all Americans can gain access and enhance their productivity.

We are again at a crossroads. The convergence of these critical technologies — nano, bio, info, and cogno — may cause deeper disruptions sooner then any prior technologies. We may not have generations or decades to foster national collaboration. We may have a brief period, perhaps a few years, to raise awareness and committed actions at the national scale before serious global competitive challenges arise.

Convergent Technologies and Human Resources

There already is a crisis of inadequate qualified human resources to manage the future opportunities that may lay before us. Already we confront low math and science test scores in our students. Most of the doctoral students in the technical sciences are from abroad. We have close to one million high-tech jobs a year that go begging. Immigration policy cannot keep pace with attracting the number of skilled knowledge workers our economy needs to grow — and this is only the beginning of the talent wars. Clearly, the emergence of radical innovations in science, such as the convergent technology paradigm described here, will accelerate the nation's need for deep science and technical human resources.

How are we as a nation to compete in the super-charged high-tech global economy of the future if we do not have the skilled human resources? Consider the stakeholders of this crisis and what we must do today to rectify this problem before it becomes the nation's Waterloo. Too long has this message been ignored or simply not addressed with the resources required to make a difference for institutions, the private sector, and individuals.

In our modern era we have seen large transformations in nations due to the globalization of trade, emergence of communications technologies, and the expansion of offshore manufacturing. Increasingly, new technology is emerging as the key driver of change where once the train, the telephone, and before that the steamship, drove economic opportunity.

Given the prospects of advanced NBIC technologies, efforts towards large-systems-managed change represent a daunting task for policymakers across all sectors of society. In some ways, the social policymaking process has lagged behind scientific and technological progress. It is time for the social policymaking process to catch up and reach further to explore the technological vectors that will shape our nation's economic future.

Preparing For the Next Economy

No society has ever had to deal with tools as massively powerful as those that are emerging today. The convergence of the NBIC technologies promise to realign the nation's economic future. These power tools are the key arbiters of the next economy, but they will seem tame compared to what is to come. It could be argued that we have passed over the threshold where it is clear that these power tools will be definitive shapers of nations, economies, and societies. How might we guide the emerging future? How might we invent the preferred future by investing in readiness on a national scale? How might we raise awareness of the radical nature of these technologies so that we can be more productive and focused on enhancing human performance?

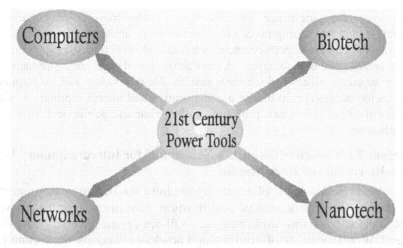

Figure A.14. 21st century power tools.

An entirely new infrastructure is emerging. This new infrastructure will need to accelerate knowledge exchange, networked markets, fast collaborative work, and workforce education. The building blocks of the next economy will be born from the convergent technologies. They represent the shift from the steel and oil of the past and point us towards a radical reshaping of the economy, now in an embryonic stage. The next economy's building blocks — bits, atoms, genes and neurons (Fig. A.15) — will be followed by photons and qubits, as well.

The nations that understand this and that support the growth and development of government and private sector collaboration will thrive. Such collaboration will enable those economies prepared to pursue new economic growth horizons. The future wealth of nations will be based on the change-management readiness we set in motion today by enabling the social adaptation to convergent technology.

How might we direct, encourage, and ultimately shape this desired future for the

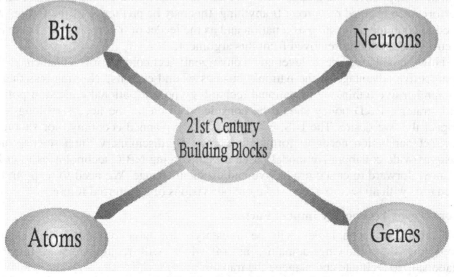

Figure A.15. 21st Century building blocks.

nation, given the emergence of convergent technologies? We can start by developing a plan and setting objectives committed to answering this question. How might we enhance human performance, as a national objective, given the emergence of these convergent technologies? A coordinated and strategic approach will be necessary to create effective long-term results. New thinking will be required that recognizes the necessity of building a collaborative and interdisciplinary strategy for integrating national policy and programs and private and public sector cooperation as never before.

Convergent Technologies: Towards a New Model for Interdisciplinary Policy, Systems-Research, and Inter-Science

Convergent technologies offer an opportunity to design a new model for policyplanners, research scientists, and business executives to consider what the probable outcomes of this work may be. Will we create longevity for the Baby Boomers? Will a new era of convergent knowledge workers be nurtured? How should the private venture community prepare to attract more capital to fuel convergent technology deals? What might healthcare do with enhanced human performance as a medical "product"? What of the ethical and social issues concerning who in our society gets enhanced? These issues and many more are waiting for us in the near future, where convergent technologies will dominate the agenda with breakthroughs too numerous to forecast with any accuracy.

Will we have ready a comprehensive and integrated science policy framework that is visionary enough to consider the development of human potential and the enhancement of human performance? This is the challenge before us, to build a framework that can nurture and experiment but that has the proper controls in place.

The central challenge may well be that we desire a higher quality of life for the nation, as well as building our competitive readiness, given the emergence of convergent technologies. Some may argue against these as non-essential. The quality of life of Americans, it could be easily argued, is influenced heavily by their easy access to leading technologies. American companies and their workers enjoy a global competitive advantage over other less tech-tool-enabled, less human performance-enabled resources. If anything, this may be predictive of the future. We need to continue innovating as a nation and as the leader of the free world. There are security issues not far removed from this argument.

How might we best leverage convergent technology for enhancing the competitive advantage of the nation's businesses and citizens? Nothing less than a comprehensive rethinking of national technology policy, national education policy, and strategic R&D policy should be considered to create the necessary long-term impact that we desire. The U.S. economy is not a planned economy, nor should it be. Yet our nation needs to formulate a new interdisciplinary, inter-science, and systems-wide collaborative model based on converging NBIC technologies in order to move forward to create productive and efficient change. We need to map out the scenarios with all sectors as we stake out our visions of a preferred future.

Convergent Technology Impact Factors

Convergent technologies will be a catalyst for large-systems social change impacting the following domains, all of which will require forward-thinking leadership to facilitate and manage the transition:

1. Workforce Training
2. Educational Curricula
3. Market and Supply Chain Infrastructure
4. Government R&D
5. Private Sector R&D
6. Private Sector Product Development
7. Next Generation Internet

Economic Readiness and Convergent Technology: Key Policy Questions

It could be argued that a nation's technological innovations shape the destiny of that nation. They certainly shape the security, economics, and social well-being of nations. The economic prosperity of the modern nation state cannot be separated from technological adaptation and leadership. But there are other less well-defined issues that we should consider in a global realtime market shaped by convergent technology. Here are some of the arenas yet to be addressed:

1. How can we use the Internet to encourage the high-level knowledge exchange and collaborative work required by convergent technology?
2. What knowledge management resources and large-scale efforts might be mission-essential to facilitate the work with convergent technology?
3. How should private and public sectors work together to facilitate change and adaptation to convergent technology?
4. What new business and economic models might we foster to better enhance productivity in convergent technology?
5. How might we best prepare the nation to compete in a global marketplace shaped by convergent technology?
6. How might we rethink social policy given the future impact of convergent technology?
7. What are the best ways to raise private sector awareness and support for convergent technologies initiatives?
8. Given the emergence of convergent technology, how might we rethink a more holistic inter-science model to better increase our understanding and enhance human performance?
9. How do we define human performance and enhanced human performance given convergent technologies?
10. What is the basis for formulating a national convergent technology initiative to foster private sector and government collaboration, increase citizens' awareness, and coordinate and conduct R&D?

A Proposal for a Convergent Technologies Enterprise Knowledge Network

A convergent technologies Enterprise Knowledge Network (EKN) could provide an online resource bank of information and jobs, a marketplace and clearinghouse for innovations in different vertical industries such as manufacturing, financial services, and entertainment. This network could coordinate information about innovations and intellectual property, and most importantly, connect people using the power of the Internet. This virtual community would be able to build upon, share, and collaborate on new developments in convergent technologies. This network would be linked to research and the marketplace to be able to quickly disperse information, available capital, breakthroughs, and communications relevant to the convergent technologies community.

Next Steps: Advancing Convergent Technologies to Enhance Human Performance

Convergent technology represents an opportunity to address the need to better share innovations, ideas, knowledge, and perhaps, as is our thesis here, to create more effective breakthroughs in enhancing human performance. This is a process that will have to untangle the silo thinking that has been at the heart of science, government, academia, and research. Given the emerging paradigm of convergent technologies, how might we conceptualize a new systems approach to science?

An adoption of a systems approach is already being explored in many areas: Information technology is considering genetic models; telecommunications is experimenting with self-healing networks; biotechnology is edging towards systems-biology; quantum computing and nanotechology are destined for a convergence.

An area that will require much policy and research work is how we define "enhancing human performance." For the physically-challenged the definition may entail gaining sight or mobility. For the aged, it may entail having access to one's memory. Even bolder, the definition of human enhancement may entail providing people with advanced capabilities of speed, language, skill, or strength beyond what humans can perform today. Just as plastic surgery and pharmacology have given new choices to human beings today, enhancement treatments will no doubt shape tomorrow.

Cybernetic Enhancement

Inevitably, the cybernetic enhancement of human performance is sneaking up on society. We already are "enhanced." We wear contact lens to see better, wear hearing aids to hear better, replace hips to improve mobility. We are already at the point of embedding devices in the heart, brain, and body to regulate behavior and promote health. From braces that straighten teeth to plastic surgery that extends youthful appearance, humans are already on the path towards human performance enhancement. Yet, the next generation of human performance enhancement will seem radical to us today .

Well beyond anticipating the sightless who will see, the lame who will walk, and the infertile couples who will be able to conceive children, we will be faced with radical choices. Who will have access to intelligence-enhancing treatments? Will we desire a genetic modification of our species? The future may hold different definitions of human enhancement that affect culture, intelligence, memory, physical performance, even longevity. Different cultures will define human performance based on their social and political values. It is for our nation to define these values and chart the future of human performance.

Summary

Research into convergent technologies may provide insight into better productivity, enhanced human performance, and opportunities to advance the betterment of individuals. No doubt the business sector will need to be a full player in the strategies to further this approach. Better collaboration within government and between government and the private sector would be a worthwhile endeavor.

The destiny of our nation and the leadership that the United States provides to the world will be influenced by how we deal with convergent technologies and the enhancement of human performance.

Convergent technologies will be a key shaper of the future economy. This will drive GDP higher while the health, prosperity, and quality of life of individuals is improved.

A national initiative that can accelerate convergent technology collaboration and innovation while fostering better inter-agency work and public or private sector work will lead to a prosperous future. Without a strategy that enable collaboration, the development of a true systems approach, and an inter-science model, future success maybe be haphazard. The future destiny of the nation as a global leader may be at risk unless a coordinated strategy is pursued to maximize the opportunity that lies inherent in convergent technologies.

References

Canton, J. 1999. *Technofutures*. Carlsbad, California: Hay House.

Christensen, C. 1997. *Innovators dilemma*. Boston, Massachusetts: Harvard Business Press.

Kurzweil, R. 1999. *Age of spiritual machines*. New York: Viking.

Paul, G., and E. Fox. 1996. *Beyond humanity*. Rockland, Massachusetts: Charles River.

de Rosnay, J. 2000. *The symbiotic man*. New York: McGraw Hill.

Tushman, M. 1997. *Winning through innovation*. Boston, Massachusetts: Harvard Business Press.

COHERENCE AND DIVERGENCE OF MEGATRENDS IN SCIENCE AND ENGINEERING

Mihail C. Roco, National Science Foundation; Chair, National Science and Technology Council's Subcommittee on Nanoscale Science, Engineering, and Technology (NSET)

Scientific discoveries and technological innovations are at the core of human endeavor, and it is expected that their role will increase over time. Such advancements evolve into coherence, with areas of temporary confluence and divergence that bring both synergism and tension for further developments. Six increasingly interconnected megatrends (Fig. A.16) are perceived as dominating the science and engineering (S&E) scene for the next several decades: (a) information and computing, (b) nanoscale science and engineering, (c) biology and bio-environmental approaches, (d) medical sciences and enhancement of human physical capabilities, (e) cognitive sciences and enhancement of intellectual abilities, and (f) collective behavior and systems approaches.

This paper presents a perspective on the process of identifying, planning, and implementing S&E megatrends, with illustration for the U.S. research initiative on nanoscale science, engineering, and technology. The interplay between coherence and divergence that leads to unifying science and converging technologies does not develop only among simultaneous scientific trends but also over time and across geopolitical boundaries. There is no single way to develop S&E: here is the value of visionary thinking, to anticipate, inspire, and guide development. Scientists with a view of societal implications should be involved from the conceptual phase of any program that responds to an S&E megatrend.

Introduction

Discoveries and advancements in science and technology evolve into coherence reflecting the trends towards unifying knowledge and global society and have areas of both enduring confluence and temporary divergence. The dynamics bring synergism and tension that stimulate further developments following, on average, an exponential growth. Besides addressing societal needs for wealth, health, and peace, a key driver for discoveries is the intrinsic human need for intellectual advancement, to creatively address challenges at the frontiers of knowledge. A few of the most relevant discoveries lead to the birth of megatrends in science and engineering after passing important scientific thresholds, then building up to a critical mass and inducing wide societal implications. After reaching this higher plateau, such discoveries spread into the mainstream of disciplines and are assimilated into general knowledge. S&E megatrends always are traceable to human development and societal needs, which are their origin and purpose (Fig. A.16). We speak about both science and engineering, because engineering skills provide the tools to implement scientific knowledge and thus the capability to transform society.

Funding a megatrend means enhancing the chance to support researchers moving into the respective field while maintaining most of the investment in the original research fields. The goals are to increase the research outcomes of the total investment, obtain the benefits sooner, and create a suitable infrastructure for the new field in the long term.

At times, groups of researchers argue, targeted funding of S&E megatrends could present a threat to open science and technology advancement. We agree that targeted funding may present a threat to the uniform distribution of R&D funding and could present a larger threat to scientific advancement if the megatrend selection were arbitrary. With proper input from the scientific community to identify the megatrend to support, the primary purpose of a focused S&E effort at the national level is the big payoff in terms of accelerated and synergistic S&E development at the frontiers of science and at the interfaces between scientific disciplines. Without such divergent developments, the entire S&E dynamics would be much slower. There is a need for synergy and cooperative efforts among the disciplines supporting a new field of science or engineering, as well as the need to focus on and fund the key contributing disciplines in a timely fashion.

How should society identify an S&E megatrend? A megatrend is usually motivated by a challenge that may appear unfeasible and even unreasonable at the beginning, as were flying, landing on the Moon, or going into the nanoworld. The goals must be sufficiently broad, the benefits sufficiently valuable, and the development time frame sufficiently long to justify national attention and expense. This paper presents an overview of what we see as key national S&E trends in the United States and illustrates the process of identifying a new megatrend in the recent "National Nanotechnology Initiative" (NNI). Finally, the paper discusses the coherence and synergism among major S&E trends and the role of macroscale management decisions.

Six Increasingly Interconnected Megatrends

The S&E communities and society at large share a mutual interest in advancing major new areas of technological focus in response to objective opportunities, with the goal of accelerating the progress of society as a whole. Six increasingly interconnected scientific megatrends, some closely followed by engineering and technology advancements, are expected to dominate the scene for the coming decades in the United States:

1. *Information and computing.* The bit-based language (0,1) has allowed us to expand communication, visualisation, and control beyond our natural intellectual power. Significant developments beginning in the 1950s have not slowed down, and it is expected that we will continue the exponential growth of opportunities in this area. The main product is in the form of software.

2. *Nanoscale science and engineering.* Working at the atomic, molecular, and supramolecular levels allows us to reach directly the building blocks of matter beyond our natural size limitation, that is, on orders of magnitude smaller than what we can see, feel, or smell. At this moment, this is the most exploratory of all megatrends identified in this list. The field was fully recognised in the 1990s and is at the beginning of the development curve. The main outcome of nanotechnology is in the form of hardware, that is, in the creation of new materials, devices, and systems. The nanoscale appears to

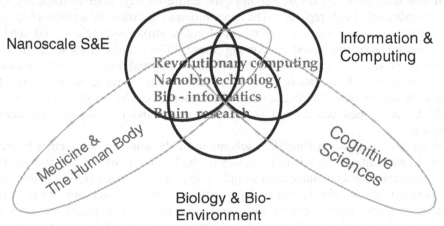

Figure A.16. Coherence and synergism at the confluence of NBIC science and engineering streams.

be the most efficient scale for manufacturing, as we understand its nature now, promising the smallest dissipation of energy, material consumption, and waste and the highest efficiency in attaining desired properties and functions.

3. *Modern biology and bioenvironmental approaches.* Studying cells, their assemblies, and their interactions with their surroundings presents uniquely challenging issues because of their unparalleled complexity. Biology introduces us to self-replicating structures of matter. It uses the investigative methods of information and nanoscale technologies. One important aspect is genetic engineering, another is the connection between life and its environment, including topics such as global warming. Modern biology began its scientific ascendance in the 1970s, and its role continues to expand.

4. *Medical sciences and enhancement of the human body.* The goals are maintaining and improving human physical capabilities. This includes monitoring health, enhancing sensorial and dynamical performance, using implant devices, and extending capabilities by using human-machine interfaces. Healthcare technology is a major area of R&D; it has general public acceptance, and its relative importance is growing as the population ages.

5. *Cognitive sciences and enhancement of intellectual abilities.* This area is concerned with exploring and improving human cognition, behavior, and intellect. Enhancing communication and group interaction are an integral part of improving collective behavior and productivity. This area has received little public recognition, even though increasing cognitive capabilities is a natural objective for a large section of the population.

6. *Collective behavior and systems approach.* This area uses concepts found in architecture, hierarchical systems, chaos theory, and various disciplines to study nature, technology, and society. It may describe a living system, cultural traits, reaction of the society to an unexpected event, or development of global communication, to name a few examples. Recognition of the value of systems approaches increased in the late 1990s.

If one were to model the evolution of the entire society, none of these six S&E megatrends could be disregarded. The nano, bio, and information megatrends extend naturally to engineering and technology, have a strong synergism, and tend to gravitate towards one another. Among these three trends, nanoscale S&E is currently the most exploratory area; however, it is a condition for the development of the other two. Information technology enhances the advancement of both the others. A mathematical formulation of the coherent evolution of research trends could be developed based on a systems approach and time-delayed correlation functions.

Figure A.16. shows a simplified schematic of the complex interaction between the main elements of the scientific system of the beginning of the 21st century. Bits (for computers and communication to satisfy the need for visualization, interaction, and control), genes and cells (for biology and biotechnology), neurons (for cognition development and brain research), and atoms and molecules (to transform materials, devices, and systems) are all interactive components (part of a system approach). But it is important to note that there is a melding of human and S&E development here: human development, from individual medical and intellectual development to collective cultures and globalization, is a key goal.

The main trends of this 21st century scientific system overlap in many ways; their coherence and synergy at the interfaces create new research fields such as bioinformatics, brain research, and neuromorphic engineering. Let's illustrate a possible path of interactions. Information technology provides insights into and visualization of the nanoworld; in turn, nanotechnology tools help measure and manipulate DNA and proteins; these contribute to uncovering brain physiology and cognition processes; and brain processes provide understanding of the entire system. Finally, the conceived system and architecture are used to design new information technology. *Four transforming tools have emerged: nanotechnology for hardware, biotechnology for dealing with living systems, information technology for communication and control, and cognition-based technologies to enhance human abilities and collective behavior.*

Unifying Science and Engineering

There are several reasons why unifying principles in science and engineering are arising now:

- Scientists have increased their depth of understanding of physical, chemical, and biological phenomena, revealing the fundamental common ground in nature.

- Significant advances exist at the interfaces among disciplines, in such a way that the disciplines are brought closer together and one can more easily identify the common principles, fractal patterns, and transforming tools.

- There is a convergence of principles and methods of investigation in various disciplines at the nanoscale, using the same building blocks of matter in analysis. Now it is possible to explore within human cell and neural systems.

- There is a need to simulate complex, simultaneous phenomena, and hierarchical processes where the known physico-chemico-biological laws are too specific for effective multiscale modeling and simulation. An obvious illustration is the requirements for modeling many-body interactions at the nanoscale, where the laws are specific for each material, and variable within bodies and at the boundaries, at different environmental parameters and for different phenomena.

The unifying science may manifest in three major ways:

- Unification of the basic understanding of various natural phenomena and bringing under the same umbrella various laws, principles, and concepts in physical, chemical, biological, and engineering sciences using cause-and-effect explanation. For example, in physics, there is an increasing awareness that weak, strong, electromagnetic, and gravitational forces may collapse into the same theory in the future (Grand Unified Theory). Mathematical language and other languages for improved communication at S&E interfaces and the system approach offer general tools for this process. Furthermore, unification of knowledge of natural sciences with social sciences and humanities forms *a continuum across levels of increasingly complex architectures and dynamics.*

- Observation of collective behavior in physics, chemistry, biology, engineering, astronomy, and society. *Integrative theories are being developed* using the concepts of self-organized systems, chaos, multi-length and time-scale organizations, and complex systems.

- *Convergence of investigative methods to describe the building blocks of matter at the nanoscale.* The nanoscale is the natural threshold from the discontinuity of atoms and molecules to the continuity of bulk behavior of materials. Averaging approaches specific to each discipline collapse in the same multibody approach.

Identifying and using unifying science and engineering has powerful transforming implications on converging technologies, education, healthcare, and the society in the long term.

National S&E Funding Trends

The foundation of major S&E trends are built up over time at the confluence of other areas of R&D and brought to the front by a catalytic development such as a scientific breakthrough or a societal need. For example, space exploration has grown at the confluence of developments in jet engines, aeronautics, astronomy, and advanced materials and has been accelerated by global competitiveness and defense challenges. Information technology advancement has grown at the confluence of developments in mathematics, manufacturing on a chip, materials sciences, media, and many other areas and has been accelerated by the economic impact of improved computing and communication. Nanotechnology development has its origins in scaling down approaches, in building up from atomic and molecular levels, and in the confluence of better understanding of chemistry, biosystems, materials, simulations, and engineering, among others; it has been accelerated by its promise to change the nature of almost all human-made products. Biotechnology development has grown at the confluence of biology, advanced computing, nanoscale tools, medicine, pharmacy, and others and has been accelerated by its obvious benefits in terms of improved healthcare and new products.

Development of initiatives for such fields of inquiry has led to additional funding for these and similar initiatives. The last two national research initiatives are the Information Technology Research (ITR) initiative, announced in 1999, and the National Nanotechnology Initiative (NNI), announced in 2000. For ITR, there is a report from the President's Information Technology Advisory Committee (PITAC), a committee with significant participation from industry, that shows new elements and expectations. According to this report, the Internet is just a small token development on the way to larger benefits.

How is a new trend recognized for funding? There is no single process for raising an S&E trend to the top of the U.S. national priorities list. One needs to explore the big picture and the long term. It is, of course, important to identify a significant trend correctly; otherwise, either a gold mine may not be exploited, or a wasteful path may be chosen. We note that major U.S. R&D initiatives are designed to receive only a relatively small fraction of the total research budget, because the country must provide support for all fields, including the seeds for future major trends. Generally, one must show a long-term, cross-cutting, high-risk/high-return R&D opportunity in order to justify funding a trend. However, this may be insufficient. Of the six major trends listed above, only the first two have led to multiagency national research initiatives, although there is *de facto* national priority on the fourth trend — that related to human health. Information technology and nanotechnology received national recognition through the National Science and Technology Council (NSTC). In another example, the driving force for support for a program for global change has been international participation.

Table A.5 summarizes the main reasons for national recognition and funding of several S&E programs. A few years ago, NSF proposed a research focus on biocomplexity in the environment, a beautiful (and actual) subject. This topic so far has not received attention from other funding agencies; a reason may be that no dramatic scientific breakthrough or surge of societal interest was evident at the date of proposal to justify reallocating funds at the national level. On the other hand, cognitive sciences are key for human development and improvement, and it is expected that this area will receive increased attention. Converging technologies starting from the nanoscale is another area for future consideration.

We could relate the S&E developments to the perception and intellectual ability of the contributing researchers. The left-brain handles the basic concepts; the right-brain looks into pictures and assemblies. "Your left-brain is your verbal and rational brain; it thinks serially and reduces its thoughts to numbers, letters, and words. Your

**Table A.5. Reasons for national recognition for funding purposes:
No unique process of identification of U.S. national R&D programs**

(PITAC: Presidential Information Technology Advisory Committee; NSTC: National Science and Technology Council)

S&E Funding Trends in U.S.	Main reasons for recognition at the national level
Information Technology Research (1999 -)	Economic implications; proposed by PITAC; promise of societal implications; recognized by NSTC
National Nanotechnology Initiative (2000 -)	Intellectual drive towards the nanoscale; promise of societal implications; recognized by NSTC
Medicine (NIH)	Public interest in health, and aging population; focus at the National Institutes of Health
Biology and bioenvironment	Distributed interest; NSF focus on biocomplexity
Cognitive	Not yet well recognized; included in education
Collective behavior	Not yet well recognized; not focused, part of others
Others in the last 50 years:	
Nuclear program	National security
Space exploration	International challenge
Global change research	International agreements
Partnerships for a new generation of vehicles	Economic competitiveness; environment

right brain is your non-verbal and intuitive brain; it thinks in patterns, or pictures, composed of "whole things" (Bergland 1985). Accordingly, the brain combines reductionist elements with assembling views into a cooperative and synergistic thinking approach. Those two representations of thinking may be identified as development steps for each S&E megatrend, as illustrated in Table A.6.

It is relevant to keep track of this connection when developing a new research program. For example, the basic concepts originating in the left brain allow individuals and groups to develop representations further from their primary perception (point of reference). Let's consider the human representation of length scale. Initially, we used our hands to measure and developed representations at our natural length scale; then we used mechanical systems, and our representation moved towards the smaller scale of exact dimensions; later, optical tools helped us move into the microscale range of length representation; and electron microscopes and surface probes have helped us move into the nanoscale range. This process continues into the arena of nuclear physics and further on. In a similar manner, abstract concepts handled by the left brain have helped humans move into larger representation scales, beginning with the representation of a building and geography of a territory; later moving to representation of the Earth (useful in sustainable development and global change R&D), then of the universe (needed in space exploration).

The left brain tends to favor reductionist analysis and depth in a single field, which may contribute to "divergent" advancements. Within finite time intervals, such advancements tend to develop faster, to diverge, to take on a life of their own. Meantime, the "whole think" approach is favored by right-brain activities. It is the role of the right brain to assemble the global vision for each initiative and see the coherence among initiatives. This coherence leads to unifying concepts and converging technologies.

Societal feedback is the essential and ultimate test for the nation to establish and assimilate S&E megatrends. There are clear imperatives: increasing wealth, improving healthcare, protecting a sustainable environment, enhancing the culture, and providing national security. When one looks from the national point of view and in the long term, scientific communities, government, and society at large all have the same goals, even if the R&D funds for a megatrend favor some S&E communities in short-term.

Table A.6. S&E megatrends as related to human representation

Left-brain focus	Right-brain focus	S&E trend
DNA, cell (from natural environment)	Biosystems, organisms	Modern biology
Atom, molecule (from natural environment)	Patterns, assemblies	Nanoscale S&E
Bits (chosen language)	Visualization, networking	Information and computing

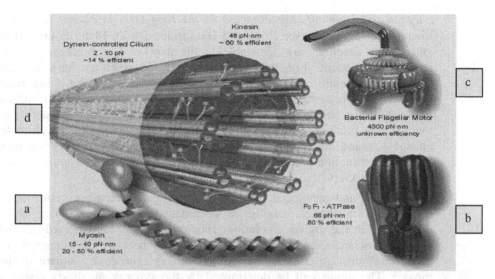

Figure A.17. All living systems work at the nanoscale: illustration of cellular nanomachines (after Montemagno 2001): (a) Myosin, the principle molecular motor responsible for muscle movement (characteristic dimension L about a few nm); (b) ATP synthase, a chemical assembler (L about 10 nm); (c) Bacterial flagella motor (L about 20 nm); (d) A dynein-microtube complex assembled to form a cilium (L about 50 nm).

Motivation, Preparation, and Approval Process of the National Nanotechnology Initiative

Four imperatives define the National Nanotechnology Initiative:

1. There is a need for long-term fundamental research leading to systematic methods of control of matter at the nanoscale. All living systems and man-made products work at this scale. This is because all basic building blocks of matter are established and their basic properties are defined in the range between one and 100 molecular diameters. The first level of organization in biosystems is in the same nanometer range. For example, our body cells typically include nanobiomotors converting energies to the forms needed, such as chemical, electrical, or mechanical. The typical size of the organelles (see Fig. A.17) in a cell is 10 nanometers, which corresponds approximately to 10 shoulder-to-shoulder molecules of water. Fundamental understanding of matter at the nanoscale may change our long-term strategies concerning healthcare, the way we manage the environment, our manufacturing practices. This is the first initiative at the national level motivated by and focused on fundamental research.

2. *Nanotechnology promises to become the most efficient length scale for manufacturing.* While we know that the weak interactions at the nanoscale would require small amounts of energy for manufacturing and that precise assembly of matter would lead to products with high performance and no waste, we do not yet have systematic, economic manufacturing methods for production at the nanoscale. Again, a focus on fundamental research is essential in this regard.

3. *Large societal pay-offs are expected in the long term in almost all major areas of the economy* (see Roco and Bainbridge 2001). Material properties and system functions are adjustable at the nanoscale, a function of size, shape, and pattern. For this reason, nanoscale sciences have created tremendous scientific interest. However, this alone would have not been sufficient to start a national research initiative. Nanotechnology has acquired national interest only in the last two years because of our increasing ability to manufacture products with structures in the nanometer range, as well as to change life and environmental ventures. This possibility promises a new industrial revolution leading to a high return on investments and to large benefits for society.

4. Nanoscience and nanotechnology development are necessary contributing components in the converging advancements in S&E, including those originating in the digital revolution, modern biology, human medical and cognitive sciences, and collective behavior theory. The creation of "hardware" through control at the nanoscale is a necessary square in the mosaic. The future will be determined by the synergy of all six research areas, although in the short term, the synergy will rely on the information, nano- and bio- sciences starting from the molecular length scale. The developments as a result of the convergent technologies will be significant, but are difficult to predict because of discontinuities.

NNI was the result of systematic preparation. It was done with a similar rigor as used for a research project, and documents were prepared with the same rigor as for a journal article. In 1996-1998, there was an intellectual drive within various science and engineering communities to reach a consensus with regard to a broad definition of nanotechnology. In the interval 1997-2000, we prepared detailed materials answering several defining questions:

- What are the research directions in the next 10 years? (See *Nanotechnology research directions. A vision for nanotechnology research and development in the next decade.* Roco, Williams, and Alivisatos 1999/2000; http://nano.gov/nsetrpts.htm.)

- What is the national and international situation? (See *Nanostructure science and technology, A worldwide study.* Siegel et al. 1999; http://nano.gov/nsetrpts.htm.)

- What are the societal implications? (See *Societal implications of nanoscience and nanotechnology.* NSF 2000; http://nano.gov/nsetrpts.htm.)

- What are the vision and implementation plans for the government agencies? (See NNI, Budget request submitted by the president to Congress. NSTC 2000; http://nano.gov.)

- How do we inform and educate the public at large about nanotechnology? (See *Nanotechnology. Reshaping the world atom by atom*, NSTC/CT 1999; http://nano.gov/nsetrpts.htm.)

The approval process began with various S&E communities, and advanced with the positive recommendations of the Presidential Council of Science Advisory and Technology and of the Office of Management and Budget. The president proposed

NNI on January 21, 2000, in a speech at the California Institute of Technology. The proposed budget was then approved by eight congressional committees, including those for basic science, defense, space, and health-related issues. Finally, the Congress appropriated $422 million for NNI in fiscal year 2001 (see Roco 2001a).

The Role of Macroscale Management Decisions

It is essential that we take time to explore the broad S&E and societal issues and that we look and plan ahead. These activities require information at the national level, including macroscale management decisions, which must be sufficiently flexible to allow creativity and imagination to manifest themselves during implementation of planning and programs. (Firm predictions are difficult because of the discontinuities in development and synergistic interactions in a large system.) Industry provides examples of the value of applying visionary ideas at the macroscale and making corresponding management decisions. At General Electric, for example, Jack Welsh both articulated a clear vision and spearheaded measures structured at the level of the whole company for ensuring long-term success. R&D activities depend on the decisions taken at the macroscale (national), S&E community (providers and users), organization (agency), and individual levels. In addition, the international situation increasingly affects results in any individual country. An international strategy would require a new set of assumptions as compared to the national ones (Roco 2001b).

a) **Strategic macroscale decisions taken at the national level.** These have broad, long-term implications. Different visions and implementation plans may lead to significantly different results. Examples and principles follow.

   – NSF collects information on the evolution of sources of R&D funding like the one shown in Fig. A.18. Federal funding is relatively constant

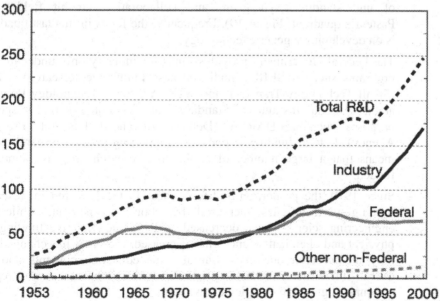

Figure A.18. National R&D funding by source (NSF 2002)

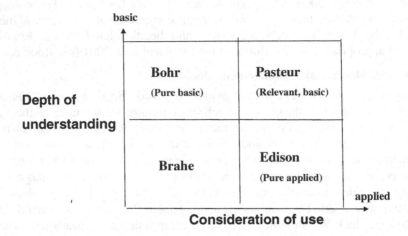

Figure A.19. Pasteur's Quadrant (schematic after Stokes 1997): Redirecting R&D investments with a new role for engineering.

from 1992 to 2000. In the same time interval, private R&D funding has increased and approximately doubled as compared to federal funding. The federal government share of support for the nation's R&D decreased from 44.9 percent in fiscal year 1988 to 26.7 percent in fiscal year 1999. Also, more funds in industry are dedicated to funding development and applied research. That is, society spends more overall for shorter-term outcomes and less for long-term outcomes. Government needs to direct its funding more on complementary aspects: fundamental research (see Bohr's quadrant, Fig. A.19) and mission-oriented projects that encourage depth of understanding, synergism, and collaboration among fields (see Pasteur's quadrant, Fig. A.19). Frequently, the focus in this last quadrant is on developing a generic technology.

– The Federal Government provides funds for industry only under limited programs such as SBIR (Small Business Innovative Research), STTR (Small Technology Transfer), and ATP (Advanced Technology Program at the National Institute of Standards and Technology), or for special purposes such as DARPA (Defense Advanced Research Program Agency). If total funding is constant, supporting applied research often means that a large number of exploratory research projects cannot be funded.

– Since 1970, the proportion of life sciences in the U.S. federal research funding portfolio has increased by about 13 percent, while the engineering sciences have decreased by the same. Relative funding for physical and chemical sciences has decreased, too. This has changed not only the research outcomes, but also the education contents and the overall infrastructure. One way to address this imbalance is to prepare national programs in complementary areas.

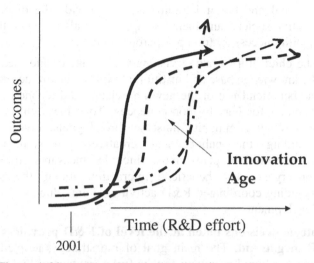

Figure A.20.   The "Innovation Age." Organizations will change the focus from repetitive to creative, innovation-based activities and transfer efforts from machines to human development.

- The measures need a collective discipline and flexibility in implementation. A bio-inspired funding approach within the major NNI research areas has been adopted. The funding agencies have issued solicitations for proposals addressing relatively broad R&D themes identified by panels of experts according to the agency missions. In their proposals, researchers respond to those solicitations with specific ideas in a manner suggesting a bottom-up assembly of projects for each theme.

- The coherence among various S&E areas should be evaluated periodically in order to create conditions for convergence and synergism. Figure A.20 suggests that the major trends identified in this paper will play an increased role, beginning with the synergism of nanoscience, modern biology, information technology, and neuro-cognitive sciences, integrated from the molecular level up, with the purpose of enhancing cognitive, human body, and social performance. This coherence will create an unprecedented transformational role for innovation. Organizations will augment creative, knowledge-based activities, with larger conceptual versus physical work components.

- Macroscale measures should address the increased role of the partnerships between the government-sponsored research providers and industry.

- The measures should encourage international collaboration based on mutual interest. The U.S. investments in the areas of nanoscience and nanotechnology represent about one-third of the global investment made by government organizations worldwide. At NSF, support is made available to investigator-initiated collaborations and through activities sponsored by the Foundation.

- National and cultural traditions will provide the diverse support for a creative society, and their role appears to also provide the continuity and stability necessary for a prosperous society.

- The chief aim of taking visionary and macroscale measures is to create the knowledge base and institutional infrastructure necessary to accelerate the beneficial use of the new knowledge and technology and reduce the potential for harmful consequences. To achieve this, the scientific and technology community must set broad goals; involve all participants, including the public; and creatively envision the future. The implementation plans must include measures for stimulating the convergence and beneficial interaction among the S&E megatrends, including coordinated R&D activities, joint education, and infrastructure development.

b) **Strategic decisions taken at the level of R&D providers and users of an S&E megatrend.** The main goal of the strategy adopted by the National Nanotechnology Initiative is to take full advantage of this new technology by coordinated and timely investment in ideas, people, and tools. A coherent approach has been developed for funding the critical areas of nanoscience and engineering, establishing a balanced and flexible infrastructure, educating and training the necessary workforce, promoting partnerships, and avoiding unnecessary duplication of efforts. Key investment strategies are as follows:

- *Focusing on fundamental research.* This strategy aims to encourage revolutionary discoveries and open a broader net of results as compared to development projects for the same resources.

- *Maintaining a policy of inclusion and partnerships.* This applies to various disciplines, areas of relevance, research providers and users, technology and societal aspects, and international integration.

- *Recognizing the importance of visionary, macroscale management measures.* This includes defining the vision of nanotechnology; establishing the R&D priorities and interagency implementation plans; integrating short-term technological developments into the broader loop of long-term R&D opportunities and societal implications, using peer review for NNI; developing a suitable legal framework; and integrating some international efforts. Work done under NSTC (the White House) has allowed us to effectively address such broader issues.

- *Preparing the nanotechnology workforce.* A main challenge is to educate and train a new generation of skilled workers in the multidisciplinary perspectives necessary for rapid progress in nanotechnology. The concepts at the nanoscale (atomic, molecular, and supramolecular levels) should penetrate the education system in the next decade in a manner similar to that of microscopic approach over the last 40 to 50 years.

- *Addressing the broad goals of humanity.* Nanoscale science and engineering must be designed to lead to better understanding of nature, improved wealth, health, sustainability, and peace. This strategy has strong roots, and, it is hoped, may bring people and countries together. An integral aspect of broader goals is increasing productivity by applying innovative nanotechnology for commerce (manufacturing, computing and communications, power systems, and energy).

- – *Identifying and exploiting coherence with other major S&E trends.* As part of an S&E trend, one may address a scientific and technological "grand challenge" at the national level.

c) **Strategic decisions taken at the organizational level.** The organization level is concerned with optimum outcome in each department, agency, national laboratory, or other organization.

d) **Strategic decisions taken at the level of the individual.** The individual level addresses issues related to education, motivation, productivity, and personal involvement.

Common Ground for the Science Community and Society at Large

a) We envision the bond of humanity driven by an **interconnected virtual brain** of the Earth's communities searching for intellectual comprehension and conquest of nature (Roco 1999), where individuals are empowered with access to collective knowledge while safeguarding privacy. In the 21st century, we estimate that scientific and technological innovation will outperform for the first time the societal output of the physical activities separated by geographical borders. Knowledge and technologies will cross multiple institutional boundaries on an accelerated path before application, in a world dominated by movement of ideas, people, and resources. National and cultural diversity will be a strength for the new creative society. The interplay between information, nano-, bio- and healthcare technologies, together with cognitive sciences and cultural continuity will determine the share of progress and prosperity of national communities, no matter what their size.

b) **Researchers need the big picture of different disciplines.** The current focus on reductionism and synthesis in research will be combined with and partially overtaken by a recognition of various aspects of unity in nature and a better understanding of complexity, crossing streams in technology, and crossing national and cultural borders. The ability to see complex systems at the molecular and atomic level will bring a new renaissance. Leonardo da Vinci, equally brilliant in the art of painting and in mechanical, hydraulic, military, and civil engineering, embodied the quintessence of the original Renaissance. The laboratory investigations that began in the 17th century led researchers to separate, reductionist pathways. Today, all disciplines share a common ability to work at the molecular and nano length scales using information technology and biology concepts. The reductionist divergence of sciences and engineering of old seems to be regrouping and finding a confluence. The collective multiphenomena and multiscale behavior of systems between single atoms and bulk have become the center of attention in order to extract new properties, phenomena, and function — like a new alchemy. For researchers to acquire a "big picture" approach requires depth in each discipline and good communication across disciplines.

c) **Visionary R&D planning pays off.** It is essential to take the time to courageously look into the future. "The best way to predict the future is to create it" according to Alan Kaye of Xerox Park. Technological progress may be accelerated by a wise structuring of science and engineering that helps the main trends (or megatrends) be realized sooner and better. Why do all of this? We cite U.S. Federal Reserve Chairman Allen Greenspan (1999), who credits our nation's "phenomenal" economic performance to technological innovation that has accelerated productivity: "Something special has happened to the American economy in recent years . . . . a remarkable run of economic growth that appears to have its roots in ongoing advances in technology."

We have seen in the last 20 years that industrial productivity has steadily increased. This is the key reason why the U.S. economy is growing, indicating the strong connection between science, engineering, and development. The productivity growth rate increased from 0.8 percent during the Carter administration, to 1.6 percent during the Reagan administration, 1.7 percent during the first Bush administration, and 2.1 percent during the Clinton administration. These increases are attributed to technological innovation. Several case studies show that investment in research at the national level also brought about 20 percent additional benefits in the private sector and 50 percent in social return.

Because there is no single or proven way to successfully develop S&E, the role of visionary R&D planning is to set priorities and provide the infrastructure for major promising projects at the national level. The coherence and synergy of various S&E trends and the rate of implementation and utilization are affected by management decisions at the macroscale. The measures must be based on good understanding of the global societal environment and on long-term trends. Professors do not leave their students to do everything they like in academic research. On the contrary, if a research project goes well, more resources are guided in that direction. This idea should be held true at the national level, where there are additional advantages such as synergistic and strategic effects.

d) **The risk of S&E developments should be evaluated in the general context of potential benefits and pitfalls in the long term.** Significant S&E developments inevitably have both desired and undesired consequences. Dramatic discoveries and innovations may create a tension between societal adoption of revolutionary new technologies in the future and our strong desire for stability and predictability in the present. Important research findings and technological developments may bring undesirable negative aspects. Bill Joy has raised such issues with the public, presenting scenarios that imply that nanoscale science and engineering may bring a new form of life, and that their confluence with biotechnology and the information revolution could even place the human species in danger.

In our opinion, raising this general issue is very important, but several of Joy's scenarios are speculative and contain unproven assumptions (see comments from Smalley 2000) and extrapolations. However, one has to treat these concerns responsibly. For this reason we have done studies and tasked coordinating offices at the national level to track and respond to unexpected developments, including public health and legal aspects. So far, we all agree that while all possible risks should be considered, the need for economic and technological progress must be counted in the balance. We underscore that the main aim of our national research initiatives is to develop the knowledge base and to create an institutional infrastructure to bring about broader benefits for society in the long term. To this end, it is essential to involve the entire community from the start, including social scientists, to maintain a broad and balanced vision.

e) **Contributions to the broader vision and its goals are essential** at any level of activity, including organizational and individual levels. Researchers and funding agencies need to recognize the broad societal vision and contribute to the respective goals in a useful and transforming manner, at the same time allowing the unusual (divergent) ideas to develop for future discoveries and innovations. The funded megatrends provide temporary drivers that seem to be part of the overall dynamics of faster advancements in S&E. The vision and goals should be inclusive and equally understandable by top researchers, the productive sector, and society at large. In a similar manner, one needs to observe the international trends and respond accordingly. Internationalization with free movement of ideas, people, and resources makes impossible long-term advances in only one country. Cultural and national diversity is an asset for creative, divergent developments in S&E.

In a system with R&D management structured at several levels as discussed above, the macroscale measures have major implications, even if they are relatively less recognized by an S&T community that tends to be more focused on specific outcomes at the organizational and individual levels and on the distribution of funds. The recognition system centered on individual projects in R&D universities and other research organizations may be part of the reason for the limited recognition of the role of macroscale measures.

f) **Maintaining a balance between continuity and new beginnings (such as funding S&E megatrends) is an important factor for progress at all levels.** Coherence and convergence are driven by both intrinsic scientific development (such as work at the interfaces) and societal needs (such as the focus on healthcare and increased productivity). The divergent tendencies are driven also by both internal stimuli (such as special breakthrough in a scientific and engineering field) and external stimuli (such as political direction). We need to stimulate the convergence and allow for temporary divergence for the optimum societal outcomes, using, for example, the mechanisms of R&D funds allocation and enhancing education based on unity in nature. Such activities need to be related to the individual capabilities, where the left-brain (new beginnings) and right-brain (coherence) have analogous dual roles as the drivers of S&E trends.

g) **The societal importance of innovation is growing**, where innovation is defined as "knowledge applied to tasks that are new and different." In many ways, science and engineering have begun to affect our lives as essential activities because of innovation that motivates, inspires, and rewards us.

While the ability to work has been a defining human quality and increasing industrial productivity was the motor of the 20th century, we see innovation being the main new engine joining other key humanity drivers in the 21st century. The coherence and divergence of major S&E trends are intrinsic processes that ensure more rapid progress in science and technology, enhancing human performance and improving the quality of life. We envision the S&E trends converging towards an *"Innovation Age"* in the first half of the 21st century, where creativity and technological innovation will become core competencies. Current changes are at the beginning of that road. They are triggered by the inroads made in understanding the unity of nature manifested equally at the nanoscale and in broad complex systems, by reaching a critical mass of knowledge in physical and biological sciences and their interfaces and by the increased ability to communicate effectively between scientific and engineering fields.

Acknowledgements

This essay is based on the keynote lecture at the Annual Meeting of the Swiss Scientific Academies (CASS), Bern, Switzerland, on November 30 - December 2, 2000. It includes personal opinions and does not necessarily represent the position of the National Science and Technology Council or the National Science Foundation.

References

Bergland, R. 1985. *The fabric of mind*. New York: Viking Penguin.

Greenspan, A. 1999 (June 14). Statement at the Joint Economic Committee, Washington, D.C.

Montemagno, C.D. 2001. Nanomachines: A roadmap for realizing the vision. *J. Nanoparticle Research*, 3:1-3.

NSF. 2002. *Science and engineering indicators*. Arlington, VA: National Science Foundation.

NSTC. 2000. *National Nanotechnology Initiative: The initiative and its implementation plan*. WH, Washington, D.C.; website http://nano.gov.

Roco, M.C. 1999. Scientific and engineering innovation in the world: A new beginning. SATW, Sept. 23, 1999, Zurich — Aula der Eidergennossischen Technischen Hochschule, Switzerland.

Roco, M.C., R.S. Williams, and P. Alivisatos, eds. 2000.Nanotechnology research directions, Boston, MA: Kluwer Academic Publishers.

Roco, M.C. 2001a. From vision to the implementation of the National Nanotechnology Initiative. *J. Nanoparticle Research* 3(1):5-11.

Roco, M.C. 2001b. International strategy for nanotechnology research and development. *J. Nanoparticle Research* 3(5-6):353-360.

Roco, M.C., and W.S. Bainbridge, eds. 2001. Societal implications of nanoscience and nanotechnology, Boston: Kluwer Academic Publishers.

Schwartz, P., P. Leyden, and J. Hyatt. 1999. *The long boom*. New York: Perseus Books.

Smalley, R. 2000. "Nanotechnology, education, and the fear of nanorobots." In *Societal implications of nanoscience and nanotechnology*, NSF Report (also Kluwer Acad. Publ., 2001, pp. 145-146).

Stokes, D.E. 1997. *Pasteur's quadrant: Basic science and technological innovation*. Washington, D.C.: Brookings Institution Press.

B. EXPANDING HUMAN COGNITION AND COMMUNICATION

THEME B SUMMARY

Panel: W.S. Bainbridge, R. Burger, J. Canton, R. Golledge, R.E. Horn, P. Kuekes, J. Loomis, C.A. Murray, P. Penz, B.M. Pierce, J. Pollack, W. Robinett, J. Spohrer, S. Turkle, L.T. Wilson

In order to chart the most profitable future directions for societal transformation and corresponding scientific research, five multidisciplinary themes focused on major goals have been identified to fulfill the overall motivating vision of convergence described in the previous pages. The first, "Expanding Human Cognition and Communication," is devoted to technological breakthroughs that have the potential to enhance individuals' mental and interaction abilities. Throughout the twentieth century, a number of purely psychological techniques were offered for strengthening human character and personality, but evaluation research has generally failed to confirm the alleged benefits of these methods (Druckman and Bjork 1992; 1994). Today, there is good reason to believe that a combination of methods, drawing upon varied branches of converging science and technology, would be more effective than attempts that rely upon mental training alone.

The convergence of nanotechnology, biotechnology, information technology, and cognitive science could create new scientific methodologies, engineering paradigms, and industrial products that would enhance human mental and interactive abilities. By uniting these disciplines, science would become ready to succeed in a rapid program to understand the structure and functions of the human mind, The Human Cognome Project. Truly, the mind is the final frontier, and unraveling its mysteries will have tremendous practical benefits. Among the most valuable spin-offs will be a host of devices that enhance human sensory capabilities. We will be able to build a vast variety of humane machines that adapt to and reflect the communication styles, social context, and personal needs of the people who use them. We will literally learn how to learn in new and more effective ways, revolutionizing education across the life span. New tools will greatly enhance creativity, industrial design, and personal productivity. Failure to invest in the necessary multidisciplinary research would delay or even prevent these benefits to the economy, to national security, and to individual well-being.

Rapid recent progress in cognitive science and related fields has brought us to the point where we could achieve several breakthroughs that would be of great value to mankind. However, we will need to make a significant special effort to bring together the often widely dispersed scientific and technical disciplines that must contribute. For example, progress in the cognitive neuroscience of the human brain has been achieved through new research methodologies, based in both biology and information science, such as functional magnetic resonance imagining (fMRI) and infrared sensors. However, we are reaching the resolution limits of current instrumentation, for example, because of concerns about the safety of human research subjects (Food and Drug Administration 1998), so progress will stall quickly unless breakthroughs in NBIC can give us research tools with much greater resolution, sensitivity, and capacity to analyze data. Many other examples could be cited in which scientific, technological, and economic progress is approaching a barrier that can be surmounted only by a vigorous program of multidisciplinary research.

The panel identified five main areas in which integration of the NBIC sciences can enhance the cognitive and communicative aspects of human performance. Each of these is a challenging field for multidisciplinary research that will lead to many beneficial applications.

1. The Human Cognome Project

It is time to launch a Human Cognome Project, comparable to the successful Human Genome Project, to chart the structure and functions of the human mind. No project would be more fundamental to progress throughout science and engineering or would require a more complete unification of NBIC sciences. Success in the Human Cognome Project would allow human beings to understand themselves far better than before and therefore would enhance performance in all areas of human life.

While the research would include a complete mapping of the connections in the human brain, it would be far more extensive than neuroscience. The archaeological record indicates that anatomically modern humans existed tens of thousands of years before the earliest examples of art, a fact that suggests that the human mind was not merely the result of brain evolution but also required substantial evolution in culture and personality. Central to the Human Cognome Project would be wholly new kinds of rigorous research on the nature of both culture and personality, in addition to fundamental advances in cognitive science.

The results would revolutionize many fields of human endeavor, including education, mental health, communications, and most of the domains of human activity covered by the social and behavioral sciences. Some participants in the human cognition and communication working group were impressed by the long-term potential for uploading aspects of individual personality to computers and robots, thereby expanding the scope of human experience, action, and longevity. But at the very least, greater understanding of the human mind would allow engineers to design technologies that are well suited to human control and able to accomplish desired goals most effectively and efficiently. Success in the Human Cognome Project would greatly facilitate success in the other four areas identified by this working group.

2. Personal Sensory Device Interfaces

Fundamental scientific and engineering work needs to be done to permit development of an array of personal sensory device interfaces to enhance human abilities to perceive and communicate. Human senses are notoriously limited. Whereas we can hear ten octaves of musical tones, we can see only one octave of the colors of light, and our ears have a poor ability to form detailed "images" from sound the way our eyes can with light. Today's communication technology has revolutionized the ability of people to communicate across large distances, but little has been done to help with small area communication, for example, between individuals in a conference room. These are only two of many areas where NBIC sensor efforts can increase human performance.

Research can develop high bandwidth interfaces between devices and the human nervous system, sensory substitution techniques that transform one type of input (visual, aural, tactile) into another, effective means for storing memory external to the brain, knowledge-based information architectures that facilitate exploration and understanding, and new kinds of sensors that can provide people with valuable data about their social and physical environments. For example, increased awareness of the chemical composition of things in our immediate environment will improve human productivity, health, and security. Artificial agents based in microelectronics, nanotechnology, and bioengineering may endow people with entirely new senses or existing senses operating in new ways, in some cases employing neural interfaces to deliver complex information directly into the human mind.

3. Enriched Community

Enlightened exploitation of discoveries in the NBIC sciences will humanize technology rather than dehumanize society. Robots, intelligent agents, and information systems need to be sensitive to human needs, which is another way of saying that they must to some extent embody human personality. Over the next two decades, as nanotechnology facilitates rapid improvement of microelectronics, personal digital assistants (PDAs) are likely to evolve into smart portals to a whole world of information sources, acting as context aware personal brokers interacting with other systems maintained by corporations, governments, educational institutions, and individuals. Today's email and conference call systems could evolve into multi-media telepresence communication environments. Global Positioning System (GPS) units could become comprehensive guides to the individual's surroundings, telling the person his or her location and also locating everything of interest in the immediate locale.

To accomplish these practical human goals, we must invest in fundamental research on how to translate human needs, feelings, beliefs, attitudes, and values into forms that can guide the myriad devices and embedded systems that will be our artificial servants of the future. We must understand how interacting with and through machines will affect our own sense of personhood as we create ever more personable machines. As they become subtle reflections of ourselves, these technologies will translate information between people who are separated by perspective, interests, and even language. Without the guidance provided by the combined NBIC sciences, technology will fail to achieve its potential for human benefit. Multidisciplinary research to humanize computing and communications technology will expand the social competence of individuals and increase the practical effectiveness of groups, social networks, and organizations.

4. Learning How to Learn

We need to explore fresh instructional approaches, based in the NBIC sciences, to help us learn how to learn. Such educational tools as interactive multimedia, graphical simulations, and game-like virtual reality will enhance learning not merely from kindergarten through graduate school but also throughout the entire life course in school, in corporations, and at home. The results of past efforts have often been disappointing, because they failed to draw upon a sufficiently broad and deep scientific base. For example, instructional software typically lacked a firm grounding in the findings of cognitive science about how people actually think and learn (Bransford, Brown, and Cocking 1999).

In the future, everyone will need to learn new skills and fundamental knowledge throughout life, often in fields connected to mathematics, engineering, and the sciences. Thus we will need new kinds of curricula, such as interactive virtual reality simulations run over the Internet that will allow a student anywhere to experience the metabolic processes that take place within a living cell, as if seeing them from a nanoscale perspective. New, dynamic ways to represent mathematical logic could be developed based on a correct understanding of how the human mind processes concepts like quantity and implication, allowing more people to learn mathematics more quickly, thoroughly, and insightfully. The social interaction resulting from multiuser video games can be harnessed as a strong learning motivator, if they are designed for the user's demographic and cultural background and can infuse the learning with mystery, action, and drama. The goal would be to revolutionize science, mathematics, and engineering education through experiences that are emotionally exciting, substantively realistic, and based on accurate cognitive science knowledge about how and why people learn.

5. Enhanced Tools for Creativity

As technology becomes ever more complex, engineering design becomes an increasingly difficult challenge. For example, it is extremely costly to create large software systems, and the major bottlenecks reducing their effectiveness are unreliability and inefficiency. Similar problems beset systems for large-scale organization administration, supply chain management, industrial design, mass media, and government policy making. We can anticipate that future industries in biotechnology and nanotechnology will present unprecedented design challenges.

Investment in research and development of wholly new industrial design methods will pay great dividends. Among these, biologically inspired techniques, such as evolutionary design methods analogous to genetic algorithms, are especially promising. Terascale and petascale computer simulations are excellent approaches for many design problems, but for the foreseeable future the cost of creating a facility to do such work would be prohibitive for universities and most companies. Therefore, a national center should be established for high-end engineering design simulations. This facility could be linked to a network of users and specialized facilities, providing a distributed design environment for advanced research in engineering. Good models for creating the National Center for Engineering Design would be the supercomputer networks established by the National Science Foundation: the National Computational Science Alliance, the National Partnership for Advanced Computational Infrastructure, and the new Terascale Computing System.

At the same time, radically new methods would enhance small-scale design activities by a wide range of individuals and teams in such fields as commercial art, entertainment, architecture, and product innovation. New developments in such areas as visual language, personalized design, designing around defects, and the cognitive science of engineering could be extremely valuable. Breakthroughs in design could become self-reinforcing, as they energize the economic and technical feedback loops that produce rapid scientific and technological progress.

Statements and Visions

Participants in the human cognition and communication panel contributed a number of *statements*, describing the current situation and suggesting strategies for building upon it, as well as transformative *visions* of what could be accomplished in 10 or 20 years through a concentrated effort. The contributions include statements about societal opportunities and challenges, sensory systems, networking architecture, spatial cognition, visual language, and "companion" computers, as well as visions on predicting social behavior, design complexity, enhancing personal area sensing, understanding the brain, stimulating innovation and accelerating technological convergence.

References

Bransford, J.D., A.L. Brown, and R.R. Cocking, eds. 1999. *How people learn: Brain, mind, experience, and school.* Washington D.C.: National Research Council.

Druckman, D. and R.A. Bjork, eds. 1992. *In the mind's eye: Enhancing human performance.* Washington, D.C.: National Research Council.

_____. 1994. *Learning, remembering, believing: Enhancing human oerformance.* Washington, D.C.: National Research Council.

Food and Drug Administration. 1998. Guidance for the submission of premarket notifications for magnetic resonance diagnostic devices. U.S. Food and Drug Administration, November 14 1998: http://www.fda.gov/cdrh/ode/mri340.pdf.

Kurzweil, R. 1999. *The age of spiritual machines: When computers exceed human intelligence.* New York: Viking.

STATEMENTS

NBICS (NANO-BIO-INFO-COGNO-SOCIO) CONVERGENCE TO IMPROVE HUMAN PERFORMANCE: OPPORTUNITIES AND CHALLENGES

Jim Spohrer, IBM, CTO Venture Capital Relations

This paper is an exploration of new opportunities and challenges for improving human performance from the perspective of rapid technological change and convergence. In the past two million years, human performance has primarily been improved in two ways: evolution (physical-cognitive-social changes to people) and technology (human-made artifacts and other changes to the environment). For example, approximately one hundred thousand generations ago, physical-cognitive-social evolution resulted in widespread spoken language communication among our ancestors. About 500 generations ago, early evidence of written language existed. Then the pace of technological progress picked up: 400 generations ago, libraries existed; 40 generations ago, universities appeared; and 24 generations ago, printing of language began to spread. Again, the pace of technological advancements picked up: 16 generations ago, accurate clocks appeared that were suitable for accurate global navigation; five generations ago, telephones were in use; four, radios; three, television; two, computers; and one generation ago, the Internet.

In the next century (or in about five more generations), breakthroughs in nanotechnology (blurring the boundaries between natural and human-made molecular systems), information sciences (leading to more autonomous, intelligent machines), biosciences or life sciences (extending human life with genomics and proteomics), cognitive and neural sciences (creating artificial neural nets and decoding the human cognome), and social sciences (understanding "memes" and harnessing collective IQ) are poised to further pick up the pace of technological progress and perhaps change our species again in as profound a way as the first spoken language learning did some one hundred thousand generations ago. NBICS (nano-bio-info-cogno-socio) technology convergence has the potential to be the driver of great change for humankind. Whether or not this is in fact desirable, reasoned speculation as to how this may come to pass and the threats posed by allowing it to come to pass are increasingly available from futurists. Currently, this technology road of human performance augmentations is at the stage of macroscopic external human-computer interfaces tied into large social networking systems that exist today. Recently, there are the tantalizing first experiments of microscopic internal interfaces to assist the elderly or others with special needs; and then there is the further speculative road, with potentially insurmountable obstacles by today's standards, that leads to the interfaces of the future.

After setting the stage with longer term visions and imaginings, this paper will focus on the nearer term opportunities and challenges afforded by NBICS research and development (R&D) over the next half a generation or so. In conclusion, while futurists may be overestimating the desirability and feasibility of achieving many of their visions, we are probably collectively underestimating the impact of many of the smaller technological steps along the way.

Introduction: Motivations and Goals

At the beginning of the NBIC workshop, the participants were challenged by Newt Gingrich to think outside the box and to ambitiously consider the possible implications of the nano-info-bio-cogno convergence over the coming decades. We were also instructed to consider human dignity as an important issue, which tempered some of the cyborg speculations and other visions of humans with technology implants and augments that might seem unappealing to most people today. Thus, while social norms can shift significantly over several generations, we were primarily concerned with the world of our children and our own old-age years. We were also treated to a number of presentations describing state-of-the-art results in areas such as nanotechnology; learning technology; social acceptance of technology; designer drugs to combat diseases and other degenerative conditions; neurological implants; advanced aircraft designs highlighting smart, polymorphic (shape-shifting) materials; reports on aging, blindness, and other challenges; evolutionary software and robots; the needs of the defense department for the military of the future; augmented reality and virtual reality; and other useful perspectives on the topic of augmenting human performance. While it would be well beyond the scope of this paper to try to summarize all of these perspectives, I have tried to integrate ideas from these presentations into my own thinking about nano-info-bio-cogno convergence. Additionally, my perspective has been shaped by interactions with Doug Engelbart, whose pioneering work in the area of human augmentation systems stresses the importance of the co-evolution of technological and social systems. Because the social sciences will strongly influence which paths humans will ultimately explore as well as help us understand why, we are really concerned here with nano-bio-info-cogno-socio convergence.

Nano-bio-info-cogno-socio convergence assumes tremendous advances in each of the component science and technology areas:

1. Nanoscience advances in the coming decade will likely set the stage for a new generation of material science, biochemistry, and molecular electronics, as well as of new tools for measuring and manipulating the world at the level of individual atoms and molecules. Nanotechnology advances are poised to give humans the capabilities that bacteria have had for billions of years, the ability to create molecular machines that solve a wide range of problems on a global scale. Ultimately, these advancements will blur the distinction between natural and human-made objects.

2. Bioscience or life sciences will expand the mapping of the human genome to the human proteome, leveraging both to create new drugs and therapies to address a host of maladies of the past and new threats on the horizon.

3. Information science advances will find many applications in the ongoing e-business transformation already underway, as well as pervasive communication and knowledge management tools to empower individuals. More importantly, information science will provide both the interlingua to knit the other technologies together and the raw computational power needed to store and manipulate mountains of new knowledge.

4. Cognitive science and neuroscience will continue to advance our understanding of the human information processing system and the way our brains work.

5. Social science advances (obtained from studies of real systems as well as simulations of complex adaptive systems composed of many interacting individuals) will provide fresh insights into the collective IQ of humans, as well as interspecies collective IQ and the spread of memes. A meme, which is a term coined by the author and zoologist Richard Dawkins, is "a habit, a technique, a twist of feeling, a sense of things, which easily flips from one brain to another." It is no coincidence that meme rhymes with gene, for one is about replicating ideas (from one brain to another brain) and the other is about replicating molecules (from one cell to another cell).

6. Thus, the central question of this paper is "how might the convergence of nano-bio-info-cogno-socio technologies be accomplished and used to improve human performance" or, in the words of one workshop participant, Sandia National Laboratory scientist Gerry Yonas, to "make us all healthier, wealthier, and wiser"?

7. To gain some traction on this question, a framework, here termed simply the Outside-Inside Framework, is proposed in the next section. This framework makes explicit four of the key ways that new technologies might be used to augment human performance: (a) outside the body (environmental); (b) outside the body (personal); (c) inside the body (temporary); (d) inside the body (permanent). This framework will be shown to be largely about how and where information is encoded and exchanged: (i) info: bits and the digital

environment, (ii) cogno-socio: brains and memes and the social environments, (iii) nano-bio: bacteria and genes and the bioenvironment, (iv) nano-cogno: bulk atoms, designed artifacts, and the physical environments. In conclusion, near-term implications of NBICS technology convergence will be discussed.

The Outside-Inside Framework and Future Imaginings

The Outside-Inside framework consists of four categories of human performance-enhancing technologies:

- Outside the body and environmental
- Outside the body and personal
- Inside the body and temporary
- Inside the body and permanent

In this section, while briefly describing the categories and subcategories, some extremely speculative visions of the future will be discussed to help stretch our imaginations before "coming back to earth" in the last section to discuss more practical and near term possibilities. Readers are encouraged to view this section as a number of imagination challenges and to create their own answers to questions like what new materials, agents, places, mediators, ingestibles, senses, and species might come to be in the next few decades. In the true spirit of brainstorming, anything goes in this section. Also, it is worth noting that while futurists may be overestimating the desirability and feasibility of how quickly, if ever, we can achieve many of their visions, we are probably collectively underestimating the impact of many of the smaller technological steps along the way. Finally, as an example of improving human performance, the task of learning will be considered, focusing on the way existing and imaginary technologies may improve our ability to learn and/or perform more intelligently.

Outside the Body and Environmental.

People perform tasks in a variety of environmental contexts or places, such as homes, offices, farms, factories, hotels, banks, schools, churches, restaurants, amusement parks, cars, submarines, aircraft, space stations, and a host of other environments that have been augmented by what is termed here environmental technologies. From the materials that are used to construct the buildings and artifacts at these locations to the agents (people, domesticated animals) that provide services in these locations to the very nature of the places themselves, environmental technologies account for most of the advances in human performance that have occurred in the past 500 generations of recorded history (most of us overlap and therefore experience only about five generations of perspectives from grandparents to grandchildren). For the task of learning, consider the important roles that the three innovations — paper (material), teachers (agents), and schools (places) — have had on education. NBICS convergence will surely lead to new materials, new agents, and new places.

Outside the body and environmental: Materials. We expect that the progression from rocks, wood, bricks, cloth, ceramics, glass, bronze, iron, cement, paper, steel, rubber, plastic, semiconductors, and so on, will be augmented with new materials, such as smart, chromatically active (change color), polymorphic (change shape) materials such as those NASA is already experimenting with. For a thought-provoking vision of where new materials could lead, the reader is directed to the currently infeasible but intriguing notion of "utility fog" developed by Rutgers computer science professor J. Storrs Hall in the early 1990s. Smaller than dust, "foglets" are speculative tiny interlocking machines that can run "programs" that make collections of billions of them work together to assume any shape, color, and texture, from flowing, clear water to fancy seat belt body suits that appear only when an accident has occurred. If utility fog were a reality, most artifacts could be made invisible until needed, making them quite portable. There would be no need to carry luggage on trips; one could simply create clothes out of utility fog. Materializing objects out of thin air (or fog), while wildly infeasible today, nevertheless provides an interesting springboard for imagining some of the ultimate human-computer interfaces (such as a second skin covering human bodies, eyes, ears, mouth, nose, and skin) that may someday exist. Perhaps these ultimate interfaces might connect us to telerobotic versions of ourselves assembled out of utility fog in distance places.

There are many reasons to be skeptical about utility fog (the Energy budget, for one), but notions like utility fog help us understand the potential of NBICS. For example, multi-cellular organisms provide a vast library of examples of the ways cells can be interlinked and grouped to produce shapes, textures, and macroscopic mechanical structures. Social insects like ants have been observed to interlink to solve problems in their environments. And while I'm unaware of any types of airborne bacteria that can spontaneously cluster into large groups, I suspect that mechanisms that bacteria and slime molds use for connecting in various arrangements may one day allow us to create new kinds of smart materials. Hopefully the notion of utility fog has served its brainstorming purpose of imagination stretching, and there are a number of related but nearer term investigations underway. For example, U.C.-Berkeley professor and microroboticist Kris Pister's Smart Dust and Micromechanical Flying Insect projects are good examples of the state-of-the-art in building microrobots, and as these microrobots get smaller, they may very well pave the way to many exciting new materials.

Outside the body and environmental: Agents. Interacting with intelligent agents, such as other people and other species (e.g., guide dogs), has clear advantages for augmenting human performance. Some of the most important agents we interact with daily are role-specialized people and businesses (organization as agent). The legal process of incorporating a business or nonprofit organization is essentially equivalent to setting up a fictitious person with specialized rights, responsibilities, and capabilities. The notion of new agents was an active area of discussion among the workshop participants: from the implications of digital personae (assumed identities on-line) to artificial intelligence and robotics, as well as the evolution of new types of organizations. The successful entrepreneur and futurist Ray Kurzweil has a website *kurzweilai.net* (see Top KurzweilAI News of 2001) that explores these and other futures and interestingly includes Kurzweil's alter-ego, Ramona!, that has been interviewed by the press to obtain Kurzweil's views on a variety of subjects. Undoubtedly, as technology evolves, more digital cloning of aspects of human interactions will occur. An army of trusted agents that can interact on our behalf has the potential to be very empowering as well as the potential to be quite difficult to update and maintain synchrony with the real you. What happens when a learning agent that is an extension of you becomes more knowledgeable about a subject than you? This is the kind of dilemma that many parents and professors have already faced.

Outside the body and environmental: Places. New places create new opportunities for people. The exploration of the physical world (trade connecting ancient civilization, New World, the Wild West, Antarctica, the oceans, the moon, etc.) and the discovery of new places allows new types of human activities and some previously constrained activities to flourish. For example, the New World enhanced the Puritans' abilities to create the kind of communities they wanted for themselves and their children. Moving beyond the physical world, science fiction writer William Gibson first defined the term *cyberspace.* The free thinking artist and futurist Jaron Lanier, who coined the term *virtual reality*, and many other people have worked to transform the science fiction notion of cyberspace into working virtual reality technologies. Undoubtedly, the digital world will be a place of many possibilities and affordances that can enhance human performance on a wide variety of tasks, including both old, constrained activities as well as new activities. The increasing demand for home game machines and combinatorial design tools used by engineers to explore design possibilities is resulting in rapid advances in the state-of-the-art creation of simulated worlds and places. Furthermore, in the context of learning, inventor and researcher Warren Robinett, who was one of the workshop participants, co-created a project that allows learners to "feel" interactions with simulated molecules and other nanostructures via virtual realities with haptic interfaces. In addition, Brandeis University professor Jordan Pollack, who was also one of the workshop participants, described his team's work in the area of combinatorial design for robot evolution, using new places (simulated worlds) to evolve new agents (robots) and then semi-automatically manifest them as real robots in the real world. Also, it is worth noting that in simulated worlds, new materials, such as utility fog, become much easier to implement or, more accurately, at least emulate.

Outside the Body and Personal

The second major category, personal technologies, are technologies that are outside of the body, but unlike environmental technologies are typically carried or worn by a person to be constantly available. Two of the earliest examples of personal technologies were of course clothing and jewelry, which both arose thousands of generations ago. For hunter gatherers as well as cowboys in the American West, weapons were another form of early personal technology. Also included in this category are money, credit cards, eyeglasses, watches, pens, cell phones, handheld game machines, and PDAs (personal digital assistants). For learners, a number of portable computing and communication devices are available, such as leapfrog, which allows students to prepare for quizzes on chapters from their school textbooks, and graphing calculators from Texas Instruments. Recently, a number of wearable biometric devices have also appeared on the market.

Outside the body and personal: Mediators. Mediators are personal technologies that include cellphones; PDAs; and handheld game machines that connect their users to people, information, and organizations and support a wide range of interactions that enhance human performance. WorldBoard is a vision of an information infrastructure and companion mediator devices for associating information with places. WorldBoard, as originally conceived in my papers in the mid-1990s, can be thought of either as a planetary augmented reality system or a sensory augment that would allow people to perceive information objects associated with locations (e.g., virtual signs and billboards). For example, on a nature walk in a national park a person could use either heads-up display glasses or a cell phone equipped with a display, camera, and GPS (Global Positioning System) to show the names of mountains, trees, and buildings virtually spliced into the scenes displayed on the glasses or cell phone. WorldBoard mediators might be able to provide a pseudo X-ray vision, allowing construction equipment operators to see below the surface to determine the location of underground buried pipes and cables rather than consulting blueprints that might not be available or might be cumbersome to properly orient and align with reality. The slogan of WorldBoard is "putting information in its place" as a first step to contextualizing and making useful the mountains of data being created by the modern day information explosion.

Human-made tools and artifacts are termed mediators, in this paper, because they help externalize knowledge in the environment and mediate the communication of information between people. Two final points are worth making before moving inside the body. First, the author and cognitive scientist Don Norman, in his book *Things that Make Us Smart* provides an excellent, in-depth discussion of the way human-made tools and artifacts augment human performance and intelligence. Furthermore, Norman's website includes a useful article on the seeming inevitability of implants and indeed cyborgs in our future, and why implants will become increasingly accepted over time for a wider and wider range of uses. A second point worth making in the context of mediators is that human performance could be significantly enhanced if people had more will power to achieve the goals that they set for themselves. Willpower enforcers can be achieved in many ways, ranging from the help of other people (e.g., mothers for children) to mediator devices that remove intentionality from the equation and allow multitasking (e.g., FastAbs electric stimulation workout devices).

Inside the Body and Temporary

The third major category, inside the body temporary technologies, includes most medicines (pills) as well as new devices such as the camera that can be swallowed to transmit pictures of a journey through a person's intestines. A number of basic natural human processes seem to align with this category, including inhaling and exhaling air; ingesting food and excreting waste; spreading infections that eventually overcome the body's immune system; as well as altered states of awareness such as sleep, reproduction, pregnancy, and childbirth.

Inside the body and temporary: Ingestibles. Researchers at Lawrence Livermore National Laboratories have used mass spectrometry equipment to help study the way that metabolisms of different people vary in their uptake of certain chemical components in various parts of the body. Eventually, this line of investigation may lead to precisely calibrating the amount of a drug that an individual should take to achieve an optimal benefit from ingesting it. For example, a number of studies show positive effects of mild stimulants, such as coffee, used by subjects who were studying material to be learned, as well as positive effects from being in the appropriate mental and physical states when performing particular tasks. However, equally clear from the data in these studies are indications that too much or too little of a good thing can result in no enhancement or detrimental side effects instead of enhanced performance.

With the exception of an Air Force 2025 study done by the Air University, I have not yet found a reference (besides jokes, science fiction plots, and graduate school quiz questions), to what I suspect is someone's ultimate vision of this ingestible enhancements subcategory, namely a learning pill or knowledge pill. Imagine that some day we are able to decode how different brains store information, and one can simply take a custom designed learning pill before going to sleep at night to induce specific learning dreams, and when morning arrives the person's wetware will have been conditioned or primed with memories of the new information. Staggeringly improbable, I know.

Nevertheless, what if someone could take a pill before falling asleep at night, and awaken in the morning knowing or being conditioned to more rapidly learn how to play, for example, a game like chess? If learning could be accelerated in this manner, every night before going to bed, people would have a "learning nightcap." Imagine an industry developing around this new learning pill technology. The process at first might require someone spending the time to actually learn something new, and monitoring and measuring specific neurological changes that occur as a result of the learning experience, and then re-encoding that information in molecular machines custom-designed for an individual to attach himself or herself to locations in the brain and interact with the brain to create dream-like patterns of activation that induce time-released learning. Businesses might then assign learning pills to their employees, schools might assign learning pills to their students, soldiers might take learning pills before being sent out on missions (per the Air Force 2025 study that mentioned a "selective knowledge pill"), and families might all take learning pills before heading out on vacations. However, perhaps like steroids, unanticipated side effects could cause more than the intended changes.

What makes the learning pill scenario seem so far-fetched and improbable? Well, first of all, we do not understand much about the way that specific bits of information are encoded in the brain. For example, what changes in my brain (short-term and then long-term memory) occur when I learn that there is a new kind of oak tree called a Live Oak that does not lose its leaves in the winter? Second, we do not know how to monitor the process of encoding information in the brain. Third, different people probably have idiosyncratic variations in the ways their brains encode information, so that one person's encoding of an event or skill is probably considerably different from another person's. So how would the sharing work, even if we did know how it was encoded in one person's brain? Fourth, how do we design so many different molecular machines, and what is the process of interaction for time-released learning? Fifth, exactly how do the molecular machines attach to the right parts of the brain? And how are they powered? We could go on and on, convincing ourselves that this fantasy is about as improbable as any that could possibly be conceived. Nevertheless, imagination-stretching warmups like these are useful to help identify subproblems that may have nearer term partial solutions with significant impacts of their own.

Inside the Body and Permanent

The fourth major category, inside the body permanent technologies, raises the human dignity flag for many people, as negative images of cyborgs from various science fiction fare leap immediately to mind. The science fact and e-life writer Chris O'Malley recently wrote a short overview of this area. Excerpts follow:

> Largely lost in the effort to downsize our digital hardware is the fact that every step forward brings us closer to an era in which computers will routinely reside within us. Fantasy? Hardly. We already implant electronics into the human body. But today's pacemakers, cochlear implants, and the like will seem crude — not to mention huge — in the coming years. And these few instances of electronic intervention will multiply dramatically... The most pervasive, if least exciting, use of inner-body computing is likely to be for monitoring our vital stats (heart rate, blood pressure, and so on) and communicating the same, wirelessly, to a home healthcare station, physician's office, or hospital. But with its ability to warn of imminent heart attacks or maybe even detect early-stage cancers, onboard monitoring will make up in saved lives what it lacks in sex appeal... More sensational will be the use of internal computers to remedy deficiencies of the senses. Blindness will, it seems reasonable to speculate, be cured through the use of electronic sensors — a technology that's already been developed. So, too, will deafness. Someday, computers may be able to mimic precisely the signal that our muscles send to our brain and vice versa, giving new mobility to paralysis victims. Indeed, tiny computers near or inside our central processing unit, the human brain, could prove a cure for conditions such as Alzheimer's, depression, schizophrenia, and mental retardation... Ethical dilemmas will follow, as always...

Inside the body and permanent: New organs (senses and effectors). This subcategory includes replacement organs, such as cochlear implants, retinal implants, and pacemakers, as well as entirely new senses. People come equipped with at least five basic senses: sight, hearing, touch, taste, and smell. Imagine if we were all blind but had the other four senses. We'd design a world optimized for our sightless species, and probably do quite well. If we asked members of that species to design a new sense, what might they suggest? How would they even begin to describe vision and sight? Perhaps they might describe a new sense in terms of echolocation, like a species of bats, that would provide a realtime multipoint model of space in the brain of the individual that could be reasoned to be capable of preventing tripping on things in hostile environments.

In our own case, because of the information explosion our species has created, I suggest that the most valuable sixth sense for our species would be a sense that would allow us to quickly understand, in one big sensory gulp, vast quantities of written information (or even better, information encoded in other people's neural nets). Author Robert Lucky has estimated that all senses give us only about 50 bits per second of information, in the Shannon sense. A new high bandwidth sense might be called a Giant UpLoad Process or the GULP Sense. Imagine a sixth sense that would allow us to take a book and gulp it down, so that the information in the book was suddenly part of our wetware, ready for inferencing, reference, etc., with some residual sense of the whole, as part of the sensory gulp experience. Just as some AI programs load ontologies and rules, the GULP sense would allow for rapid knowledge uptake. A GULP sense would have a result not unlike the imaginary learning pill above. What makes the information-gulping sixth sense and the learning pill seem so fantastic has to do in part with how difficult it is for us to transform information encoded in one format for one set of processes into information encoded in another format for a different set of processes — especially when one of those formats is idiosyncratic human encoding of information in our brains. Perhaps the closest analogy today to the complexity of transforming information in one encoding to another is the ongoing transformation of businesses into e-businesses, which requires linking idiosyncratic legacy systems in one company to state-of-the-art information systems in another company.

The process of creating new sensory organs that work in tandem with our own brains is truly in a nascent state, though the cochlear implant and retinal implant directions seem promising. University of Texas researcher Larry Cauller, who was one of the workshop participants, grabbed the bull by the horns and discussed ways to attack the problem of building an artificial brain as well as recent technology improvements in the area of direct neural interfaces. As neural interface chips get smaller, with finer and more numerous pins, and leveraging RF ID tag technology advances, the day is rapidly approaching where these types of implants can be done in a way that does minimal damage to a brain receiving a modern neural interface implant chip. Improved neural interface chips are apparently already paying dividends in deepening the understanding of the so-called mirror neurons that are tied in with the "monkey see, monkey do" behaviors familiar in higher primates. One final point on this somewhat uncomfortable topic, MIT researcher and author Sherry Turkle, who was also a workshop participant, presented a wealth of information on the topic of sociable technologies as well as empirical data concerning people's attitudes about different technologies. While much of the discussion centered on the human acceptance of new agents such as household entertainment robots (e.g., Sony's AIBO dog), there was unanimous agreement among all the participants that as certain NBICS technologies find their way into more universally available products, attitudes will be shaped, positively as well as negatively, and evolve rapidly, often in unexpected ways for unexpected reasons.

Tokyo University's Professor Isao Shimoyama has created a robo-roach or cyborg roach that can be controlled with the same kind of remote that children use to control radio-controlled cars. Neural interfaces to insects are still crude, as can be seen by going to Google and searching for images of "robo-roach." Nevertheless, projecting the miniaturization of devices that will be possible over the next decade, one can imagine tools that will help us understand the behaviors of other species at a fine level of detail. Ultimately, as our ability to rapidly map genes improves, neural interface tools may even be valuable for studying the relationship between genes and behaviors in various species. NBICS convergence will accelerate as the linkages between genes, cellular development, nervous systems, and behavior are mapped.

Inside the body and permanent: New skills (new uses of old sensors and effectors). Senses allow us to extract information from the world, exchange information between individuals, and encode and remember relevant aspects of the information in our brains (neural networks, wetware). Sometimes physical, cognitive, and social evolution of a species allows an old sense to be used in a new way. Take, for example, verbal language communication or speech. Long before our ancestors could effectively listen to and understand spoken language, they could hear. A lion crashing through the jungle at them registered a sound pattern in their prehuman brains and caused action. However, over time, a set of sound associations with meaning and abstractions, as well as an ability to create sounds, along with increased brain capacity for creating associations with symbols and stringing them together via grammars to create complex spoken languages, occurred. Over time, large groups of people shared and evolved language to include more sounds and more symbolic, abstract representations of things, events, and feelings in their world. An important point about acquiring new skills, such as sounds in a language, is that infants and young children have certain advantages. Evidence indicates that brains come prewired at the neural level for many more possibilities than actually get used, and if those connections are not needed, they go away. Once the connections go away, learning can still occur, but the infant brain advantage is no longer available. Essentially, the infant brain comes prewired to facilitate the development of new uses of old sensors and effectors.

Entrepreneur and author Bob Horn, who was also a participant at the workshop, argues that visual languages have already evolved and can be further evolved — perhaps, dramatically so for certain important categories of complex information, and thus progressing towards the information gulp-like sense alluded to above. In addition, researchers at IBM's Knowledge Management Institute and elsewhere offer stories and story languages as a highly evolved, and yet mostly untapped — except for entertainment purposes — way to rapidly convey large volumes of information. For example, when I mention the names of two television shows, The Honeymooners and The Flintstones, many TV-literate Americans in their forties and fifties will understand that these have in fact the same basic story formula, and will immediately draw on a wealth of abstractions and experience to interpret new data in terms of these stories. They may even be reminded of a Honeymooner episode when watching a Flintstone cartoon — this is powerful stuff for conveying information. The generation of television and videogame enthusiasts have a wealth of new cognitive constructs that can be leveraged in the evolution of a new sense for rapid, high volume information communication. Certainly, new notations and languages (e.g., musical notation, programming languages, and mathematics) offer many opportunities for empowering people and enhancing their performance on particular tasks. All these approaches to new uses of old senses are primarily limited by our learning abilities, both individually and collectively. Like the evolution of speech, perhaps new portions of the brain with particular capabilities could accelerate our ability to learn to use old senses in new ways. An ability to assimilate large amounts of information more rapidly could be an important next step in human evolution, potentially as important as the evolution of the first language spoken between our ancestors.

Inside the body and permanent: New genes. If the notion of "computers inside" or cyborgs raise certain ethical dilemmas, then tinkering with our own genetic code is certain to raise eyebrows as well. After all, this is shocking and frightening stuff to contemplate, especially in light of our inability to fully foresee the consequences of our actions. Nevertheless, for several reasons, including, for the sake of completeness in describing the Outside-Inside Framework, this is an area worth mentioning. While selective breeding of crops, animals, and people (as in ancient Sparta) is many hundreds of generations old, only recently have gene therapies become possible as the inner working of the billion-year-old molecular tools of bacteria for slicing and splicing DNA have been harnessed by the medical and research communities. Just as better understanding of the inner working of memory of rodents and its genetic underpinnings have allowed researchers to boost the IQs of rodents on certain maze running tasks, soon we can expect other researchers building on these results to suggest ways of increasing the IQs of humans.

University of Washington researcher and medical doctor Jeffrey Bonadio (Bonadio 2002), who was a workshop participant, discussed emerging technologies in the area of gene therapies. Gene therapy is the use of recombinant DNA as a biologic substance for therapeutic purposes, using viruses and other means to modify cellular DNA and proteins for a desired purpose.

In sum, the Outside-Inside Framework includes four main categories and a few subcategories for the ways that technology might be used to enhance human performance:

- Outside the body and environmental
  - new materials
  - new agents
  - new places
  - new mediators (tools and artifacts)
- Outside the body, personal
  - new mediators (tools and artifacts)
- Inside the body, temporary
  - new ingestibles
- Inside the body, permanent
  - new organs (new sensors and effectors)
  - new skills (new uses of old sensors and effectors)
  - new genes

The four categories progress from external to internal changes, and span a range of acceptable versus questionable changes. In the next section, we'll consider these categories from the perspective of information encoding and exchange processes in complex dynamic systems.

Information Encoding and Exchange: Towards a Unified Information Theoretic Underpinning

The Outside-Inside Framework provides one way to organize several of the key issues and ideas surrounding the use of NBICS technology advances to enhance human performance ("make us all healthier, wealthier, and wiser"). This simple framework can be shown to be largely about understanding and controlling how, where, and what information is encoded and exchanged. For example, consider the following four loosely defined systems and the way information is encoded differently, and interdependently, in each: (a) bits and the digital environment (information), (b) brains and memes and the social environment (cogno-socio), (c) bacteria and genes and the bioenvironment (nano-bio), (d) bulk atoms, raw materials, designed artifacts, and the physical environment (nano-based).

At this point, a brief digression is in order to appreciate the scale of successful information encoding and evolution in each of these loosely defined systems. People have existed in one form or another for about 2 million years, which is a few hundred thousand generations (to an order of magnitude). Today, there are about six billion people on Earth. The human body is made up of about 10^{13} cells, the human brain about 10^{10} cells (10^{27} atoms), and the human genome is about 10^9 base pairs. Humans have been good problem solvers over the generations, creating successful civilizations and businesses as well as creating a growing body of knowledge to draw on to solve increasingly complex and urgent problems. However, in some ways, even more impressive than humans are bacteria, according to author Howard Bloom (2001). Bacteria have existed on Earth for about 3.5 billion years, which is an estimated 10^{14} bacteria generations ago. Today, there are an estimated 10^{30} bacteria (or about one hundred million bacteria for every human cell) on Earth living inside people, insects, soil, deep below the surface of the Earth, in geothermal hot springs in the depths of the ocean, and in nearly every other imaginable place. Bacteria have been successful "problem-solvers," as is evidenced by their diversity and ever-growing bag of genetic tricks for solving new problems. People have made use of bacteria for thousands of generations (though electronic digital computers only recently) in producing bread, wine, and cheese, but only in the past couple of generations have bacteria become both a tool kit and a road map for purposeful gene manipulation. Bacteria and viruses are both an ally and a threat to humans. For example, bacterial or viral plagues like the influenza outbreak of 1917 are still a major threat today. Among our best new allies in this fight are the advances in life sciences technologies enabled by more powerful digital technology. Most recently, electronic transistors have been around for less than a century, and at best, we have only a few dozen generations of manufacturing technology. Today, there are more than 10^{18} transistors on Earth, and very roughly 10 million transistors per microprocessor, 100 million PCs manufactured per year, and 10 billion embedded processors.

Returning to the issue of understanding and controlling how, where, and what information is encoded and exchanged, consider the following milestones in human history (where GA is human generations ago), as seen through the lens of the Outside-Inside Framework:

- Speech (100,000 GA): A new skill (new use of old sensors and effectors, requires learning a new audible language), encoding information in sounds, for exchanging information among people. Probably coincides with the evolution of new brain centers, new organs.

- Writing (500 GA): A new mediator and new skill (new use of old sensors and effectors, requires learning a new visual language), encoding information in visual symbols on materials from the environment for recording, storing, and exchanging information between people. Did not require new brain centers beyond those required for spoken language.

- Libraries (400 GA): A new place and agent (organization) for collecting, storing and distributing written information.
- Universities (40 GA): A new place and agent (organization) for collecting, storing, and distributing information as social capital.
- Printing (14 GA): A new mediator (tool) for distributing information by making many physical copies of written and pictorial information.
- Accurate clocks (16 GA): A new mediator (tool) for temporal information and spatial information (accurate global navigation).
- Telephone (5 GA): A new mediator (tool) for exchanging audio information encoded electrically and transported via wires over great distances.
- Radio (4 GA): A new mediator (tool) for distributing audio information encoded electromagnetically and transported wirelessly over great distances.
- Television (3 GA): A new mediator (tool) for distributing audiovisual information encoded electromagnetically, transported wirelessly over great distances.
- Computers (2 GA): A new mediator and agent for storing, processing, creating, and manipulating information encodable in a binary language.
- Internet (1 GA): A new mediator for distributing information encodable in a binary language.
- Global Positioning System or GPS (0 GA): A new mediator for spatial and temporal (atomic clock accuracy) information.

Stepping back even further for a moment (per Bloom 2001), we can identify six fundamental systems for encoding and accumulating information: matter, genes, brains, memes, language, and bits:

- Big Bang (12 billion years ago): New place and new material - the Universe and matter
- Earth (4.5 billion years ago): New place and new materials - the Earth and its natural resources
- Bacteria (3.5 billion years ago): New species and agent, encoding information in primitive genome (DNA) in cells
- Multicellular (2.5 billion years ago): New species with multicellular chains and films
- Clams (720 million years ago): New species with multiple internal organs with primitive nervous systems
- Trilobites (500 million years ago): New species with simple brains for storing information (memes possible)
- Bees (220 million years ago): New species and agent; social insect with memes, collective IQ

- Humans and Speech (2 million years ago): New species and agent, with primitive spoken language and tools, extensive memes, collective IQ

- Writing (about 10 thousand years ago): New mediator, recordable natural language and symbolic representations

- Computers (about 50 years ago): New mediator and agent, binary language and predictable improvement curve through miniaturization

Of course, all these dates are very approximate. The important point is simply this: if the past is the best predictor of the future, then we can expect NBICS convergence to shed light on all of these key systems for encoding, exchanging, and evolving information. If (and this is a big if) we can (1) truly understand (from an information processing standpoint) the working of material interactions, genes and proteins, nervous systems and brains, memes and social systems, and natural language, and translate all this into appropriate computational models, and (2) use this deep model-based understanding to control and directly manipulate their inner workings to short-cut the normal processes of evolution, then perhaps we can create improvements (solve complex urgent problems) even faster. Of course, accelerating evolution in this way is both staggeringly difficult to do in reality as well as potentially very empowering and dangerous if we should succeed.

Again, the point here is simply that NBICS convergence has zeroed in on the key, few separate information systems that drive enhancements not only to human performance, but to the universe as we know it: matter, genes, brains, memes, language, and bits. Does this mean that we have bitten off too much? Perhaps, but it does seem to be time to ask these kinds of convergence questions, much as physicists in the late 1800s began a quest to unify the known forces. In essence, the quest for NBICS convergence is looking for the Maxwell's equations or, better yet, the "unified field theory" for complex dynamic systems that evolve, but in terms of models of information encoding, and exchange instead of models of particle and energy exchange. Author and scientist Richard Dawkins in his book *The Selfish Gene* foreshadows some of this thinking with his notion of a computational zoology to better understand why certain animal behaviors and not others make sense from a selfish gene perspective. Author and scientist Stuart Kaufman in his book *At Home in the Universe: The Search for the Laws of Self-Organization and Complexity* is searching for additional mechanisms beyond evolution's natural selection mechanism that could be at work in nature. Testing and applying these theories will ultimately require enormous computational resources.

It is interesting to note that computational power may become the limiting factor to enhancing human performance in many of the scenarios described above. What happens when Moore's Law runs out of steam? To throw one more highly speculative claim into the hopper, perhaps quantum computing will be the answer. Recently, IBM researchers and collaborators controlled a vial of a billion-billion (10^{18}) molecules designed to possess seven nuclear spins. This seven qubit quantum computer correctly factored the number 15 via Shor's algorithm and had its input programmed by radio frequency pulses and output detected by a nuclear magnetic resonance instrument. Certainly, there is no shortage of candidates for the next big thing in the world of more computing power.

Concluding Remarks: Near-Term Opportunities for e-Business Infrastructure

So what are the near-term opportunities? The R&D community is engaged. From an R&D perspective, the five innovation ecosystems (university labs, government labs, corporate labs, venture capital backed start-ups, and nonprofit/nongovernment organizations) have already geared up initiatives in all the separate NBICS (nano-bio-info-cogno-socio) areas, somewhat less in socio, and cogno is perhaps secondary to neuro. However, what about real products and services coming to market and the converged NBICS as opposed to separate threads?

From a business perspective, a number of existing technology trends generally align with and are supportive of NBICS directions. One of the major forces driving the economy these days is the transformation of businesses into e-businesses. The e-business evolution (new agent) is really about leveraging technology to enhance all of the connections that make businesses run: connections to customers, connections to suppliers, connections between employees and the different organizations inside a business, and connections to government agencies, for example. Some aspects of the NBICS convergence can not only make people healthier, wealthier, and wiser, but can make e-businesses healthier, wealthier, and wiser as well. I suspect that while many futurists are describing the big impact of NBICS convergence on augmenting human performance, they are overlooking the potentially larger and nearer term impacts of NBICS convergence on transforming businesses into more complete e-businesses. The area of overlap between what is good for business and what is good for people is in my mind one of the first big, near term areas of opportunity for NBICS convergence. Improving human performance, like improving business performance will increasingly involve new interfaces to new infrastructures.

a) *Communication infrastructure*: The shift from circuits to packets and electronics to photonics and the roll out of broadband and wireless will benefit both businesses and individuals.

b) *Knowledge infrastructure*: Knowledge management, semantic search, and natural language tools will make businesses and people act smarter.

c) *Sensor infrastructure*: Realtime access to vital information about the health of a person or business will be provided.

d) *Simulation infrastructure*: There will be a shift from *in vitro* to *in silico* biology for the design and screening of new drugs for people and new products for businesses.

e) *Intellectual property, valuations and pricing, human capital infrastructure*: Inefficiencies in these areas are a major drag on the economy overall.

f) *Miniaturization, micromanipulation, microsensing infrastructure*: Shrinking scales drive chip businesses and open new medical applications.

g) *Computing infrastructure (grid - social)*: This is still emerging, but ultimately, computer utility grids will be an enormous source of computing power for NBICS efforts.

h) *Computing infrastructure (autonomic - biological)*: The cost of managing complex technology is high; the autonomic borrows ideas from biological systems.

Already, IBM Research has begun to articulate some of the challenges and the promise of autonomic computing (http://www.research.ibm.com/autonomic/), which seeks to build a new generation of self-managing, self-regulating, and self-repairing information technology that has some of the advantages of living systems. As NBICS convergence happens, our information technology infrastructure will benefit, making many businesses more efficient and more viable.

Ultimately, NBIC convergence will lead to complete computational models of materials, genes, brains, and populations and how they evolve, forever improving and adapting to the demands of changing environments. A first step is to understand the way information is encoded and exchanged in each of these complex dynamic systems and to apply that new understanding to enhance each system. While this is an exciting undertaking, especially in light of recent advances in mapping the human genome, nanotechnology advances, and 30 some years of unabated miniaturization (Moore's Law) driving up computational capabilities, it is also a time to admit that this is still a multi-decade undertaking with lots of twists and turns in the road ahead. Better frameworks that help us inventory and organize the possibilities, as well as glimpse the ultimate goal of NBICS convergence, are still needed.

References

Bloom, H. 2001. Global brain: The evolution of mass mind from the big bang to the 21st century. John Wiley & Sons.
Bonadio, J. 2002. Gene therapy: Reinventing the wheel or useful adjunct to existing paradigms? In this volume.

SENSOR SYSTEM ENGINEERING INSIGHTS ON IMPROVING HUMAN COGNITION AND COMMUNICATION

Brian M. Pierce, Raytheon Company

The improvement of human cognition and communication can benefit from insights provided by top-down systems engineering used by Raytheon and other aerospace and defense companies to design and develop their products. Systems engineering is fundamental to the successful realization of complex systems such as multifunction radar sensors for high performance aircraft or the Army's Objective Force Warrior concept for the dismounted soldier. Systems engineering is very adept at exploring a wide trade space with many solutions that involve a multitude of technologies. Thus, when challenged by the theme of the workshop to evaluate and explore convergent technologies (nanoscience and nanotechnology, biotechnology and biomedicine, information technology, and cognitive science) for improving human cognition and communication, it was natural to start with a top-down systems engineering approach.

One of the first questions to be asked is what is meant by improvement. In sensor systems engineering, improvement covers a wide range of issues such as performance, cost, power and cooling, weight and volume, reliability, and supportability. The ranking of these issues depends on the mission for the system in question and on the sensor platform. For example, a surveillance radar system on an aircraft has much more emphasis on the power and cooling issues than does a ground-based radar system.

Improvement has many facets in the context of the Army's Objective Force Warrior system for the dismounted soldier: enhanced fightability without impeding movement or action; minimal weight; efficient, reliable, and safe power; integratability; graceful degradation; trainability; minimal and easy maintenance (ultra-reliability); minimal logistics footprint; interoperability; and affordability. The prioritization of these requirements could change depending on whether the warrior is based on space, airborne, surface ship, or undersea platforms. Ideally, the adaptability of the system is high enough to cover a wide range of missions and platforms, but issues like cost can constrain this goal.

Improvement is a relative term, and improvement objectives in the case of human cognition depend on the definition of the baseline system to be improved, e.g., healthy versus injured brain. Furthermore, does one focus solely on cognition in the waking conscious state, or is the Rapid Eye Movement (REM) sleeping conscious state also included? Although recent memory, attention, orientation, self-reflective awareness, insight, and judgment are impaired in the REM sleep state, J. Allen Hobson suggests that this state may be the most creative one, in which the chaotic, spontaneous recombination of cognitive elements produces novel configurations of new information resulting in new ideas (Hobson 1999).

Improvement objectives for human communication include enhancements in the following:

a) communication equipment *external* to the individual, e.g., smaller, lighter cell phones operable over more frequencies at lower power

b) information transfer *between* equipment and individual, i.e., through human-machine interface

c) communication and cognitive capabilities *internal* to the individual, e.g., communication outside the normal frequency bands for human vision and hearing.

If one reviews the evolution of cognitive and communication enhancement for the dismounted soldier during the last several decades, improvements in equipment *external* to the soldier and the human-machine interface predominate. For example, Raytheon is developing uncooled infrared imagers for enhanced night vision, a tactical visualization module to enable the visualization of a tactical situation by providing realtime video, imagery, maps, floor plans, and "fly-through" video on demand, and GPS and antenna systems integrated with the helmet or body armor. Other external improvements being developed by the Department of Defense include wearable computers, ballistic and laser eye protection, sensors for detection of chemical and biological warfare agents, and smaller, lighter, and more efficient power sources. Improvements that would be inside the individual have been investigated as well, including a study to enhance night vision by replacing the visual chromophores of the human eye with ones that absorb in the infrared, as well as the use of various drugs to achieve particular states of consciousness.

The convergent technologies of nanoscience and nanotechnology, biotechnology and biomedicine, information technology, and cognitive science have the potential to accelerate evolutionary improvements in cognition and communication *external* to the individual and the human-machine interface, as well as enable revolutionary improvements *internal* to the individual. The recent workshop on nanoscience for the soldier identified several potential internal improvements to enhance soldier performance and to increase soldier survivability: molecular internal computer, sensory, and mechanical enhancement; active water reclamation; short-term metabolic enhancement; and regeneration/self-healing (Army Research Laboratory 2001).

The trend in sensor systems is towards the integrated, wide band, multifunction sensor suite, in which processor/computer functions are extended into the sensing elements so that digitization occurs as early as possible in the sensing process. This type of sensor architecture enables a very high degree of adaptability and performance. However, one still has to trade the pros and cons of handling the increasing torrent of bits that results from digitizing closer to the sensor's front-end. For example, the power consumption associated with digitization can be an important consideration for a given platform and mission.

Improvements in human cognition and communication will also follow a path of higher integration and increased functionality. The exciting prospect is that the convergent technologies encompass the three major improvement paths: external, human-machine interface, and internal. This breadth should make it possible to pursue a more complete system solution to a particular need. If better night vision is desired, the convergent technologies could make it possible to trade a biological/chemical approach of modifying the photoreceptors in the eye, a micro/nano-optoelectronic imager external to the eye, or a hybrid of the two. Memory enhancement is an important element of improving human cognition, and perhaps convergent technologies could be used to build on work that reports using external electrical stimulation (Jiang, Racine, and Turnbull 1997) or infusion of nerve growth factor (Frick et al. 1997) to improve/restore memory in aged rats.

Sensor systems have benefited enormously from architectures inspired by the understanding of human cognition and communication. The possibility exists for sensor system engineering to return the favor by working in concert with the convergent technologies of nanoscience and nanotechnology, biotechnology and biomedicine, information technology, and cognitive science.

References

Army Research Laboratory/Army Research Office. 2001. Workshop on Nanoscience for the Soldier. February 8-9.

Frick, K.M., D.L. Price, V.E. Koliatsos, and A.L. Markowska. 1997. The effects of nerve growth factor on spatial recent memory in aged rats persist after discontinuation of treatment. *Journal of Neuroscience.* 17(7):2543-50 (Apr 1).

Hobson, J.A. 1999. *Consciousness.* NY: Scientific American Library. pg. 45.

Jiang, F., R. Racine, and J. Turnbull. 1997. Electrical stimulation of the septal region of aged rats improves performance in an open-field maze. *Physiology and Behavior* 62(6):1279-82 (Dec.).

CAN NANOTECHNOLOGY DRAMATICALLY AFFECT THE ARCHITECTURE OF FUTURE COMMUNICATIONS NETWORKS?

Cherry A. Murray, Lucent Technologies

We live in an era of astounding technological transformation — the Information Revolution — that is as profound as the two great technological revolutions of the past — the Agricultural and Industrial Revolutions. All around us are now familiar technologies whose very existence would have seemed extraordinary just a generation ago: cellular telephones, the optical fiber telecommunications backbone, the Internet, and the World Wide Web. All of the underlying technologies of the Information Age are experiencing exponential growth in functionality due to decreasing size and cost of physical components — similar to Moore's Law in silicon-integrated electronics technology. In the next decade, the size scale of many communications and computing devices — such as individual transistors — is predicted to decrease to the dimension of nanometers; where fundamental limits may slow down, single device functionality will increase. Before these fundamental limits are even attained, however, we must address the difficult assembly and interconnection problems with a network of millions of small devices tied together to provide the increased functionality at lower cost. If the interconnection problem is solved and if the cost of physical elements is dramatically reduced, the architectures of future communications networks — and the Internet itself — can be dramatically changed. In order for this to happen, however, we must have a breakthrough in our ability to deal with the statistical nature of devices in the simulation and design of complex networks on several levels.

Fundamental Limits to Individual Devices

Communications and computing rely ultimately on individual devices such as transistors, optical switching elements, memory elements, and detectors of electrical, optical, and radio signals. These devices are linked into physical modules like integrated circuits that perform the necessary functions or computations needed for the communications network or computer. For the last two decades, the trends of technology have dramatically decreased the size and power requirements of individual elements so that they can be integrated into a single complex package, thus reducing parts needed, space, and cost of functional modules such as communications receivers. I expect that these trends will continue, using what we have learned from nanotechnology research, until either fundamental physical limits to the individual devices are reached, or which is more likely, until we hit a new bottleneck of how to design and achieve the interconnection of these devices. When devices approach the nanometer scale, they will no longer be identical but will have a statistical distribution of characteristics: for example, in a 10 nm channel length transistor, the number of dopant atoms will be in the tens or hundreds and vary from transistor to transistor produced in an identical way, due to the random nature of the dopant diffusion process. This means that there will necessarily be a statistical variation of turn-on voltages, break-down voltages, channel conductivity, and so forth.

The Interconnection Problem

The engineering research of the next decade will most likely bring to fruition the integration of different functionalities on a single small platform, such as compact electro-photonic modules using engineered photonic bandgap structures to reduce the size of optical modulation-demodulation elements. I expect that the newest integration architectures will necessarily include fully three-dimensional circuits and up to a million or so "zero cost" single devices. These integrated modules will, in themselves, be complex networks of individual elements, each element type described by a statistical distribution of characteristics. We will need a breakthrough in the methods of integrated circuit simulation and design in order to deal with the complexity of designing these modules and to deal with the latency of signals travelling across long paths or through many connections. Right now, the simulation and design software for merely pure electronic integrated circuits, assuming that each element type is identical, is a major bottleneck in the production of application-specific integrated circuits (ASICS). One possibility in the far future is to harness the methods of directed self-assembly of the network of devices, much as our brain manages to learn from its environment how to assemble the synapses between neurons. We are not even close today.

Future Communications Network Architectures

As extremely small and low-cost communications modules are developed, certainly personal access networks — the equipment used by an individual to communicate with his or her near surroundings and to gain access to larger area local area networks and ultimately to the global wide area communications networks of the future — will become ubiquitous. These will mostly be wireless ad hoc networks, since people are mobile. Local area networks, for example, campus or in-building networks with range below one km, will be ubiquitous as well, whether wireless or wireline, depending on deployment costs. But how will the dramatic reduction of cost of the physical infrastructure for communications equipment affect the major communication long haul or wide area networks? Currently, the architectures of cross-continental or undersea or satellite communications systems are determined not only by the cost of components but by the costs associated with deployment, provisioning, reconfiguration, protection, security, and maintenance. The simulation and design tools used for complex wide area networks are in their infancy, as are the simulation and design tools for the integrated modules of which they are comprised. We need a breakthrough in simulation and design techniques. As the costs of the physical hardware components for wide area networks come down, the deployment costs will not fall as much, due to the power requirements needed in wide area systems, and this and the complexity of network management will probably determine network architectures. For example, the complexity of managing security and quality of service in a nationwide ad hoc wireless network comprised of billions of only small, low power base stations is enormous. Thus it is much more likely to have hierarchies of scale in networks, first personal, then local, and then medium range, culminating in a backbone network similar to what we have today. Again, we may be able to learn much from how biological networks configure themselves as we develop self-configuring, self-protecting, and self-monitoring networks.

SPATIAL COGNITION AND CONVERGING TECHNOLOGIES

Reginald G. Golledge, University of California at Santa Barbara

This paper explores aspects of spatial cognition and converging technologies following five themes:

1. Nano-Bio-Info-Cognitive technology (NBIC) and improving learning
2. Enhancing sensory and cognitive capabilities in the spatial domain
3. NBIC and improving human-machine interfaces
4. Suggestions about what should be done
5. Expected outcomes

NBIC and Improving Learning

What will NBIC allow us to achieve in the learning domain that we cannot achieve now?
The effects of NBIC may be

- improved knowledge of brain functioning and capabilities
- new learning domains such as immersive virtual environments
- more widespread use of nonvisual experiences for solving spatial problems
- examining sensory substitution as a way to enhance learning .

Let us briefly examine how these might occur.

Improving Knowledge of Brain Functioning and Capabilities: Place Cell Analysis.

Advances in Magnetic Resonance Imagery (MRI) have given some promise for tracking what parts of the brain are used for what functions. Opinions differ regarding the value of this technology, but much of the negative criticism is directed towards identifying which parts of the brain appear to be used for emotions such as love or hate, or for aesthetic reactions to concepts of beauty, danger, and fear. Somewhat less controversy is present in the spatial domain, where the 25-year-old hypothesis of O'Keefe and Nadel (1978) that the hippocampus is one's "cognitive map" (or place where spatial information is stored) is being actively investigated. Neurobiologists may be able to determine which neurons "fire" (or are excited) when spatial information relating to objects and their locations are sensed and stored. If NBIC can develop reliable place cell analysis, the process of mapping the human brain could be transformed into examining the geography of the brain. To do this in a thorough manner, we need to know more about spatial cognition, including understanding spatial concepts, spatial relations, spatial thinking, and spatial reasoning.

Within the domains of spatial thinking and reasoning — domains that span all scales of science and technology from the nano scale to a universe-wide scale — there is enormous potential for improving our understanding of all facets of the spatial domain. Spatial thinking and reasoning are dominated by perceptualizations, which are the multisensory expansion of visualization. The major processes of information processing include encoding of sensed experiences, the internal manipulation of sensed information in working memory, the decoding of manipulated information, and the use of the results in the decision-making and choice processes involved in problem-solving and spatial behavior. According to Golledge (2002), thinking and reasoning spatially involves

- Understanding the effects of scale
- Competently mentally transforming perceptions and representations among different geometric dimensions (e.g., mentally expanding 1-dimensional traverses or profiles to 2-D or 3-D configurations similar to that involved in geological mapping, or reducing 3-D or 4-D static or dynamic observations to 2-D formats for purposes of simplification or generalization (as when creating graphs, maps, or images)
- Comprehending different frames of reference for location, distance estimation, determining density gradients, calculating direction and orientation, and other referencing purposes (e.g., defining coordinates, vectors, rasters, grids, and topologies)
- Being capable of distinguishing spatial associations among point, line, area, and surface distributions or configurations
- Exercising the ability to perform spatial classification (e.g., regionalization)
- Discerning patterns in processes of change or spread (e.g., recognizing patterns in observations of the spatial spread of AIDS or city growth over time)
- Revealing the presence of spatial and nonspatial hierarchies

Each of the above involves sensing of phenomena and cognitive processing to unpack embedded detail. It should also be obvious that these perceptual and cognitive processes have their equivalents in information technology (IT), particularly with respect to creating, managing, and analyzing datasets. While we are creating multiple terabytes of data each day from satellites, from LIght Detection And Ranging (LIDAR), from cameras, and from visualizations, our technology for dealing with this data — particularly for dynamic updating and realtime analysis — lags somewhat, even in the most advanced systems currently invented. Even in the case of the most efficient data collector and analyzer ever developed, the human mind, there is still a need to simplify, summarize, generalize, and represent information to make it legible. The activities required to undertake this knowledge acquisition process are called education, and the knowledge accumulation resulting from this exposure is called learning. Thus, if NBIC can empower spatial thinking and reasoning, it will promote learning and knowledge accumulation among individuals and societies, and the results will have impact the entire spatial domain. (Note, there is a National Research Council committee on spatial thinking whose report is due at the end of 2002.)

To summarize, spatial thinking is an important part of the process of acquiring knowledge. In particular, spatial knowledge, defined as the product of spatial thinking and reasoning (i.e., defined as cognitive processes) can be characterized as follows:

- Spatial thinking and reasoning do not require perfect information because of the closure power of cognitive processes such as imaging, imagining, interpolating, generalizing, perceptual closure, gestalt integration, and learning

- Spatial metaphors are being used — particularly in IT related database development and operation — but it is uncertain whether they may or may not be in congruence with equivalent cognitive functioning.

- Spatial thinking has become an important component of IT. IT has focused on visualization as a dominant theme in information representation but has paid less attention to other sensory modalities for its input and output architectures; more emphasis needs to be given to sound, touch, smell, gaze, gesture, emotion, etc. (i.e., changing emphasis from visualizations to perceptualizations).

New Learning Domains

One specific way that NBIC developments may promote learning is by enhancement of virtual systems. In geography and other spatial sciences, learning about places other than one's immediate environment is achieved by accessing secondary information, as in books, maps, images, and tables. In the future, one may conceive of the possibility that all place knowledge could be learned by primary experience in immersive virtual environments. In fact, within 20 years, much geospatial knowledge could be taught in immersive virtual environments (VE) labs. This will require

- solution of the space sickness or motion sickness problems sometimes associated with immersion in VE

- quick and immediate access to huge volumes of data — as in terabytes of data on a chip — so that suitably real environments can be created

- adoption of the educational practice of "learning by doing"

- major new development of hardware and virtual reality language (VRL) software

- conviction of teachers that use of VE labs would be a natural consequence of the educational premise that humans learn to think and reason best in the spatial domain by directly experiencing environments.

- Investigation of which types of learning experiences are best facilitated by use of VE.

Using More Nonvisual Methods

Because of the absence of geography in many school curricula in the United States, many people have severely restricted access to (and understanding of) representations of the environment (for example, maps and images) and more abstract concepts (including spatial concepts of hierarchy and association or adjacency displayed by maps or data represented only in tables and graphs) that are fundamental in education and daily life. Representations of the geographic world (maps, charts, models, graphs, images, tables, and pictures) have the potential to provide a rich array of information about the modern world. Learning from spatialized representations provides insights into layout, association, adjacency, and other characteristics that are not provided by other learning modes. But, electronic spatial representations (maps and images) are not accessible to many groups who lack sight, training, or experience with computerized visualizations, thus contributing to an ever-widening digital divide. With new technological developments, such as the evolution from textual interfaces to graphically based Windows environments, and the increasing tendencies for website information to be restricted to those who can access visualizations and images, many people are being frustrated in their attempts to access necessary information — even that relevant to daily life, such as weather forecasts.

When viewing representations of the geographic world, such as a map on a computer screen, sight provides a gestalt-like view of information, allowing the perception of the synoptic whole and almost simultaneously recognizing and integrating its constituent parts. However, interacting with a natural environment is in fact a multi-modal experience. Humans engage nearly all of their sensory modalities when traversing space. Jacobson, Rice, Golledge and Hegarty (2002) summarize recent literature relating to non-visual interfaces. They suggest that, in order to attend to some of this multisensory experience and to provide access to information for individuals with restricted senses, several research threads can be identified for exploring the presentation of information multimodally. For example, information in science and mathematics (such as formulae, equations, and graphs) has been presented through auditory display (e.g., hearing a sine wave) and through audio-guided keyboard input (Gardner et al. 1998; Stevens et al. 1997). Mynatt (1977) has developed a tonal interface that allows users without vision to access Windows-style graphical user interfaces. Multimodal interfaces are usually developed for specialist situations where external vision is not necessarily available, such as for piloting and operating military aircraft (Cohen and Wenzel 1995; Cohen and Oviatt 1995; Rhyne and Wolf 1993).

Jacobson et al. also point out that abstract sound variables have been used successfully for the presentation of complex multivariate data. Parkes and Dear (1990) incorporated "sound painting" into their tactual-auditory information system (NOMAD) to identify gradients in slope, temperature, and rainfall. Yeung (1980) showed that seven chemistry variables could be presented through abstract sound and reported a 90% correct classification rate prior to training and a 98% correct response rate after training. Lunney and Morrison (1981) have shown that sound graphs can convey scientific data to visually impaired students. Sound graphs have also been compared to equivalent tactual graphs; for example, Mansur et al. (1985) found comparable information communication capabilities between the two media, with the auditory displays having the added benefit of being easier to create and quicker to read. Recent research has represented graphs by combining sound and brailled images with the mathematical formula for each graph being verbally presented while a user reads the brailled shape. Researchers have investigated navigating the Internet World Wide Web through audio (Albers 1996; Metois and Back 1996) and as a tool to access the structure of a document (Portigal and Carey 1994). Data sonification has been used to investigate the structure of multivariate and geometric data (Axen and Choi 1994; Axen and Choi 1996; Flowers et al. 1996), and auditory interfaces have been used in aircraft cockpits and to aid satellite ground control stations (Albers 1994; Ballas and Kieras 1996; Begault and Wenzel 1996). But while hardware and software developments have shown "proof of concept," *there appear to be few successful implementations of the results for general use (except for some gaming contexts) and no conclusive behavioral experiments to evaluate the ability of the general public or specialty groups (e.g., the vision-impaired) to use these innovations to interpret on screen maps, graphics, and images.*

Thus, while Jacobson et al. (2002) have illustrated that multimodal interfaces have been explored within computer science and related disciplines (e.g., Delclos and Hartman 1993; Haga and Nishino 1995; Ladewski 1996; Mayer and Anderson 1992; Merlet, Nadalin, Soutou, Lapujade, and Ravat 1993; Morozov 1996; Phillips 1994; Stemler 1997; Hui et al. 1995; and others), and a number of researchers have looked at innovative interface mediums such as gesture, speech, sketching, and eye tracking (e.g., Ballas and Kieras 1996; Briffault and Denis 1996; Dufresne et al. 1995; Schomaker et al. 1995; Taylor et al. 1991), they also claim that only recently are such findings beginning to have an impact upon technology for general education, a view shared by Hardwick et al. (1996; 1997).

In summary, extrapolating from this example, one can assume that developments in NBIC will impact the learning activities of many disciplines by providing new environments for experience, by providing dynamic realtime data to explore with innovative teaching methods, and (if biotechnology continues to unpack the secrets of the brain and how it stores information as in place cell theory), the possibility of direct human-computer interaction for learning purposes may all be possible. Such developments could

- enhance the process of spatial learning by earlier development of the ability to reason abstractly or to more readily comprehend metric and nonmetric relations in simple and complex environments

- assist learning by discovering the biotechnological signatures of phenomena and discovering the place cells where different kinds of information are stored, and in this way enhance the encoding and storage of sensed information

- where functional loss in the brain occurs (e.g., if loss of sight leaves parts of the brain relatively inactive), to find ways to use the cells allocated to sight to be reallocated to other sensory organs, thus improving their functioning capabilities.

- Representations of the geographic world (maps, charts, models, graphs, images, tables, and pictures) have the potential to provide a rich array of information about the modern world.

- Learning from spatialized representations provides insights into layout, association, adjacency, and other spatial characteristics that are not provided by other learning modes.

- However, interacting with a natural environment is in fact a multimodal experience. Humans engage nearly all of their sensory modalities when traversing or experiencing space.

Given the dominance of computer platforms for representing information and the overwhelming use of flat screens to display such information, there is reason to believe that multimodal representations may not be possible until alternatives to 2-D screen surfaces have been developed for everyday use. The reasons for moving beyond visualization on flat screens are compelling and are elaborated on later in this chapter.

Enhancing Sensory and Cognitive Capabilities in the Spatial Domain

How can we exploit developments in NBIC to enhance perceptual and cognitive capabilities across the life span, and what will be the types of developments needed to achieve this goal?

To enhance sensory and cognitive capabilities, a functional change in the way we encode information, store it, decode it, represent it, and use it may be needed. Much of the effort in Information Technology has been directed towards developing bigger and bigger databases that can be used on smaller and smaller computers. From satellites above we get terabytes of data (digitized records of the occurrence of phenomena), and we have perhaps outgrown our ability to examine this data. As nanotechnology and IT come into congruence, the terabytes of data being stored in boxes will be stored on chips and made accessible in real time via wearable and mobile computers, and even may be fed into smart fabrics woven into the clothes we wear. But just how well can we absorb, access, or use this data? How much do we need to access? And how best *can* we access it and use it? The question arises as to how we can exploit human perception and cognition to best help in this process, and the answer is to find out more about these processes so that they can be enhanced. Examples of questions to be pursued include the following:

- How can we enhance the sensory and cognitive aspects of human wayfinding for use in navigating in cyberspace?

- What particular sensory and cognitive capabilities are used in the field, and how do we enhance them for more effective fieldwork with wearable and mobile computers (e.g., for disaster responses)?

- How do we solve problems of filtering information for purposes of representation and analysis (e.g., enhance visualizations)?

- How do we solve the problem of resolution, particularly on the tiny screens typical of wearable and field computers?

- What alternatives to visualization may be needed to promote ease of access, representation, and use of information?

- What is the best mode for data retrieval in field settings (e.g., how do we get the information we need now)?

- How can we build technology to handle realtime dynamic input from several sources, as is done by human sensory organs and the human brain?

- Will we need a totally new approach to computer design and interface architecture (e.g., abandon keyboards and mice) that will allow use of the full range of sensory and cognitive capabilities, such as audition, touch, gaze, and gesture (e.g., the use of Talking Signs® and Internet connections to access websites tied to specific locations)?

Visualization is the dominant form of human-IT interaction. This is partly because the visual sense is so dominant, particularly in the spatial domain. It is also the dominant mode for representation of analyzed data (on-screen). But visualization is but a subset of spatialization, which goes beyond the visual domain by using everyday multimodal situations (from desktops and file cabinets to overlay and digital worlds) to organize and facilitate access to stored information. These establish a linking by analogy and metaphor between an information domain and familiar elements of everyday experience. Spatial (and specifically geographic) metaphors have been used as database organizing systems. But even everyday geospatial experiences are biased, and to enhance our sensory and cognitive abilities we need to recognize those biases and mediate them if successful initiation of everyday knowledge and experience (including natural languages) are to be used to increase human-IT interactions.

The main problem arising from these usages is simply that an assumption of general geospatial awareness is false. Basic geographic knowledge (at least in the United States) is minimal, and knowledge of even rudimentary spatial concepts like distance, orientation, adjacency, and hierarchy is flawed. Recent research in spatial cognition has revealed a series of biases that permeate naïve spatial thinking. Partly because of a result of cognitive filtering of sensed information and partly because of inevitable technical errors in data capture and representation, biases occur. Golledge (2002) has suggested that these include the following:

- conceptual bias due to improper thinking and reasoning (e.g., applying metric principles to nonmetric situations)

- perceptual biases, including misunderstandings and misconceptions of notions of symmetry, alignment, clustering, classification, closure, and so on (e.g., assuming Miami, Florida, MUST be east of Santiago, Chile, because Miami is on the east cost of North America and Santiago is on the west coast of South America) (Fig. B.1)

Map by
S. Baumgart

Figure B.1.   Cognitive East/ West alignment effects.

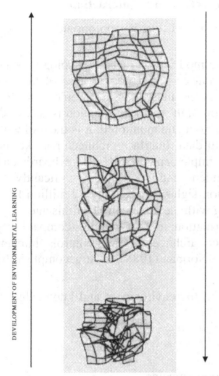

DEVELOPMENT OF ENVIRONMENTAL LEARNING

DEVELOPMENT OF ENVIRONMENTAL FORGETTING

Figure B.2. Three examples of cognitive maps, of long-term residents (top), mid-term residents, (middle), and newcomers (bottom), recovered using non-metric multidimensional scaling of cognitive interpoint distances. (The exact parallel reversals for memory loss is speculative.)

- violating topological features of inclusion and exclusion when grouping (spatial) data

- assuming distance asymmetry when distance symmetry actually exists, and vice versa (e.g., different perceptions of trips to and from work)

- inappropriate use of cognitive concepts of rotation and alignment (e.g., misreading map orientation)

- cognitively overestimating shorter distances and underestimating longer distances (Stevens' Law or regression towards the mean)

- distortions in externalized spatial products (e.g., distorted cognitive maps) (Liben 1982; Fig. B.2)

- bias that results from using imprecise natural language (e.g., fuzzy spatial prepositions like "near" and "behind" that are perspective dependent). (Landau and Jackendoff 1993)

Golledge has argued that these (and other storage, processing, and externalizing biases) result in perceptual and cognitive errors in encoding, internally manipulating, decoding, and using stored information. The following are examples of the accommodations humans make to deal with these biases (incidentally developing new ones):

- making naturally occurring irregular shapes and areas regular for purposes of simplification, representation, and generalization

- mentally rotating features or distributions to fit preconceptions (e.g., vertically aligning North and South America, as shown in Figure B.1)

- imposing hierarchical orderings to clarify distributions (e.g., systems of landmarks)

- making effective rational decisions without perfect information

- cognitively generalizing from one scale to another without appropriate empirical evidence (e.g., from laboratory to real world scales)

- realizing that data collected for machine use has to be more perfect than data collected for human use.

NBIC and Improving Human-Computer Interfaces and Interactions

A key question is why won't existing interface architecture be appropriate for human-computer interaction in the future?

Existing interface architecture is still being modeled on dated technology — the typewriter keyboard and the cursor driven mouse — and not for ease of human-computer interaction. The interface concern is the most pressing problem of HCI and is its most critical part. It is the medium through which information is accessed, questions are posed, and solution paths are laid out and monitored. It is the tool with which the user manipulates and interacts with data. Interface architectures like the desktop, filing cabinet, and digital world are implemented (still) via keyboards and mice. Today's interfaces are cursor dependent and contribute significantly to creating a digital divide that impedes 8 million sight-impaired and 82 million low-vision (potential) users from freely interacting with the dominant IT of this age.

Communicating involves transferring information; to do so requires compatibility between sender and receiver. The interface architecture that controls human-computer information exchange, according to Norman (1988), must accomplish the following:

- facilitate the exchange of knowledge in the environment and knowledge in the head
- keep the interaction task simple
- ensure that operations are easy to do
- ensure correct transfer among information domains
- understand real and artificial restraints on interaction
- acknowledge existence of error and bias due to modal difficulties
- eventually standardize procedures

Thus, the interface must maximize the needs of both human user and computer.

These needs raise the question of what cutting edge hardware (e.g., rendering engines, motion tracking by head mounted display units, gaze tracking, holographic images, avatars complete with gestures, and auditory, tactual, and kinesthetic interface devices), adds to information processing? Besides the emphasis on historic input devices (keyboard and mouse), there is a similar emphasis on a dated output device, the limited domain of the flat computer screen (inherited from the TV screen of the 1930s), which is suited primarily for visualization procedures for output representation. While there is little doubt that the visual senses are the most versatile mode for the display of geospatial data and data analysis (e.g., in graph, table, map, and image mode), it is also argued that multiple modality interfaces could enrich the type, scale, and immediacy of displayed information. One of the most critical interface problems relates to the size and resolution of data displays. This will be of increasing importance as micro-scale mobile and wearable computers have to find alternatives to 2-inch square LED displays for output presentation. The reasons for moving beyond visualization on flat screens are compelling. Examples include the following:

- multimodal access to data and representations provide a cognitively and perceptually rich form of interaction

- multimodal input and output interfaces allow HC interaction when sight is not available (e.g., for blind or sight-impaired users) or when sight is an inappropriate medium (e.g., accessing onscreen computer information when driving a vehicle at high speeds)
- when absence of light or low precludes the use of sight
- when visual information needs to be augmented
- when a sense other than vision may be necessary (e.g., for recording and identifying bird calls in the field)

Nonvisual technology allows people with little or no sight to interact (e.g., using sound, touch, and force-feedback) with computers. Not only is there a need for text to speech conversion, but there is also a need to investigate the potential use of nonvisual modalities for accessing cursor-driven information displays, icons, graphs, tables, maps, images, photos, windows, menus, or other common data representations. Without such access, sight-disabled and low-sight populations are at an immense disadvantage, particularly when trying to access spatial data. This need is paramount today as home pages on the World Wide Web encapsulate so much important information in graphic format, and as digital libraries (including the Alexandria Digital Map and Image Library at the University of California, Santa Barbara) become the major storage places for multidimensional representations of spatial information.

In the near future, one can imagine a variety of new interfaces, some of which exist in part now but which need significant experimentation to evaluate human usability in different circumstances before being widely adopted. Examples of underutilized and underinvestigated technologies include the following:

- a force-feedback mouse that requires building virtual walls around on-screen features, including windows, icons, objects, maps, diagrams, charts, and graphs. The pressure-sensitive mouse allows users to trace the shape of objects or features and uses the concept of a gravity well to slip inside a virtual wall (e.g., a building entrance) to explore the information contained therein (Jacobson et al. 2002).
- vibrotactile devices (mice) that allow sensing of different surfaces (dots, lines, grates, and hachures) to explore flat, on-screen features (e.g., density shading maps and meteorological or isoline temperature maps) (O'Modhrain and Gillespie 1995; Jacobson, et al. 2002)
- use of real, digitized, or virtual sounds including speech to identify on-screen phenomena (e.g., Loomis, Golledge, and Klatzky 2001)
- avatars to express emotions or give directions by gesturing or gazing
- smart clothing that can process nearby spatial information and provide information on nearby objects or give details of ambient temperature, humidity, pollution levels, UV levels, etc.

Currently, the use of abstract sound appears to have significant potential, although problems of spatial localization of sound appear to offer a significant barrier to further immediate use. Some uses (e.g., combinations of sound and touch — NOMAD — and sound and Braille lettering — GPS Talk — are examples of useful multimodal interfaces (e.g., Parkes and Dear 1990; Brabyn and Brabyn 1983; Sendero Group 2002). Some maps (e.g., isotherms/density shading) have proven amenable to sound painting, and researchers in several countries have been trying to equate sound and color. At present, much of the experimentation with multimodal interfaces is concentrated in the areas of video games and cartoon-like movies. Researchers such as Krygier (1994) and Golledge, Loomis, and Klatzky (1994) have argued that auditory maps may be more useful than tactual maps and may, in circumstances such as navigating in vision-obstructed environments, even prove more useful than visual maps because they don't require map-reading ability but rely on normal sensory experiences to indicate spatial information such as direction.

What Needs to be Done to Help NBIC Make Contributions in the Spatial Domain?

- If space is to be used as a metaphor for database construction and management, and if human wayfinding/navigation practices are to be used as models for Internet search engines, there are a host of spatial cognition research activities that need to be pursued. First there is a need for a concept-based common vocabulary. There must be a sound ontology, an understanding of spatial primitives and their derivatives, and a meaningful way to communicate with a computer using natural language and its fuzzy spatial prepositions (i.e., a common base of spatial linguistics, including a grammar).

- We need to find matches between information types and the best sensory modalities for representing and using each type of information.

- We need an educated and IT-enlightened science and engineering community that understands spatial thinking and reasoning processes.

- We need to change educational and learning practices to produce an NBIC-enlightened public and an IT-enlightened set of decision makers. Part of this need can be achieved by producing spatially aware professionals who understand and use actual or enhanced sensory and cognitive capabilities to understand and react to different situations and settings.

- We need to explore the cognitive processes used in risky decision making and use innovative IT practices to develop databases, management systems, and analytical techniques that are cognitively compatible with these processes (Montello 2001).

- We need to develop new realtime dynamic human-computer interfaces (both input and output) that facilitate collaborative decision making. This may involve building virtual environments suited for realtime collaborative image exchange and simultaneous use, analysis, modification, and representation of data, even when researchers are continents apart.

- We need to determine what dimensions of cyberspace are compatible with perceptualization and visualization, particularly in the spatial domain.

- We need to define the impacts of selecting specific scales and levels of resolution for visual or perceptual representation of information.

- We need to explore the value of changing network representations and displays of information in cyberspace to grid layout or configurational displays — the expansion from 1- to 2- or 3-dimensional information representations would facilitate a higher level of abstract thinking and reasoning to be implemented in analyzing configurational displays.

- The explosion of interfaces built upon visualization has produced too many graphic interfaces that do not maximize cognitive capabilities of users and have further disadvantaged disabled groups such as the blind or sight-impaired. This latter fact is continuing the computer alienation of aged populations, where over 70% have low vision or other sight problems. There are, according to census estimates, over 52 million disabled people in the United States. Approximately 3-4 million of these are blind, legally blind, or severely vision-impaired. A further 80+ million people have low vision. We cannot ignore these groups or exclude them from use of future technology.

- We need to determine optimal output interfaces for wearable computers that do not limit the user to visually reading complex displays (e.g., maps) on tiny screens. This carries with it the various cartographic representation problems of choosing scale, resolution, degree of simplification, generalization, and accuracy. This is not just a computer graphics problem, but a problem for cartographic theorists, empirical researchers, and researchers in spatial perception and spatial cognition, and it may involve innovative nanotechnology to build "fold-out" or "expandable" screens.

- There is a need to explore interfaces that can meaningfully display dynamic data at various scales and degrees of resolution.

- There is a need to examine whether nano- or biotechnology can alter the senses and cognitive capabilities of humans to enhance HCI. In particular, can nano-biotechnology enhance our tactual and auditory capabilities (e.g., sensing gloves and ear implants) to ensure that information processing becomes perceptually and cognitively less biased and error ridden?

- There is a need for distributed national learning and research networks to be developed to encourage timely transfer of information from the research to the educational domains; otherwise, the current 3-5 year lags needed for much of this transfer to take place will continue.

- As we learn more about how the mind stores data, there is a need to examine whether we can use the mind as a model to enhance efforts to build a national network of digital libraries.

- There is a need for solving problems associated with using immersive virtual environments (e.g., motion sickness) so that their real potential in research and decision making can be exploited and evaluated.

- There is a need to explore ways to increase the effectiveness of human-environment relations. This may involve

  - developing personal guidance and spatial information systems that allow people to carry with them in a wearable computer all the local

environmental infor-
mation that they need
to undertake daily
activities (Fig. B.3)

- developing smart en-
vironments that allow
people to access
wireless information
(e.g., infrared-based
auditory signage or
locally distributed
servers that allow
immediate access to
the Internet and web
pages) (Fig. B.4).

Figure B.3. Personal guidance system.

- Since environmental in-
formation is filtered
through our senses and consequently is biased, individually selective, and
related to stage of cognitive development, we need to know to what extent
human sensing is dependent on perspective or point of view for encoding
spatial relations. Attention must be paid to the roles of alignment, frames of
reference, and scale or resolution (e.g., asymmetries of distance, orientation
error, or locational inaccuracy), which produce information not always
consistent with metric geometries and logically based algebras used to unpack
information from data about the real
world. Perhaps a new subjective
mathematics is needed to interpret our
cognitive maps.

- We need to determine if knowledge of
wayfinding in the real world can help
us find our way in cyberspace. Spatial
knowledge in humans develops from
landmark route configurational
understanding. Much high-order spatial
knowledge in humans concerns
understanding spatial relations
embedded in configurational or layout
knowledge, whereas much of the
knowledge in IT is link- and network-
based, potentially reducing its
information potential by requiring
human ability to integrate information
obtained from specific routes in
cyberspace.

Figure B.4. "Smart environments."

There are two dominant ways for NBIC to impact the 52+ million disabled people in the United States:

1. free them from the tyranny of print and other "inaccessible" visual representations
2. help them obtain independence of travel

Enacting measures like the following will increase mobility, employability, and quality of life:

- changing computer interface architecture so that disabled groups (e.g., blind, sight impaired, dyslexic, arthritic, immobile) can access the Internet and its webpages as transparently and quickly as able-bodied people
- enabling wearable computers for use in everyday living (e.g., finding when the next bus is due or where it is now) (Fig. B.4)
- developing voice-activated personal guidance systems using GPS, GIS, and multimodal interfaces that will enable people to travel in unfamiliar environments (Fig. B.4)
- improve speech recognition for input to computers
- use infrared-based remote auditory signage systems (RASS) (e.g., talking sign technology) to facilitate wayfinding, business or object location identification, recognition of mass transit services and promotion of intermodal transfer, and to define other location-based services and information systems

Outcomes

Following are some outcomes of the integration of spatial cognition and converging NBI technologies:

- Expanding sensory and cognitive capabilities should improve learning and result in a more NBIC-enlightened public, scientists, engineers, and public policymakers.
- Developing multimodal input and output interfaces will enrich human ability to process and analyze information, covering all types of spatial information required for microscopic, global, or extraterrestrial research. It will also help to remove the rapidly growing effects of the digital divide by allowing more disabled (or otherwise disadvantaged) people to join the computer-literate population, thus improving employment possibilities and improving quality of life.

Converging NBIC technology will broaden our abilities to think "outside the box" in a variety of sensory domains, such as the following examples of convergence of NBI and spatial cognition methods:

- Natural language-driven mobile and wearable computers
- Internet search engines based on human wayfinding practices
- Smart fabrics that sense the environment and warn us of pollution levels, etc.
- Smart environments (e.g., remote auditory signage systems) that talk to us as we travel through them

Figure B.5. Talking maps.

- GPS-based personal guidance systems that facilitate travel (e.g., tourism) in unfamiliar places

- Smart maps that explain themselves at the touch of a stylus or as a result of gaze or gesture (e.g., "You are here" maps or on-screen computer representations of data) (Fig. B.5)

- Robotic guide dogs that carry large environmental databases and can develop routes to unfamiliar places

- Smart buildings that inform about their contents and inhabitants, e.g., transit terminals (Fig. B.6).

Of particular interest are NBIC-based knowledge and devices that enhance spatial cognition used in wayfinding performance:

- Remote auditory signage (Talking Signs/Remote Infrared Auditory Signage) (at places or on vehicles, including mass transit)

- Talking fluorescent lights inside buildings such as shopping centers and transit terminals (Fig. B.7)

GPS-based guidance systems with Pointlink capabilities to locations and websites for place-based information.

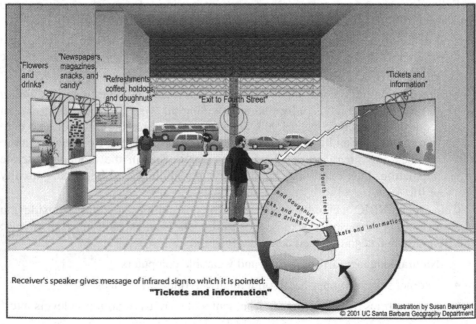

Figure B.6. Transit terminal with remote auditory signage.

Figure B.7. Talking neon lights in airport terminal.

Conclusion

The convergence of nano-, bio-, info- technology and spatial cognition research will

- broaden our ability to think outside the box
- ensure that NBI technologies are compatible with ways humans think and reason
- facilitate new product development
- help remove barriers to the natural integration of disabled and disadvantaged groups into the community, thus improving their quality of life
- provide new environments for learning
- enhance cognitive functioning by improving perceptual and cognitive capabilities
- help create less abstract and more "naturally human" computer interface architecture
- once we have learned how and where spatial information is stored in the brain (place cell analysis), this may prompt new ideas about how we think and reason

For example, eventually, the most powerful computer interface will rely on an architecture that combines geospatial metaphors with spatialization principles and multimodal input and output devices that provide access to text, maps, images, tables, and gestures.

But there is the inevitable downside, such as the thorny ethical and legal issues of defining and maintaining appropriate levels of individual privacy and security of public or business information. But developments in NBIC are the future of humankind, and these and other unrealized problems, must — in the way of humankind — be faced and solved.

Finally, if VE can be developed in an effective way, humans will have many of the capabilities of the Star Trek holodeck. They will stroll through the Amazon jungles, trek to the North or South Pole, explore an active volcano, avalanche, or hurricane, redesign cities or parts of them, change transport systems to maximize the benefits of intelligent highways, visit drought areas, explore areas of poverty or crime, all within the safety of VE. The contribution of such systems to education, research, and decision making in the policy arena could be immense. As long as we can solve the cognition and technical problems of building and using VE, these goals may be achievable.

References

Albers, M.C. 1994. The Varese System, hybrid auditory interfaces and satellite-ground control: Using auditory icons and sonification in a complex, supervisory control system. In G. Kramer and S. Smith (Eds.), *Proceedings of the Second International Conference on Auditory Display* (pp. 3-13). Sante Fe, NM.

Albers, M.C. 1996. Auditory cues for browsing, surfing, and navigating the WWW: The audible web. In S.P. Frysinger and G. Kramer (Eds.), *Proceedings of the Third International Conference on Auditory Display* (pp. 85-90). Palo Alto, Ca.

Axen, U., and I. Choi. 1994. Using additive sound synthesis to analyze simple complexes. In G. Kramer and S. Smith (Eds.), *Proceedings of the Second International Conference on Auditory Display* (pp. 31-44). Sante Fe, NM.

Axen, U., and I. Choi. 1996. Investigating geometric data with sound. In S.P. Frysinger and G. Kramer (Eds.), *Proceedings of the Third International Conference on Auditory Display* (pp. 25-28). Palo Alto, Ca.

Ballas, J., and D.E. Kieras. 1996. Computational modeling of multimodal I/O in simulated cockpits. In S.P. Frysinger and G. Kramer (Eds.), *Proceedings of the Third International Conference on Auditory Display* (pp. 135-136). Palo Alto, Ca.

Begault, D.R., and E.M. Wenzel. 1996. A virtual audio guidance and alert system for commercial aircraft operations. In S.P. Frysinger and G. Kramer (Eds.), *Proceedings of the Third International Conference on Auditory Display* (pp. 117-122). Palo Alto, Ca.

Brabyn, L.A., and J.A. Brabyn. 1983. An evaluation of "Talking Signs" for the blind. *Human Factors, 25*(1), 49-53.

Briffault, X., and M. Denis. 1996 (August). *Multimodal interactions between drivers and co-drivers: An analysis of on-board navigational dialogues.* Paper presented at the Proceedings of the Twelfth European Conference on Artificial Intelligence, Budapest.

Cohen, M., and E.M. Wenzel. 1995. The design of multidimensional sound interfaces. In T. A. F. I. I. I. E. Woodrow Barfield (Ed.), *Virtual environments and advanced interface design.* (pp. 291-346): Oxford University Press, New York, NY, US.

Cohen, P.R., and S.L. Oviatt. 1995. The role of voice input for human-machine communication. *Proceedings of the National Academy of Sciences, 92*(22), 9921-9927.

Delclos, V.R., and A. Hartman. 1993. The impact of an interactive multimedia system on the quality of learning in educational psychology: an exploratory study. *Journal of Research on Technology in Education, 26*(1), 83-93.

Dufresne, A., O. Martial, and C. Ramstein. 1995. *Multimodal user interface system for blind and "visually occupied" users: Ergonomic evaluation of the haptic and auditive dimensions.*

Flowers, J.H., D.C. Buhman, and K.D. Turnage. 1996. Data sonification from the desktop: Should sound be a part of standard data analysis software. In S.P. Frysinger and G. Kramer (Eds.), *Proceedings of the Third International Conference on Auditory Display* (pp. 1-8). Palo Alto, Ca.

Gardner, J.A., R. Lundquist, and S. Sahyun. 1998 (March). *TRIANGLE: A Tri-Modal Access Program for Reading, Writing and Doing Math.* Paper presented at the Proceedings of the CSUN International Conference on Technology and Persons with Disabilities, Los Angeles.

Golledge, R.G. 2002. The nature of geographic knowledge. *Annals of the Association of American Geographers, 92*(1), In Press.

Golledge, R.G., J.M. Loomis, and R.L. Klatzky. 1994 (February 20-26). *Auditory maps as alternatives to tactual maps.* Paper presented at the 4th International Symposium on Maps and Graphics for the Visually Impaired, Sao Paulo, Brazil.

Haga, H., and M. Nishino. 1995 (17-21 June). *Guidelines for designing hypermedia teaching materials.* Paper presented at the Educational Multimedia and Hypermedia 1995. Proceedings of ED-MEDIA 95 - World Conference on Educational Multimedia and Hypermedia, Graz, Austria.

Hardwick, A., S. Furner, and J. Rush. 1996 (21 Jan). *Tactile access for blind people to virtual reality on the World Wide Web.* Paper presented at the IEE Colloquium on Developments in Tactile Displays (Digest No.1997/012) IEE Colloquium on Developments in Tactile Displays, London, UK.

Hardwick, A., J. Rush, S. Furner, and J. Seton. 1997. Feeling it as well as seeing it-haptic display within gestural HCI for multimedia telematics services. In P.A. Harling and A.D. N. Edwards (Eds.), *Progress in Gestural Interaction. Proceedings of Gesture Workshop '96, York, UK* (pp. 105-16). Berlin, Germany: Springer-Verlag.

Hui, R., A. Ouellet, A. Wang, P. Kry, S. Williams, G. Vukovich, and W. Perussini. 1995. Mechanisms for haptic feedback, *IEEE International Conference on Robotics and Automation* (pp. 2138-2143).

Jacobson, R.D., M. Rice, R.G. Golledge, and M. Hegarty. 2002. *Force feedback and auditory interfaces for interpreting on-screen graphics and maps by blind users* (Technical Paper funded by the UCSB Research Across the Disciplines (RAD) Program): Geography Departments of Florida State University and University of California Santa Barbara.

Krygier, J.B. 1994. Sound and Geographic Visualisation. In A.M. MacEachren and D.R. Fraser-Taylor (Eds.), *Visualisation in Modern Cartography* (pp. 149-166): Pergamon.

Ladewski, B.G. 1996. Interactive multimedia learning environments for teacher education: comparing and contrasting four systems. *Journal of Computers in Mathematics and Science Teaching 15*(1-2) 173-97.

Landau, B., and R. Jackendoff. 1993. "What" and "where" in spatial language and spatial cognition. *Behavioral and Brain Sciences 16*, 217-238.

Loomis, J.M., R.G. Golledge, and R.L. Klatzky. 2001. GPS-based navigation systems for the visually impaired. In W. Barfield and T. Caudell (Eds.), *Fundamentals of Wearable Computers and Augmented Reality* (pp. 429-446). Mahway, NJ: Erlbaum.

Lunney, D., and R. Morrison. 1981. High Technology Laboratory Aids for Visually Handicapped Chemistry Students. *Journal of Chemical Education, 8*(3), 228-231.

Mansur, D., M. Blattner, and K. Joy. 1985. Sound graphs: A numerical data analysis for the blind. *Journal of Medical Systems, 9*(3) 163-174.

Marston, J. 2002. *Towards an Accessible City: Empirical Measurement and Modeling of Access to Urban Opportunities for those with Vision Impairments Using Remote Infrared Audible Signage.* , UCSB, Santa Barbara. Unpublished Ph.D. Dissertation.

Mayer, R.E., and R.B. Anderson. 1992. The instructive animation: Helping students build connections between words and pictures in multimedia learning. *Journal of Educational Psychology, 84*, 444-452.

Merlet, J.F., C. Nadalin, C. Soutou, A. Lapujade, and F. Ravat. 1993 (17-20 Oct.). *Toward a design method of a multimedia information system.* Paper presented at the 1993 International Conference on Systems, Man and Cybernetics. Systems Engineering in the Service of Humans (Cat. No.93CH3242-5) Proceedings of IEEE Systems Man and Cybernetics Conference - SMC, Le Touquet, France.

Metois, E., and M. Back. 1996. BROWeb: An interactive collaborative auditory environment on the world wide web. In S.P. Frysinger and G. Kramer (Eds.), *Proceedings of the Third International Conference on Auditory Display* (pp. 105-110). Palo Alto, Ca.

Montello, D.R., ed. 2001. *Spatial Information Theory: Foundations of Geographic Information Science. Proceedings, International Conference, COSIT 2001, Morro Bay, CA, September.* New York: Springer.

Morozov, M. 1996 (14-16 Sept.). *Multimedia lecture room: a new tool for education.* Paper presented at the Multimedia, Hypermedia, and Virtual Reality. Models, Systems, and Applications. First International Conference, Moscow, Russia.

Mynatt, E.D. 1997. Transforming graphical interfaces into auditory interfaces for blind users. *Human-Computer Interaction 12*(1-2), 7-45.

Norman, D.A. (1988). *The Psychology of Everyday Things.* New York: Basic Books.

O'Keefe, J., and L. Nadel. 1978. *The Hippocampus as a Cognitive Map.* Oxford: Clarendon Press.

O'Modharain, M., and B. Gillespie. 1995. *The Moose: A Haptic User Interface for Blind Persons* (Stan-M95; CCRMA). Stanford, CA: Stanford.

Parkes, D., and R. Dear. 1990. *NOMAD: AN interacting audio-tactile graphics interpreter.* Paper presented at the Reference Manual, Version 2.0, NSW Australia: Institute of Behavior Science, University of Newcastle.

Phillips, R. 1994. Producing interactive multimedia computer-based learning projects. *Computer Graphics, 28*(1), 20-4.

Portigal, S., and T. Carey. 1994. Auralization of document structure. In G. Kramer and S. Smith (Eds.), *Proceedings of the Second International Conference on Auditory Display* (pp. 45-54). Sante Fe, NM.

Rhyne, J.A., and C. Wolf. 1993. Recognition based user interfaces. In H.R. Harston and D. Hix (Eds.), *Advances in Human-Computer Interaction: Vol. 4* (pp. 191-250). Norwood, NJ: Ablex.

Schomaker, L., J. Nijtmans, A. Camurri, F. Lavagetto, P. Morasso, C. Benoit, T. Guiard-Marginy, B. Le Goff, J. Robert-Ribes, A. Adjoudani, I. Defee, S. Munch, K. Hartnung, and J. Blauert. 1995. *A taxonomy of mulitmodal interaction in the human information processing system* : Esprit Project 8579.

Sendero Group. 2002.. Available: www.senderogroup.com.

Stemler, L.K. 1997. Educational characteristics of multimedia: a literature review. *Journal of Educational Multimedia and Hypermedia, 6*(3-4), 339-59.

Stevens, R.D., A.D.N. Edwards, and P.A. Harling. 1997. Access to mathematics for visually disabled students through multimodal interaction. *Human-Computer Interaction 12*, 47-92.

Taylor, M.M., F. Neel, and D.G. Bouwhuis, eds. 1991. *The structure of multimodal dialogue.* Amsterdam: North-Holland.

Yeung, E. 1980. Pattern recognition by audio representation of multivariate analytical data. *Analytical Chemistry, 52*(7) 1120-1123.

VISUAL LANGUAGE AND CONVERGING TECHNOLOGIES IN THE NEXT 10-15 YEARS (AND BEYOND)

Robert E. Horn, Visiting Scholar, Stanford University

Visual language is one of the more promising avenues to the improvement of human performance in the short run (the next 10 to 15 years) (Horn 2000b, 2000c). The current situation is one of considerable diversity and confusion as a new form of communication arises. But visual language also represents many great opportunities. People think visually. People think in language. When words and visual elements are closely intertwined, we create something new and we augment our communal intelligence.

Today, human beings work and think in fragmented ways, but visual language has the potential to integrate our existing skills to make them tremendously more effective. With support from developments in information technology, visual language has the potential for increasing human "bandwidth," the capacity to take in, comprehend, and more efficiently synthesize large amounts of new information. It has this capacity on the individual, group, and organizational levels. As this convergence occurs, visual language will enhance our ability to communicate, teach, and work in fields such as nanotechnology and biotechnology.

Definition

Visual language is defined as the tight integration of words and visual elements and has characteristics that distinguish it from natural languages as a separate communication tool as well as a distinctive subject of research. It has been called visual language, although it might well have been called visual-verbal language.

A preliminary syntax, semantics, and pragmatics of visual language have been described. (Horn 1998) Description of, understanding of, and research on visual language overlap with investigations of scientific visualization and multimedia.

History

The tight integration of words and visual elements has a long history (Horn 1998, Chapter 2). Only in the last 50 years, with the coming together of component visual

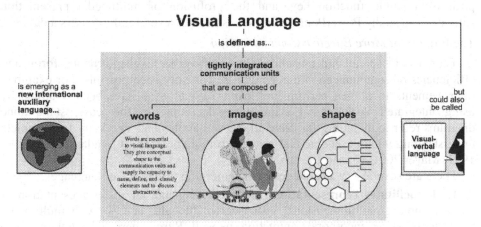

Figure B.8. Defining visual language.

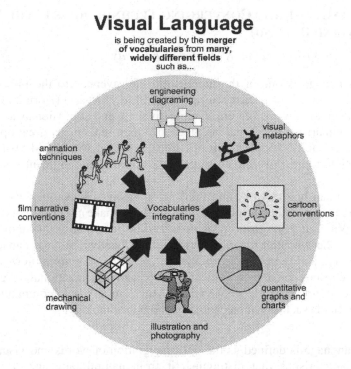

Figure B.9. Creation of visual language.

vocabularies from such widely separate domains as engineering diagramming technologies developed in medical illustration and hundreds of expressive visual conventions from the world of cartooning, has something resembling a full, robust visual-verbal language appeared (Tufte 1983, 1990).

Its evolution has been rapid in the past 10 years, especially with the confluence of scientific visualization software; widespread use of other quantitative software that permits the creation of over one hundred quantitative graphs and charts with the push of a single function key; and the profusion of multimedia presentation software, especially PowerPoint which, it is said, has several million users a day.

The Promise of *More Effective Communication*

There is widespread understanding that visual-verbal language enables forms and efficiencies of communication that heretofore have not been possible. For example, improvements in human performance from 23% to 89% have been obtained by using integrated visual-verbal stand-alone diagrams. In this case, stand-alone diagrams refer to diagrams that have all the verbal elements necessary for complete understanding without reading text elsewhere in a document (Chandler and Sweller 1991; Mayer 2001; Horton 1991).

There are several key advantages of the emerging visual-verbal language:

1. **It facilitates representation.** This new language facilitates presentation of complex, multidimensional visual-verbal thought, and — with multimedia tools — can incorporate animation, as well. Researchers and scholars are no longer constrained by the scroll-like thinking of endless paragraphs of text.

2. **It facilitates big, complex thoughts**. Human cognitive effectiveness and efficiency is constrained by the well-known limitations of working memory that George Miller identified in 1957 (Miller 1957). Large visual displays have for some time been known to help us overcome this bandwidth constraint. But only since the recent advances in visual language have we been able to imagine a major prosthesis for this human limitation. The prosthesis consists of a suite of visual language maps. This visual-verbal language (together with computer-based tools) may eliminate the major roadblocks to thinking and communicating big, complex thoughts, i.e., the problem of representing and communicating mental models of these thoughts efficiently and effectively.

This especially includes the so-called "messy" (or "wicked" or "ill-structured") problems (Horn 2001a). Problems have straightforward solutions; messy problems do not. They are

– more than complicated and complex; they are ambiguous

– filled with considerable uncertainty — even as to what the conditions are, let alone what the appropriate actions might be

– bounded by great constraints and tightly interconnected economically, socially, politically, and technologically

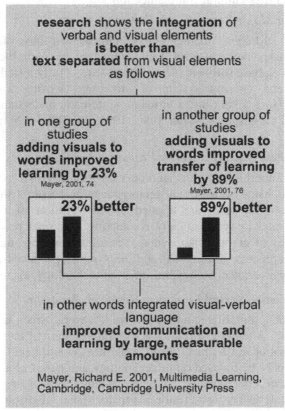

Figure B.10. Enhancing learning through visual language.

- seen differently from different points of view and quite different worldviews
- comprised of many value conflicts
- often alogical or illogical

These kinds of problems are among the most pressing for our country, for the advancement of civilization, and for humanity; hence, the promise of better representation and communication of complex ideas using visual-verbal language constructs has added significance.

Premises Regarding Visual Language

A deep understanding of the patterns of visual language will permit the following:

- more rapid, more effective interdisciplinary communication
- more complex thinking, leading to a new era of thought
- facilitation of business, government, scientific, and technical productivity
- potential breakthroughs in education and training productivity
- greater efficiency and effectiveness in all areas of knowledge production and distribution
- better cross-cultural communication

Readiness for Major Research and Development

A number of major jumping-off research platforms have already been created for the rapid future development of visual language: the Web; the ability to tag content with XML; database software; drawing software; a fully tested, widely used content-organizing and tagging system of structured writing known as Information Mapping® (Horn 1989); and a growing, systematic understanding of the patterns of visual-verbal language (Kosslyn 1989, 1994; McCloud 1993; Horton 1991; Bertin 1983).

Rationale for the Visual Language Projects

A virtual superhighway for rapid development in visual language can be opened, and the goals listed above in the premises can be accomplished, if sufficient funds over the next 15 years are applied to the creation of tools, techniques, and taxonomies, and to systematically conducting empirical research on effectiveness and efficiency of components, syntax, semantics, and pragmatics of this language. These developments, in turn, will aid the synergy produced in the convergence of biotechnology, nanotechnology, information technology, and cognitive science.

Goals of a Visual-Verbal Language Research Program

A research program requires both bold, general goals and specific landmarks along the way. A major effort to deal with the problem of increasing complexity and the limitations of our human cognitive abilities would benefit all human endeavors and could easily be focused on biotechnology and nanotechnology as prototype test beds. We can contemplate, thus, the steady, incremental achievement of the following goals as a realistic result of a major visual language program:

1. **Provide policymakers with comprehensive visual-verbal models.** The combination of the ability to represent complex mental models and the ability to collect realtime data will provide sophisticated decision-making tools for social policy. Highly visual cognitive maps will facilitate the management of and navigation through major public policy issues. These maps provide patterned abstractions of policy landscapes that permit the decisionmakers and their advisors to consider which roads to take within the wider policy context. Like the hundreds of different projections of geographic maps (e.g., polar or Mercator), they provide different ways of viewing issues and their backgrounds. They enable policymakers to drill down to the appropriate level of detail. In short, they provide an invaluable information management tool.

2. **Provide world-class, worldwide education for children.** Our children will inherit the results of this research. It is imperative that they receive the increased benefits of visual language communication research as soon as it is developed. The continued growth of the Internet and the convergence of intelligent visual-verbal representation of mental models and computer-enhanced tutoring programs will enable children everywhere to learn the content and skills needed to live in the 21st century. But this will take place only if these advances are incorporated into educational programs as soon as they are developed.

3. **Achieved large breakthroughs in scientific research.** The convergence of more competent computers, computer-based collaborative tools, visual representation breakthroughs, and large databases provided by sensors will enable major improvements in scientific research. Many of the advances that we can imagine will come from interdisciplinary teams of scientists, engineers, and technicians who will need to become familiar rapidly with fields that are outside of their backgrounds and competencies. Visual language resources (such as the diagram project described below) will be required at all levels to make this cross-disciplinary learning possible. This could be the single most important factor in increasing the effectiveness of nano-bio-info teams working together at their various points of convergence.

4. **Enrich the art of the 21st century.** Human beings do not live by information alone. We make meaning with our entire beings: emotional, kinesthetic, and somatic. Visual art has always fed the human spirit in this respect. And we can confidently predict that artistic communication and aesthetic enjoyment in the 21st century will be enhanced significantly by the scientific and technical developments in visual language. Dynamic visual-verbal murals and art pieces will become one of the predominant contemporary art forms of the century, as such complex, intense representation of meaning joins abstract and expressionistic art as a major artistic genre. This has already begun to happen, with artists creating the first generation of large visual language murals (Horn 2000).

5. **Develop smart, visual-verbal thought software.** The convergence of massive computing power, thorough mapping of visual-verbal language patterns, and advances in other branches of cognitive science will provide for an evolutionary leap in capacity and in multidimensionality of thought processes. Scientific visualization software in the past 15 years has led the

way in demonstrating the necessity of visualization in the scientific process. We could not have made advances in scientific understanding in many fields without software that helps us convert "firehoses of data" (in the vivid metaphor of the 1987 National Science Foundation report on scientific visualization) into visually comprehensible depictions of *quantitative* phenomena and simulations. Similarly, every scientific field is overwhelmed with *tsunamis* of new *qualitative* concepts, procedures, techniques, and tools. Visual language offers the most immediate way to address these new, highly demanding requirements.

6. **Open wide the doors of creativity.** Visualization in scientific creativity has been frequently cited. Einstein often spoke of using visualization on his *gedanken* experiments. He saw in his imagination first and created equations later. This is a common occurrence for scientists, even those without special training. Visual-verbal expression will facilitate new ways of thinking about human problems, dilemmas, predicaments, emotions, tragedy, and comedy. "The limits of my language are the limits of my world," said Wittgenstein. But it is in the very nature of creativity for us to be unable to specify what the limits will be. Indeed, it is not always possible to identify the limits of our worlds until some creative scientist has stepped across the limit and illuminated it from the other side.

Researchers in biotechnology and nanotechnology will not have to wait for the final achievement of these goals to begin to benefit from advances in visual language research and development. Policymakers, researchers, and scholars will be confronting many scientific, social, ethical, and organizational issues; each leap in our understanding and competence in visual language will increase our ability to deal with these kinds of complex issues. As the language advances in its ability to handle complex representation and communication, each advance can be widely disseminated because of the modular nature of the technology.

Major Objectives Towards Meeting Overall Goals of Visual-Verbal Language Research

The achievement of the six goals described above will obviously require intermediate advances on a number of fronts to achieve specific objectives:

1. **Diagram an entire branch of science with stand-alone diagrams.** In many of the newer introductory textbooks in science, up to one-third of the total space consists of diagrams and illustrations. But often, the function of scientific diagrams in synthesizing and representing scientific processes has been taken for granted. However, recent research cited above (Mayer 2001, Chandler and Sweller 1991) has shown how stand-alone diagrams can significantly enhance learning. Stand-alone diagrams do what the term indicates: everything the viewer needs to understand the subject under consideration is incorporated into one diagram or into a series of linked diagrams. The implication of the research is that the text in the other two thirds of the textbooks mentioned above should be distributed into diagrams.

    "Stand-alone" is obviously a relative term, because it depends on previous learning. One should note here that automatic prerequisite linkage is one of the easier functions to imagine being created in software packages designed to handle linked diagrams. One doesn't actually have to take too large a leap of imagination to see this as achievable, as scientists are already exchanging PowerPoint slides that contain many diagrams. However, this practice frequently does not take advantage of either the stand-alone or linked property.

Stand-alones can be done at a variety of styles and levels of illustration. They can be abstract or detailed, heavily illustrated or merely shapes, arrows, and words. They can contain photographs and icons as well as aesthetically pleasing color.

Imagine a series of interlinked diagrams for an entire field of science. Imagine zooming in and out — always having the relevant text immediately accessible. The total number of diagrams could reach into the tens of thousands. The hypothesis of this idea is that such a project could provide an extraordinary tool for cross-disciplinary learning. This prospect directly impacts the ability of interdisciplinary teams to learn enough of each other's fields in order to collaborate effectively. And collaboration is certainly the key to benefiting from converging technologies.

Imagine, further, that using and sharing these diagrams were *not* dependent on obtaining permission to reproduce them, which is one of the least computerized, most time-consuming tasks a communicator has to accomplish these days. Making permission automatic would remove one of the major roadblocks to the progress of visual language and a visual language project.

Then, imagine a scientist being able to send a group of linked, stand-alone diagrams to fellow scientists.

2. **Create "periodic" table(s) of types of stand-alone diagrams.** Once we had tens of thousands of interlinked diagrams in a branch of science, we could analyze and characterize all the components, structures, and functions of all of the types of diagrams. This would advance the understanding of "chunks of thinking" at a fine-grained level. This meta understanding of diagrams would also be a jumping-off point for building software tools to support further investigations and to support diagramming of other branches of science and the humanities.

3. **Automatically create diagrams from text.** At the present moment, we do not know how to develop software that enables the construction from text of a wide variety of kinds of elaborate diagrams. But if the stand-alone diagrams prove as useful as they appear, then an automatic process to create diagrams, or even just first drafts of diagrams, from verbal descriptions will turn out to be extremely beneficial. Imagine scientists with new ideas of how processes work speaking to their computers and the computers immediately turning the idea into the draft of a stand-alone diagram.

4. **Launch a project to map the human cognome.** In the Converging Technologies workshop I suggested that we launch a project that might be named "Mapping the Human Cognome." If properly conceived, such a project would certainly be the project of the century. If the stand-alone diagram project succeeds, then we would have a different view of human thought chunks. Since human thought-chunks can be understood as fundamental building blocks of the human cognome, the rapid achievement

of stand-alone diagrams for a branch of science could, thus, be regarded as a starting point for at least one major thrust of the Human Cognome Project (Horn 2002c).

5. **Create tools for collaborative mental models based on diagramming.** Ability to come to rapid agreement at various stages of group analysis and decision-making with support from complex, multidimensional, visual-verbal murals is becoming a central component of effective organizations. This collaborative problem-solving, perhaps first envisioned by Douglas Engelbart (1962) as augmenting human intellect, has launched a vibrant new field of computer-supported collaborative work (CSCW). The CSCW community has been facilitating virtual teams working around the globe on the same project in a 24/7 asynchronous timeframe. Integration of (1) the resources of visual language display, (2) both visual display hardware and software, and (3) the interactive potential of CSCW offers possibilities of great leaps forward in group efficiency and effectiveness.

6. **Crack the unique address dilemma with fuzzy ontologies.** The semantic web project is proceeding on the basis of creating unique addresses for individual chunks of knowledge. Researchers are struggling to create "ontologies," by which they mean hierarchical category schemes, similar to the Dewey system in libraries. But researchers haven't yet figured out really good ways to handle the fact that most words have multiple meanings. There has been quite a bit of progress in resolving such ambiguities in machine language translation, so there is hope for further incremental progress and major breakthroughs. An important goal for cognitive scientists will be to produce breakthroughs for managing the multiple and changing meanings of visual-verbal communication units on the Web in real time.

7. **Understand computerized visual-verbal linkages.** Getting computers to understand the linkage between visual and verbal thought and their integration is still a major obstacle to building computer software competent to undertake the automatic creation of diagrams. This is likely to be less of a problem as the stand-alone diagram project described above (objective #1) progresses.

8. **Crack the "context" problem.** In meeting after meeting on the subject of visual-verbal language, people remark at some point that "it all depends on the context." Researchers must conduct an interdisciplinary assault on the major problem of carrying context and meaning along with local meaning in various representation systems. This may well be accomplished to a certain degree by providing pretty good, computerized common sense. To achieve the goal of automatically creating diagrams from text, there will have to be improvements in the understanding of common sense by computers. The CYC project, the attempt to code all of human common sense knowledge into a single database — or something like it — will have to demonstrate the ability to reason with almost any subject matter from a base of 50 million or more coded facts and ideas. This common-sense database must somehow be integrally linked to visual elements.

Conclusion

It is essential to the accelerating research in the fields of nanotechnology, biotechnology, information technology, and cognitive science that we increase our understanding of visual language. In the next decade, we must develop visual language research centers, fund individual researchers, and ensure that these developments are rapidly integrated into education and into the support of the other converging technologies.

References

Bertin, J. 1983. *Semiology of graphics: Diagrams, networks, and maps.* Madison, WI: Univ. of Wisconsin Press.

Chandler, P., and J. Sweller. 1991. Cognitive load theory and the format of instruction. *Cognition and Instruction* 8(4): 293-332.

Engelbart, D.C. 1962. *Augmenting human intellect: A conceptual framework.* Stanford Research Institute, Washington, D.C.: Air Force Office Of Scientific Research, AFOSR-3233, Contract AF49(638)-1024 , SRI Project No. 3578. October.

Horn, R.E. 1989. *Mapping hypertext*, Lexington, MA: The Lexington Institute (http://www.stanford.edu/ ~rhorn/MHContents.html).

Horn, R.E. 1998a. *Mapping great debates: Can computers think?* Bainbridge Island, WA: MacroVU, Inc. (http://www.stanford.edu/~rhorn/CCTGeneralInfo.html).

Horn, R.E. 1998b. *Visual language: Global communication for the 21st century.* Bainbridge Island, WA: MacroVU, Inc. (http://www.stanford.edu/~rhorn/VLBkDescription.html).

Horn, R.E. 2000. *The representation of meaning—Information design as a practical art and a fine art.* A speech at the Stroom Center for the Visual Arts, The Hague (http://www.stanford.edu/~rhorn/ VLbkSpeechMuralsTheHague.html).

Horn, R.E. 2001a. *Knowledge mapping for complex social messes.* A speech to the Packard Foundation Conference on Knowledge Management (http://www.stanford.edu/~rhorn/ SpchPackard.html).

Horn, R.E. 2001b. *What kinds of writing have a future?* A speech prepared in connection with receiving Lifetime Achievement Award by the Association of Computing Machinery SIGDOC, October 22.

Horn, R.E. 2002a. *Think link, invent, implement, and collaborate! Think open! Think change! Think big!* Keynote Speech at Doug Engelbart Day in the State of Oregon, Oregon State University, Corvalis OR, January 24.

Horn, R.E. 2002b. *Conceptual map of a vision of the future of visual language research.* (To download PDF file http://www.stanford.edu/~rhorn/MapFutureVisualLang.html).

Horn, R.E. 2002c. *Beginning to conceptualize the human cognome project.* A paper prepared for the National Science Foundation Conference on Converging Technologies (Nano-Bio-Info-Cogno) (To download PDF file: http://www.stanford.edu/~rhorn/ArtclCognome.html).

Horton, W. 1991. *Illustrating computer documentation: The art of presenting information graphically in paper and online,* N.Y.: Wiley.

Kosslyn, S.M. 1989. Understanding charts and graphs. *Applied Cognitive Psychology* 3: 185-226.

Kosslyn, S.M. 1994. *Elements of graph design.* N.Y.: W.H. Freeman.

McCloud, S. 1993. *Understanding comics: The invisible art.* Northampton, MA: Kitchen Sink Press.

Mayer, R.E. 2001. *Multimedia learning.* Cambridge: Cambridge Univ. Press.

Tufte, E. 1983. *The visual display of quantitative information.* Cheshire, CT: Graphics Press.

Tufte, E. 1990. *Envisioning information.* Cheshire, CT: Graphics Press.

SOCIABLE TECHNOLOGIES: ENHANCING HUMAN PERFORMANCE WHEN THE COMPUTER IS NOT A TOOL BUT A COMPANION

Sherry Turkle, Massachusetts Institute of Technology

> *"Replacing human contact [with a machine] is an awful idea. But some people have no contact [with caregivers] at all. If the choice is going to a nursing home or staying at home with a robot, we think people will choose the robot."* Sebastian Thrun, Assistant Professor of Computer Science, Carnegie Mellon University

> *"AIBO [Sony's household entertainment robot] is better than a real dog. It won't do dangerous things, and it won't betray you. ...Also, it won't die suddenly and make you feel very sad."'* A 32-year -old woman on the experience of playing with AIBO

> *"Well, the Furby is alive for a Furby. And you know, something this smart should have arms. It might want to pick up something or to hug me."* Ron, age six, answering the question, "Is the Furby alive?"

Artificial intelligence has historically aimed at creating objects that might improve human performance by offering people intellectual complements. In a first stage, these objects took the form of tools, instruments to enhance human reasoning, such as programs used for medical diagnosis. In a second stage, the boundary between the machine and the person became less marked. Artificial intelligence technology functioned more as a prosthetic, an extension of human mind. In recent years, even the image of a program as prosthetic does not capture the intimacy people have with computational technology. With "wearable" computing, the machine comes closer to the body, ultimately continuous with the body, and the human person is redefined as a cyborg. In recent years, there has been an increased emphasis on a fourth model of enhancing human performance through the use of computation: technologies that would improve people by offering new forms of social relationships. The emphasis in this line of research is less on how to make machines "really" intelligent (Turkle 1984, 1995) than on how to design artifacts that would cause people to experience them as having subjectivities that are worth engaging with.

The new kind of object can be thought of as a relational artifact or as a sociable technology. It presents itself as having affective states that are influenced by the object's interactions with human beings. Today's relational artifacts include children's playthings (such as Furbies, Tamagotchis, and My Real Baby dolls); digital dolls and robots that double as health monitoring systems for the elderly (Matsushita's forthcoming Tama, Carnegie Mellon University's Flo and Pearl); and pet robots aimed at the adult (Sony's AIBO, MIT's Cog and Kismet). These objects are harbingers of a new paradigm for computer-human interaction.

In the past, I have often described the computer as a Rorschach. When I used this metaphor I was trying to present the computer as a relatively neutral screen onto which people were able to project their thoughts and feelings, a mirror of mind and self. But today's relational artifacts make the Rorschach metaphor far less useful. The computational object is no longer affectively "neutral." Relational artifacts do not so much invite projection as demand engagement. People are learning to interact with computers through conversation and gesture. People are learning that to relate successfully to a computer you do not have to know how it works but can take it "at interface value," that is, assess its emotional "state," much as you would if you were relating to another person. Through their experiences with virtual pets and digital dolls, which present themselves as loving and responsive to care, a generation of children is learning that some objects require emotional nurturing and some even promise it in return. Adults, too, are encountering technology that attempts to offer advice, care, and companionship in the guise of help-software-embedded wizards, intelligent agents, and household entertainment robots such as the AIBO "dog."

New Objects are Changing Our Minds

Winston Churchill once said, "We make our buildings and then they make us." We make our technologies, and they in turn shape us. Indeed, there is an unstated question that lies behind much of our historic preoccupation with the computer's capabilities. That question is not what can computers do or what will computers be like in the future, but instead, what will we be like? What kind of people are we becoming as we develop more and more intimate relationships with machines? The new technological genre of relational, sociable artifacts is changing the way we think. Relational artifacts are new elements in the categories people use for thinking about life, mind, consciousness, and relationship. These artifacts are well positioned to affect people's way of thinking about themselves, about identity, and about what makes people special, influencing how we understand such "human" qualities as emotion, love, and care. We will not be taking the adequate measure of these artifacts if we only consider what they do *for* us in an *instrumental* sense. We must explore what they do not just for us but to us as people, to our relationships, to the way our children develop, to the way we view our place in the world.

There has been a great deal of work on how to create relational artifacts and maximize their ability to evoke responses from people. Too little attention, however, has gone into understanding the human implications of this new computational paradigm, both in terms of how we relate to the world and in terms of how humans construct their sense of what it means to be human and alive. The language for assessing these human implications is enriched by several major traditions of thinking about the role of objects in human life.

Objects as Transitional to Relationship

Social scientists Claude Levi-Strauss (1963), Mary Douglas (1960), Donald Norman (1988), Mihaly Csikzentmihalyi (1981), and Eugene Rochberg-Halton (1981) have explored how objects carry ideas, serving as enablers of new individual and cultural meanings. In the psychoanalytic tradition Winnicott (1971) has discussed how objects mediate between the child's earliest bond with the mother, who the infant experiences as inseparable from the self, and the child's growing capacity to develop relationships with other people, who will be experienced as separate beings.

In the past, the power of objects to act in this transitional role has been tied to the ways in which they enabled the child to project meanings onto them. The doll or the teddy bear presented an unchanging and passive presence. Relational artifacts take a more active stance. With them, children's expectations that their dolls want to be hugged, dressed, or lulled to sleep don't come from the child's projection of fantasy or desire onto inert playthings, but from such things as a digital doll's crying inconsolably or even saying, "Hug me!" "It's time for me to get dressed for school!" The psychology of the playroom turns from projection to social engagement, in which data from an active and unpredictable object of affection helps to shape the nature of the relationship. On the simplest level, when a robotic creature makes eye contact, follows your gaze, and gestures towards you, what you feel is the evolutionary button being pushed to respond to that creature as a sentient and even caring other.

Objects as Transitional to Theories of Life

The Swiss psychologist Jean Piaget addressed some of the many ways in which objects carry ideas (1960). For Piaget, interacting with objects affects how the child comes to think about space, time, the concept of number, and the concept of life. While for Winnicott and the object relations school of psychoanalysis, objects bring a world of people and relationships inside the self, for Piaget objects enable the child to construct categories in order to make sense of the outer world. Piaget, studying children in the context of non-computational objects, found that as children matured, they homed in on a definition of life that centered around "moving of one's own accord." First, everything that moved was taken to be alive, then only those things that moved without an outside push or pull. Gradually, children refined the notion of "moving of one's own accord" to mean the "life motions" of breathing and metabolism.

In the past two decades, I have followed how computational objects change the ways children engage with classic developmental questions such as thinking about the property of "aliveness." From the first generation of children who met computers and electronic toys and games (the children of the late 1970s and early 1980s), I found a disruption in this classical story. Whether or not children thought their computers were alive, they were sure that how the toys moved was not at the heart of the matter. Children's discussions about the computer's aliveness came to center on what the children perceived as the computer's psychological rather than physical properties (Turkle 1984). Did the computer know things on its own or did it have to be programmed? Did it have intentions, consciousness, feelings? Did it cheat? Did it know it was cheating? Faced with intelligent machines, children took a new world of objects and imposed a new world order. To put it too simply, motion gave way to emotion, and physics gave way to psychology as criteria for aliveness.

By the 1990s, that order had been strained to the breaking point. Children spoke about computers as just machines but then described them as sentient and intentional. They talked about biology, evolution. They said things like, "the robots are in control but not alive, would be alive if they had bodies, are alive because they have bodies, would be alive if they had feelings, are alive the way insects are alive but not the way people are alive; the simulated creatures are not alive because they are just in the computer, are alive until you turn off the computer, are not alive because nothing in the computer is real; the Sim creatures are not alive but almost-alive, they would be alive if they spoke, they would be alive if they traveled, they're not alive because they don't have bodies, they are alive because they can have babies and would be alive if they could get out of the game and onto America Online."

There was a striking heterogeneity of theory. Children cycled through different theories to far more fluid ways of thinking about life and reality, to the point that my daughter upon seeing a jellyfish in the Mediterranean said, "Look, Mommy, a jellyfish; it looks so realistic!" Likewise, visitors to Disney's Animal Kingdom in Orlando have complained that the biological animals that populated the theme park were not "realistic" compared to the animatronic creatures across the way at Disneyworld.

By the 1990s, children were playing with computational objects that demonstrated properties of evolution. In the presence of these objects, children's discussions of the aliveness question became more complex. Now, children talked about computers as "just machines" but described them as sentient and intentional as well. Faced with ever more sophisticated computational objects, children were in the position of theoretical tinkerers, "making do" with whatever materials were at hand, "making do" with whatever theory could be made to fit a prevailing circumstance (Turkle 1995).

Relational artifacts provide children with a new challenge for classification. As an example, consider the very simple relational artifact, the "Furby." The Furby is an owl-like interactive doll, activated by sensors and a pre-programmed computer chip, which engages and responds to their owners with sounds and movement. Children playing with Furbies are inspired to compare and contrast their understanding of how the Furby works to how they "work." In the process, the line between artifact and biology softens. Consider this response to the question, "Is the Furby alive?"

> Jen (age 9): I really like to take care of it. So, I guess it is alive, but it doesn't need to really eat, so it is as alive as you can be if you don't eat. A Furby is like an owl. But it is more alive than an owl because it knows more and you can talk to it. But it needs batteries so it is not an animal. It's not like an animal kind of alive.

Jen's response, like many others provoked by playing with Furbies, suggests that today's children are learning to distinguish between an "animal kind of alive" and a "Furby kind of alive." In my conversations with a wide range of people who have interacted with relational artifacts — from five year olds to educated adults — an emergent common denominator has been the increasingly frequent use of "sort of alive" as a way of dealing with the category confusion posed by relational artifacts. It is a category shared by the robots' designers, who have questions about the ways in which their objects are moving toward a kind of consciousness that might grant them a new moral status.

Human-Computer Interaction

The tendency for people to attribute personality, intelligence, and emotion to computational objects has been widely documented in the field of human-computer interaction (HCI) (Weizenbaum 1976; Nass, Moon, et al. 1997, Kiesler and Sproull 1997; Reeves and Nass 1999). In most HCI work, however, this "attribution effect" is considered in the context of trying to build "better" technology.

In *Computers are Social Actors: A Review of Current Research*, Clifford Nass, Youngme Moon, and their coauthors (1997) review a set of laboratory experiments in which "individuals engage in social behavior towards technologies even when such behavior is entirely inconsistent with their beliefs about machines" (p. 138). Even when computer-based tasks contained only a few human-like characteristics, the authors found that subjects attributed personality traits and gender to computers and adjusted their responses to avoid hurting the machines' "feelings." The authors suggest that "when we are confronted with an entity that [behaves in human-like ways, such as using language and responding based on prior inputs] our brains' default response is to unconsciously treat the entity as human" (p. 158). From this, they suggest design criteria: technologies should be made more "likeable":

> ... "liking" leads to various secondary consequences in interpersonal relationships (e.g., trust, sustained friendship, etc.), we suspect that it also leads to various consequences in human-computer interactions (e.g., increased likelihood of purchase, use, productivity, etc.) (p. 138).

Nass et al. prescribe "likeability" for computational design. Several researchers are pursuing this direction. At the MIT Media Lab, for example, Rosalind Picard's Affective Computing research group develops technologies that are programmed to assess their users'' emotional states and respond with emotional states of their own. This research has dual agendas. On the one hand, affective software is supposed to be compelling to users — "friendlier," easier to use. On the other hand, there is an increasing scientific commitment to the idea that objects need affect in order to be intelligent. As Rosalind Picard writes in *Affective Computing* (1997, x),

> I have come to the conclusion that if we want computers to be genuinely intelligent, to adapt to us, and to interact naturally with us, then they will need the ability to recognize and express emotions, to have emotions, and to have what has come to be called "emotional intelligence."

Similarly, at MIT's Artificial Intelligence Lab, Cynthia Breazeal has incorporated both the "attribution effect" and a sort of "emotional intelligence" in Kismet. Kismet is a disembodied robotic head with behavior and capabilities modeled on those of a pre-verbal infant (see, for example, Breazeal and Scassellati 2000). Like Cog, a humanoid robot torso in the same lab, Kismet learns through interaction with its environment, especially contact with human caretakers. Kismet uses facial expressions and vocal cues to engage caretakers in behaviors that satisfy its "drives" and its "emotional" needs. The robot "wants" to be happy, and people are motivated to help it achieve this goal. Its evocative design seems to help, Breazeal reports: "When people see Cog they tend to say, 'That's interesting.' But with Kismet they tend to say, 'It smiled at me!' or 'I made it happy!'" (Whynott 1999). I have seen similar reactions between children and simpler digital pets (both on the screen, such as neopets and in robotic form, such as Furbies and AIBOs).

When children play with Furbies, they want to know the objects' "state," not to get something "right," but to make the Furbies happy. Children want to understand Furby language, not to "win" in a game over the Furbies, but to have a feeling of mutual recognition. When I asked her if her Furby was alive, Katherine, age five, answered in a way that typifies this response:

> "Is it alive? Well, I love it. It's more alive than a Tamagotchi because it sleeps with me. It likes to sleep with me."

Children do not ask how the Furbies "work" in terms of underlying process; they take the affectively charged toys "at interface value."

With the advent of relational artifacts and their uses of emotion, we are in a different world from the old AI debates of the 1960s to 1980s, in which researchers argued about whether machines could be "really" intelligent. The old debate was essentialist; these new objects allow researchers and their public to sidestep such arguments about what is inherent in the computer. Instead, they focus attention on what the objects evoke in us. When we are asked to care for an object (the robot Kismet or the plaything Furby), and when the cared-for object thrives and offers us its attention and concern, we experience that object as intelligent. Beyond this, we feel a connection to it. So the issue here is not whether objects "really" have emotions, but what is happening when relational artifacts evoke emotional responses in the users.

People's relationships with relational artifacts have implications for technological design (i.e., how to make the objects better, more compelling), and they have implications that are the focus of this research: they complicate people's ways of thinking about themselves, as individuals, as learners and in relationships, and within communities. To augment human potential, any discussion of how to make "better" relational artifacts must be in terms of how they can best enhance people in their human purposes. It cannot be discussed in terms of any absolute notions defined solely in terms of the objects.

The questions raised by relational artifacts speak to people's longstanding fears and hopes about technology, and to the question of what is special about being human, what is the nature of "personhood." In the case of relational technology, there is a need for examination of these questions, beginning with how these objects are experienced in the everyday lives of the individuals and groups who are closest to them.

Human Performance

When people learn that AIBO, the Sony robot dog, is being introduced into nursing homes as companions to the elderly, the first question asked is usually, "Does it work?" By this the person means, "Are the old people happier when they have a robot pet? Are they easier to take care of?" My vision of the future is that we are going to have increasingly intimate relationships with sociable technologies, and we are going to need to ask increasingly complex questions about the kinds of relationships we form with them. The gold standard cannot be whether these objects keep babies and/or the elderly "amused" or "quiet" or "easier to care for." Human performance needs to be defined in a much more complex way, beginning with a set of new questions that take the new genre of objects seriously. Taking them seriously means addressing them as new social interlocutors that will bring together biology, information science, and nanoscience. Human performance needs to take into account the way we feel about ourselves as people, in our relationships and in our social groups. From this point of view, the question for the future is not going to be whether children love their robots more than their parents, but what loving itself comes to mean. From this perspective on human enhancement, some of the questions are

- How are children adapting ideas about aliveness, intentionality, and emotion to accommodate relational artifacts?

- How are designers and early adopters adapting ideas about personhood, intentionality, and relationship to accommodate relational artifacts? How do these artifacts influence the way people think about human minds?

- How are people thinking about the ethical issues raised by relational artifacts? Is a moral code for the treatment of this new type of artifacts being developed?

- How are people using relational artifacts to address needs traditionally met by other humans and animal pets, such as companionship and nurturing?

A Vision Statement

Computational objects are "evocative objects." They raise new questions and provoke new discourse about the nature of mind, about what it means to be alive, about what is special about being a person, about free will and intentionality. Computation brings philosophy into everyday life. Objects as simple as computer toys and games raise such questions as "What is intelligence? What does it mean to be alive? Or to die? What is the nature of the self? What is special about being a person?" In the next 10 to 20 years, research that will marry biology, information science, cognitive science, and nanoscience is going to produce increasingly sophisticated relational, sociable artifacts that will have the potential to profoundly influence how people think about learning, human development, intelligence, and relationships.

- As research on relational and sociable technology progresses, there will be parallel investigations of how these objects affect the people who use them, how they influence psychological development, human relationships, and additionally, how they enter into people's thinking about themselves, including about such questions as the nature of intention, the self, and the soul.

- The development of sociable technologies will require a renaissance in the sciences that study human development and personality. There will be an increasing virtuous cycle of research to understand human personality and to create person-enhancing machines. Indeed, the notion of personable machines will come to mean person-enhancing machines.

- In the past, it has been argued that technology dehumanized life, but as we become committed to person-enhancing objects, this argument will need to be revisited. Making technology personable will entail learning about ourselves. In order to make technology enhance humans, we will humanize technology.

- Historically, when technology has been designed without human fulfillment in mind, but purely in terms of the instrumental capabilities of the machine, there has been a great deal of resistance to technology. This resistance needs to be taken seriously, because it points to the ways in which people associate technology with human loss. The development of sociable technology will require that there be a flourishing of research that takes resistance to technology as a symptom of something important that needs to be studied rather than a problem that needs to be overcome. An understanding of human psychology is essential for the development of sociable technologies. This latter will proceed with vigilance and with the participation of humanists and scientists. Sociable technology will enhance human emotional as well as cognitive performance, not only giving us more satisfactory relationships with our machines but also potentially vitalizing our relationships with each other, because in order to build better sociable objects we will have learned more about what makes us social with each other.

References

Baltus, G., D. Fox, F. Gemperle, J. Goetz, T. Hirsch, D. Magaritis, M. Montemerlo, J. Pineau, N. Roy, J. Schulte, and S. Thrun. 2000. Towards personal service robots for the elderly. Published online at http://www.cs.cmu.edu/~thrun/papers/thrun.nursebot-early.html.

Breazeal (Ferrell), C., and B. Scassellati. 2000. Infant-like social interactions between a robot and a human caretaker. In *Adaptive behavior on simulation models of social agents*, Kerstin Dautenhahn, ed. (Special issue.)

Brooks, R., C. Breazeal, M. Marjanovic, B. Scassellati, and M. Williamson. 1998. The Cog project: Building a humanoid robot. In *Computation for metaphors, analogy and agents*, C. Nehaniv, ed. *Springer Lecture Notes in Artificial Intelligence*, Vol. 1562.

Brooks, R. 2000. Human-robot interaction and the future of robots in society. Presentation at the STS Colloquium, February 7, Massachusetts Institute of Technology, Cambridge, MA.

Csikzentmihalyi, M., and E. Rochberg-Halton. 1981. *The meaning of things: Domestic symbols and the self*. Cambridge: Cambridge University Press.

Dennett, D. [1987] 1998. *The intentional stance*. Reprint. Cambridge, MA: MIT Press.

Douglas, M. [1960] 1993. *Purity and danger: An analysis of the concepts of pollution and taboo*. Reprint. London: Routledge.

Drexler, M. 1999. Pet robots considered therapy for the elderly. *CNN Online*, March 12. http://www.cnn.com/TECH/ptech/9903/25/robocat.idg/.

Ito, M. 1997. Inhabiting multiple worlds: Making sense of SimCity 2000 in the fifth dimension. In *Cyborg Babies*, R.D. Floyd and J. Dumit, eds. New York: Routledge.

Kiesler, S. and L. Sproull. 1997. 'Social' human-computer interaction? In *Human Values and the Design of Computer Technology*, B. Friedman, ed. Stanford, CA: CSLI Publications.

Knorr-Cetina, K. 1997. Sociality with objects: Social relations in postsocial knowledge societies. *Theory, Culture and Society* 14(4): 1-30.

Latour, B. 1992. Where are the missing masses? The sociology of a few mundane artifacts. In *Shaping Technology/Building Society: Studies in Sociotechnical Change*, W. Bijker and J. Law, eds. Cambridge MA: MIT Press.

Levi-Strauss, C. 1963. *Structural anthropology*. New York: Basic Books.

Nass, C., Y. Moon, J. Morkes, E. Kim, and B.J. Fogg. 1997. Computers are social actors: A review of current research. In *Human values and the design of computer technology*, B. Friedman, ed. Stanford, CA: CSLI Publ..

Norman, D. 1988. *The design of everyday things*. New York: Currency/Double Day.

Ornstein, P.H., ed. 1978. *The search for the self: Selected writings of Heinz Kohut: 1950-1978, Vol. 2*. New York: International Universities Press, Inc.

Piaget, J. 1960. *The child's conception of the world*. Joan and Andrew Tomlinson, trans. Totowa, N.J.: Littlefield, Adams.

Picard, R.W. 1997. *Affective computing*. Cambridge, MA: MIT Press.

Reeves, B. and C. Nass. 1999. *The media equation: How people treat computers, television, and new media like real people and places*. Cambridge: Cambridge University Press.

Resnick, M. 1998. Technologies for lifelong kindergarten. *Educational technology research and development* 46:4.

Turkle, S. 1995. *Life on the screen: Identity in the age of the Internet.* New York: Simon and Schuster.

_____. 1984. *The second self: Computers and the human spirit.* New York: Simon and Schuster.

Weizenbaum, J. 1976. *Computer power and human reason: From judgement to calculation.* San Francisco: W.H. Freeman.

Whynott, D. 1999. The robot that loves people. *Discover,* October, online edition.

Winner, L. 1986. *The whale and the reactor: A search for limits in an age of high technology.* Chicago: University of Chicago Press.

Winnicott, D.W. [1971] 1989. *Playing and Reality.* Reprint. London: Routledge.

Zambrowski, J. 2000. CMU, Pitt developing "nursebot." *Tribune Review,* October 27, online edition.

VISIONARY PROJECTS

SOCIO-TECH: THE PREDICTIVE SCIENCE OF SOCIETAL BEHAVIOR

Gerold Yonas, Sandia National Laboratories,[1] and Jessica Glicken Turnley, Galisteo Consulting Group, Inc.

Socio-tech is the predictive — not descriptive — science of the behavior of societies. It is the convergence of information from the life sciences, the behavioral sciences (including psychology and the study of cognition), and the social sciences. Its data gathering and analysis approaches come from these fields and are significantly augmented by new tools from fields such as nanotechnology, engineering, and the information sciences. Agent-based simulations, models incorporating genetic algorithms, evolutionary computing techniques, and brain-machine interfaces provide new ways to gather data and to analyze the results.

Why Do We Care?

Most immediately, socio-tech can help us win the war on terrorism. It can help us to understand the motivations of the terrorists and so eliminate them. It also can help us to manage ourselves, to orchestrate our own country's response to a potential or real attack. In the longer term, as a predictive science, socio-tech can help us identify possible drivers for a wide range of socially disruptive events and allow us to put mitigating or preventative strategies in place before the fact.

What Is New?

The multiple drivers of human behavior have long been known. What have been missing are the theoretical paradigm and associated tools to integrate what we know about these drivers into an overarching understanding of human activity.

Currently, most of the data related to understanding human behavior has remained field-specific. The life sciences focus on the biological impacts of humans

[1] Sandia National Laboratories is a multiprogram laboratory operated by Sandia Corporation, a Lockheed Martin Company, for the United States Department of Energy under Contract DE-AC04-94AL85000.

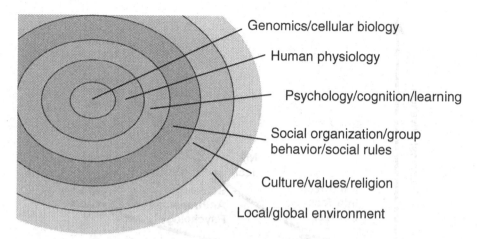

Figure B.11. Integrated studies of human behavior: Socio-tech.

functioning in physical spaces. The social sciences focus on the organizing principles of groups (rule of law, social hierarchies) and the different values groups place on behaviors (e.g., through culture or religion). The behavioral sciences are concerned with the functioning of the brain and the impact of individual experience on decision-making. The tools of science, engineering, and the information and computational sciences generally are not well integrated into these fields. C.P. Snow's 1959 Rede lecture captured this divide between the sciences on one hand and the arts and humanities on the other by the term "the two cultures."

There is little dialogue among practitioners from these different areas. They are separated by barriers of jargon, by conceptual frameworks that are difficult to translate from one field to another, and by traditional institutional compartmentalization of intellectual disciplines. Efforts such as Lewis Mumford's *Techniques and Human Development* (1989) to socially contextualize technology or E.O. Wilson's more recent and ambitious *Concilience* (1999) are the exceptions rather than the rule. We thus have no true study of human behavior, for there is no field or discipline with the interest or the tools to integrate data from these different fields. The challenge before us is to devise a way to understand data and information from each field in the context of all others. If genomics can be practiced with an awareness of human physiology, behavior, values, and environment, and, conversely, if information from genomics can be incorporated in a meaningful way into studies in these other fields, we will have made a significant leap in our understanding of human behavior (Figure B.11).

Why Now?

The time is ripe to begin such integration — to use the tremendous computing power we now have to integrate data across these fields to create new models and hence new understanding of the behavior of individuals. The ultimate goal is acquiring the ability to predict the behavior of an individual and, by extension, of groups. Recent advances in brain imaging, neuropsychology, and other sciences of the brain have significantly contributed to our knowledge of brain functioning. Genomics, molecular biology, and contributions from other areas in the life sciences

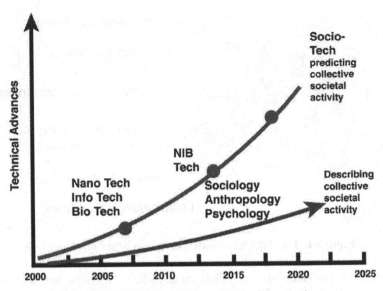

Figure B.12. Socio-tech: A qualitatively new science.

have greatly advanced our knowledge of the human body, its genetic core, and its response to various environmental stimuli. The increasing body of knowledge in the social sciences, combined with the tremendous computing (analysis) power available at affordable prices and new tools for communication and expression, have given us new ways of looking at social relationships such as social network theory, and new ways of understanding different ways of life. Incorporating these advances in a wide range of fields of study into overarching and integrating conceptual models should give us significant insights into human behavior.

Figure B.12 shows two possible trajectories for the development of knowledge. The upper trajectory combines the "two cultures," using technology to leverage the behavioral and social sciences and leads to a predictive science of behavior. The lower trajectory illustrates improvements in the behavioral and social sciences, with little incorporation of theory and tools from science and technology. It leads to greater descriptive but no predictive capabilities.

Socio-tech — the accumulation, manipulation, and integration of data from the life, social, and behavioral sciences, using tools and approaches provided by science and technology — will raise our ability to predict behaviors. It will allow us to interdict undesirable behaviors before they cause significant harm to others and to support and encourage behaviors leading to greater social goods.

References

Mumford, L. 1999. *Techniques and human development*. 2 vols. New York, NY: Harcourt Brace Jovanovich.

Snow, C.P. 1959. *The two cultures*. Introduction by Stefan Collini. New York, NY: Cambridge University Press.

Wilson, E.O. 1999. *Concilience: The unity of knowledge*. New York, NY: Random House.

BREAKING THE LIMITS ON DESIGN COMPLEXITY

Jordan Pollack, Brandeis University

As we contemplate microelectromechanical systems (MEMS) and nanotech-nologies (nano), we must study the history of the design of circuits and software, especially software that is supposed to have cognitive function, or artificial intelligence (AI). Having been working in the field of AI for 25 years, I can say with some authority that nanotechnology will not solve the AI problem. In fact, the repeated failures of AI artifacts to live up to claims made by their proponents can shed light on human expectations of nano and on the capacity of human teams to design complex objects.

We think that in order to design products "of biological complexity" that could make use of the fantastic fabrication abilities of new nano and MEMS factories, we must first liberate design by discovering and exploiting the principles of automatic self-organization that are seen in nature. A brain has 10^{11} connections. Chemistry often works with 10^{23} molecules. Advanced software is the most complex (and profitable) of all of human artifacts, yet each application only comprises between 10 million and 100 million lines of code, or a maximum of around 10^8 moving parts. Suppose an animal brain, rather than requiring the specifying over time of the bonds for every molecule, ONLY required the equivalent of 10^{10} uniquely programmed parts. Why can't we engineer that?

In circuits, achieving even a function as lowly as the "bit" of memory creates a means to replication. Now we have 32 million bits on a single chip, and that is an achievement. Building blocks that can be replicated via manufacturing in hardware make things like memory chips and CPUs faster and more capable. This replication capacity and speedup of hardware enables Moore's law, a doubling of computer power, and even disk space, every 18 months. However, this periodic doubling of computer power has not led to equivalent doubling of human capacity to manufacture significantly more complex software. Moore's law does not solve the problem of engineering 10 billion lines of code!

The simple reason we haven't witnessed Moore's law operate for software is that 32 million copies of the same line of code is just one more line of code — the DO loop. Thus today's supercomputers run the same sized programs as the supercomputers of the 1970s, which are the desktops of today. The applications can use lots of floating point multiplication, but the complexity of the tasks hasn't grown beyond word processing, spreadsheets, and animations. Faster and faster computers seem to encourage software companies to write less and less efficient code for the same essential functionality — Windows is just DOS with wallpaper.

We've learned this hard lesson from the field of software — which isn't even constrained by material cost or by physical reality: *there are limits on the complexity of achievable design.* This is true even when throwing larger and larger teams of humans at a problem, even with the best groupware CAD software, even with bigger computers. Therefore, assumptions that new fabrication methodologies will lead to a breakthrough in design complexity ought to be taken with a grain of salt.

Yet many nano pundits expect that smaller-scale manufacturing, rather than leading to homogenous materials competitive with wood and plastic, will automatically lead to artificial objects of extraordinary complexity and near life-like capacity. They seem to ignore the technical challenges of understanding and modeling cognition, plugging portals into our brains, and programming Utility Fogs of nanobots that are intelligent enough to swarm and perform coordinated missions. The reality is that making life-sized artifacts out of molecules may require the arranging of 10^{30} parts.

AI is stalled because it is starved of the much more complex blueprints than anyone has any clue how to build. Software engineering seems to have reached a complexity limit well below what computers can actually execute. Despite new programming languages and various movements to revolutionize the field, the size of programs today is about the same as it has been for 40 years: 10-100 million lines of code. Old code finally collapses under the cost of its own maintenance.

The high-level languages, object-oriented programming systems, and computer-assisted software engineering (CASE) breakthroughs have all seemed promising, yet each new breakthrough devolves back into the same old thing in new clothes: the Fortran compiler plus vast scientific libraries. The power of each new programming tool, be it PL/1, Turbo Pascal, Visual Basic, Perl, or Java, is located in the bundled collections of subroutine libraries, which eventually grow to surpass our merely human cognitive ability to remember or even look them up in burgeoning encyclopedias.

The problem illustrated here is still Brooks' Mythical Man Month: We can't get bigger and better software systems by putting more humans on the job. The best original software, whether DOS, Lotus 123, or Wordstar, have been written by one or two good programmers; large teams extend, integrate, copy, and maintain, but they do not create. The more programmers on a task, the more bugs they create for each other.

The opportunity available today is that the way out of this tarpit, the path to achieving both software and nano devices of biological complexity with tens of billions of moving parts, is very clear: it is through increasing our scientific understanding of the processes by which biologically complex objects arose. As we understand these processes, we will be able to replicate them in software and electronics. The principles of automatic design and of self-organizing systems are a grand challenge to unravel. Fortunately, remarkable progress has been shown since the computer has been available to refine the theory of evolution. Software is being used to model life itself, which has been best defined as that "chemical reaction, far from equilibrium, which dissipates energy and locally reverses entropy."

Much as logic was unconstrained philosophy before computer automation, and as psychological and linguistic theories that could not be computerized were outgunned by formalizable models, theories on the origin of life, its intrinsic metabolic and gene regulation processes, and the mechanisms underlying major transitions in evolution, are being sharpened and refuted through formalization and detailed computer simulation.

Beyond the basic idea of a genetic algorithm, the variety of studies on artificial life, the mathematical and computational bases for understanding learning, growth, and evolution, are rapidly expanding our knowledge and our know-how.

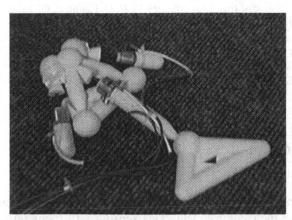

Figure B.13. Semi- and fully automatic design can help design complex systems like robot hardware and software.

My laboratory, which studies machine learning and evolutionary computation, has focused on how semi- and fully-automatic design can help design complex systems like robot hardware and software. We have used a collection of methods called "co-evolution," in which the idea is to create a sustained "arms-race" amongst or between populations of simple learning systems in order to achieve automatic design of various structures such as sorting nets, cellular automata rules, game players, and robot bodies and brains (Fig. B.13).

The field of evolutionary design, which aims at the creation of artifacts with less human engineering involvement, is in full force, documented by the books edited by Peter Bentley, as well as a NASA-sponsored annual conference on evolutionary hardware. Evolutionary robotics is a related field that started with Karl Sims' virtual robots and has grown significantly in the last five years.

So far, few artificial evolutionary processes have produced software or systems beyond those that can be designed by teams of humans. But they are competitive, and they are much cheaper than human designs. More importantly, thus far, they have not hit a barrier to complexity as seen in software engineering. Automatic design converts surplus computer time into complex design, and this will be aided by Moore's law. As inexpensive one-of-a-kind fabrication becomes possible, mass manufacture will no longer be necessary to amortize the fixed costs of engineering design, and automatic design will become necessary to generate complex designs with low cost. Success in this field holds keys to surpassing today's limits on complexity.

References

Bentley, P.J., ed.. 1999. *Evolutionary design by computers*. San Francisco: Morgan Kaufman.

Funes, P. and J.B. Pollack. 1998. Evolutionary body building: Adaptive physical designs for robots. *Artificial Life* 4: 337-357.

Holland, J.H. 1975. *Adaptation in natural and artificial systems*. Ann Arbor: University of Michigan Press.

Hornby, G.S., and J.B. Pollack. 2000. Evolving L-systems to generate virtual creatures. *Computers and Graphics* 25(6): 1041-1048.

Lipson, H., and J.B. Pollack. 2000. Automatic design and manufacture of robotic lifeforms. *Nature* 406:974-978.

Nolfi, S. and D. Floreano. 2000. Evolutionary robotics. The biology, intelligence, and technology of self-organizing machines. Cambridge, MA: MIT Press.

Pollack, J.B. and A.D. Blair. 1998. Co-evolution in the successful learning of backgammon strategy. *Machine Learning* 32(3): 225-240 (September).

Ray, T. 1992. An approach to the synthesis of life. In *Artificial Life II*, C. Langton, C. Taylor, J.F. and S. Rasmussen, eds. Reading, MA: Addison-Wesley.

Sims, K. 1994. Evolving virtual creatures. In SIGGRAPH 94 Conference Proceedings, Annual Conference Series, pages 15-22.

Thompson, A. 1997. "Artificial evolution in the physical world." In *Evolutionary Robotics: From intelligent robotics to artificial life*, T. Gomi, ed. Elsevier.

ENHANCING PERSONAL AREA SENSORY AND SOCIAL COMMUNICATION THROUGH CONVERGING TECHNOLOGIES

Rudy Burger, MIT Media Lab Europe

The next decade will see great strides in personal wearable technologies that enhance people's ability to sense their environment. This sensing will focus on at least two different areas:

a) *social sensing*, in which we may augment our ability to be aware of people in our immediate vicinity with whom we may wish to connect (or possibly avoid!)

b) *environmental sensing*, in which we may augment our ability to sense aspects of our environment (for example, the quality of the air we are breathing) that may be hazardous to us but that our normal senses cannot detect

Social Sensing

Few would question the remarkable extent to which the two pillars of modern day business communication — cell phones and email — enable us to effortlessly stay in touch with people on the other side of the planet. The paradox lurking behind this revolution is that these same technologies are steadily eroding the time and attention we devote to communicating with people in our immediate vicinity. The cost to the sender of sending an email or placing a cellular call is rapidly approaching zero. Unchecked, the cost to the recipient may rapidly become unmanageable, not in terms of financial cost, but rather in terms of demands on our time and attention. Witness the now common scene in airports and other public spaces — hundreds of people milling around in what appears to be animated conversation; on closer inspection, it turns out that they are not with each other, but rather with people connected to them via the near-invisible ear bud microphones they are wearing. Similarly, it is common to observe business colleagues in offices sitting just a few feet away from each other engaged in passionate debate. But the debate is often not verbal; rather, the only sound is the click-clack of keyboards as email flies back and forth. In desperation, some companies have resorted to the draconian measure if banning emails on certain days of the week ("email-free Fridays") or certain core hours of the day as the only way to pry their employees away from the email inboxes to engage in face-to-face dialog.

Why is it that so many people seem to find communication through email or cell phone more compelling than face-to-face dialog? Many value the fact that email permits asynchronous communication, enabling the recipient to respond only when it is convenient for them. Email also enables people to reinvent their personalities in ways that would be difficult or impossible for them socially. Cell phone technology has conquered geographical separation — anytime, anywhere communication. Rather than pursuing a chance encounter with the stranger standing next to me, it seems easier to talk to someone I know over a cell phone.

In contrast, technology has done little or nothing to enhance face-to-face communication. As we move from the current era of computing (so-called "personal" computers) to the next era (described variously as ambient intelligence or ubiquitous computing), help for social dialog will arrive in the form of next-generation personal information managers (PIMs) connected via wireless Personal Area Networks (PANs). PANs operate over a distance of just a few feet, connecting an individual to just those people within their immediate vicinity — their dinner companions, for example. First-generation PAN devices will be based on Bluetooth wireless technology.

Next generation PIMs arriving on the market over the next 24-36 months will store their owner's personal profile that will contain whatever information the owner may wish to share with others in their immediate vicinity. The information a user may wish to exchange in this way will obviously depend on the social context that the user is in at any given moment. In contrast to today's PIMs (where a lot of fumbling around will eventually result in a digital business card being exchanged between two devices), rich personal information will flow automatically and transparently between devices. It is quite likely that these PIMs will evolve to look nothing like today's devices. They may be incorporated into a pair of eyeglasses, or even in the clothes that we wear.

Widespread use of such devices will, of course, require that issues of personal privacy be resolved. However, peer-to-peer ad hoc networks of this type are inherently more respectful of individual privacy than client server systems. Users of PAN devices can specify either the exact names or the profiles of the people with whom they want their devices to communicate. They may also choose to have any information about themselves that is sent to another device time-expire after a few hours. This seems relatively benign compared to the information that can be collected about us (usually without our knowledge or consent) every time we browse the Web.

Many of us attend conferences every year for the purpose of professional networking. At any given conference of a hundred people or more, it is likely that there are a handful of potentially life-transforming encounters that could happen within the group. But such encounters are reliant on a chain of chance meetings that likely will not happen, due to the inefficiencies of the social network. Personal Area Network devices could dramatically improve our ability to identify the people in a crowd with whom we may wish to talk. Of course, we will want sophisticated software agents acting on our behalf to match our interests with the profiles of the people standing around us. We could even imagine a peer-to-peer Ebay in which my profile indicates that I am in the market to buy a certain type of car and I am alerted if anyone around me is trying to sell such a car. In Japan, it is already possible to buy a clear plastic key chain device that can be programmed to glow brightly when I encounter someone at a party whose interests are similar to mine. A high tech icebreaker!

The most profound technologies are the ones that "disappear" with use. Personal Area Network devices may enable nothing fundamentally new — they may just simplify what we already do

Environmental Sensing

We rely heavily on our natural senses (touch, sight, sound, smell) to keep us out of danger. Recent events are likely to have a lasting impact on the public's awareness that there are an increasing number of hazards that our biological senses do not help us avoid. This desire for enhanced personal area environmental awareness is not simply a function of the anthrax scare. We will increasingly want to know more about the safety of air we breathe, the water that we drink, and the things we touch. This must be accomplished without bulky instrumentation and provide realtime feedback. I expect considerable commercial effort to be devoted towards transparent technology for personal environmental sensing. This may take the form of clothing that contains chemicals that change color in the presence of certain biohazards. Equally, we can expect a new generation of nano-sensors, custom-built to detect the presence of specific molecules, to be built into our clothing. Wearable technology presents great design challenges given the need to fold and wash the fabrics, maintain wearability, fashion, and light weight. For this reason, we should expect development in this arena to focus on chemical and nano-scale sensing. We have long expected our clothing to protect us from our surroundings — whether it be from the cold, UV radiation, or industrial hazards. Designing clothes that provide protection (through awareness) from other environmental hazards is a logical extension of the function of clothing to date.

THE CONSEQUENCES OF FULLY UNDERSTANDING THE BRAIN

Warren Robinett, Consultant

We start with questions:

- How does memory work?
- How does learning work?
- How does recognition work?
- What is knowledge?
- What is language?
- How does emotion work?
- What is thought?

In short, how does the brain work?

We have nothing better than vague, approximate answers to any of these questions at the present time, but we have good reason to believe that they all have detailed, specific, scientific answers, and that we are capable of discovering and understanding them.

We want the questions answered in full detail — at the molecular level, at the protein level, at the cellular level, and at the whole-organism level. A complete answer must necessarily include an understanding of the developmental processes that build the brain and body. A complete answer amounts to a wiring diagram of the brain, with a detailed functional understanding of how the components work at every level, from whole brain down to ion channels in cell walls. *These are questions of cognitive science, but to get detailed, satisfying, hard answers, we need the tools of nanotechnology, biochemistry, and information technology.*

How important would it be if we did achieve full understanding of the brain? What could we do that we can't do now? How would it make our lives better? Unfortunately, scientific advances don't always improve the quality of life. Nevertheless, let's look at some possibilities opened up by a full understanding of how the brain works.

New Capabilities Enabled by Full Understanding of the Brain

We understand the input systems to the brain — the sensory systems — better than the rest of the brain at this time. Therefore, we start with ways of fooling the senses by means of electronic media, which can be done now, using our present understanding of the senses.

Virtual Presence

The telephone, a familiar tool for all of us, enables auditory-only virtual presence. In effect, your ears and mouth are projected to a distant location (where someone else's ears and mouth are), and you have a conversation *as if you were both in the same place.* Visual and haptic (touch) telepresence are harder to do, but nevertheless it will soon be possible to electronically project oneself to other physical locations and have the perceptions you would have if you were actually there — visually, haptically, and aurally, with near-perfect fidelity.

Tasks that could be accomplished with virtual presence include the following:

- meeting with one or more other people; this will be an alternative to business travel but will take the time of a telephone call rather than the time of a cross-country airplane flight

- interacting with physical objects in a distant location, perhaps a hazardous environment such as a nuclear power plant interior or battlefield, where actual human presence is impossible or undesirable

- interacting with objects in microscopic environments, such as in the interior of a human body (I have worked on a prototype system for doing this, the NanoManipulator; see http://www.WarrenRobinett.com/nano/)

Better Senses

Non-invasive, removable sensory enhancements (eyeglasses and contact lenses) are used now and are a useful first step. But why not go the second step and surgically correct the eyeball? Even better, replace the eyeball. As with artificial hips and artificial hearts, people are happy to get a new, better component; artificial sensory organs will follow. We can look at binoculars, night-vision goggles, and Geiger counters (all currently external to the body) to get an idea of what is possible: better resolution, better sensitivity, and the ability to see phenomena (such as radioactivity) that are normally imperceptible to humans. Electronic technology can be expected to provide artificial sensory organs that are small, lightweight, and self-powered. An understanding of the sensory systems and neural channels will enable, for example, hooking up the new high-resolution electronic eyeball to the optic nerve. By the time we have a full understanding of all human sensory systems, it is likely we will have a means of performing the necessary microsurgery to link electronic signals to nerves.

Better Memory

What is the storage mechanism for human memory? What is its architecture? What is the data structure for human memory? Where are the bits? What is the capacity of the human memory system in gigabytes (or petabytes)? Once we have answers to questions such as these, we can design additional memory units that are compatible with the architecture of human memory. A detailed understanding of how human memory works, where the bits are stored, and how it is wired will enable capacity to be increased, just as you now plug additional memory cards into your PC. For installation, a means of doing microsurgery is required, as discussed above. If your brain comes with 20 petabytes factory-installed, wouldn't 200 petabytes be better?

Another way of thinking about technologically-enhanced memory is to imagine that for your entire life you have worn a pair of eyeglasses with built-in, lightweight, high-resolution video cameras which have continuously transmitted to a tape library somewhere, so that every hour of everything you have ever seen (or heard) is recorded on one of the tapes. The one-hour tapes (10,000 or so for every year of your life) are arranged chronologically on shelves. So your fuzzy, vague memory of past events is enhanced with the ability to replay the tape for any hour and date you choose. Your native memory is augmented by the ability to reexperience a recorded past. Assuming nanotechnology-based memory densities in a few decades (1 bit per 300 nm^3), a lifetime (3 x 10^9 seconds) of video (10^9 bits/second) fits into 1 cubic centimeter. Thus, someday you may carry with you a lifetime of perfect, unfading memories.

Better Imagination

One purpose of imagination is to be able to predict what will happen or what might happen in certain situations in order to make decisions about what to do. But human imagination is very limited in the complexity it can handle. This inside-the-head ability to simulate the future has served us very well up to now, but we now have computer-based simulation tools that far outstrip the brain's ability to predict what can happen (at least in certain well-defined situations). Consider learning how to handle engine flameouts in a flight simulator: you can't do this with unaugmented human imagination. Consider being able to predict tomorrow's weather based on data from a continent-wide network of sensors and a weather simulation program. This is far beyond the amount of data and detail that human imagination can handle. Yet it is still the same kind of use of imagination with which we are familiar: predicting what might happen in certain circumstances. Thus, our native imagination may be augmented by the ability to experience a simulated future. At present, you can dissociate yourself from the flight simulator — you can get out. In future decades, with enormous computing power available in cubic micron-sized packages, we may find personal simulation capability built-in, along with memory enhancement, and improved sensory organs.

Now the Really Crazy Ones

Download Yourself into New Hardware

Imagine that the brain is fully understood, and therefore the mechanisms and data structures for knowledge, personality, character traits, habits, and so on are known. Imagine further that, for an individual, the data describing that person's knowledge, personality, and so forth, could be extracted from his brain. In that case, his mind could be "run" on different hardware, just as old video games are today run in emulation on faster processors. This, of course, raises lots of questions. What is it that makes you *you*? (Is it more than your knowledge and personality?) Is having the traditional body necessary to being human? Nevertheless, if you accept the above premises, *it could be done*. Having made the leap to new hardware for yourself, many staggering options open up:

- No death of the representation of your brain. You back "yourself" up. You get new hardware as needed.

- Turn up the clock speed. The brain representation may work faster. Goodbye, millisecond-speed neurons; hello, nanosecond-speed electronics.

- Choose space-friendly hardware. The brain representation may be transmitted to various locations.

Instant Learning

If the structure of knowledge were fully understood, and if we controlled the "hardware and software environment" of the mind, then presumably we would understand how new knowledge gets integrated with old knowledge. The quaint old-fashioned techniques of "books" and "school" would be reenacted sometimes for fun, but the efficient way would be to just get the knowledge file and run the integrate procedure. Get a Ph.D. in Mathematics with "one click."

Hive Mind

If we can easily exchange large chunks of knowledge and are connected by high-bandwidth communication paths, the function and purpose served by individuals becomes unclear. Individuals have served to keep the gene pool stirred up and healthy via sexual reproduction, but this data-handling process would no longer necessarily be linked to individuals. With knowledge no longer encapsulated in individuals, the distinction between individuals and the entirety of humanity would blur. Think Vulcan mind-meld. We would perhaps become more of a hive mind — an enormous, single, intelligent entity.

Speed-of-Light Travel

If a mind is data that runs on a processor (and its sensors and actuators), then that data — that mind — can travel at the speed of light as bits in a communication path. Thus, Mars is less than an hour away at light speed. (We needed a rocket to get the first receiver there.) You could go there, have experiences (in a body you reserved), and then bring the experience-data back with you on return.

Self-Directed Evolution

If mind could be represented by a program and data, and we control the hardware and the software, then we could make changes as we see fit. What will human-like intelligence evolve into if it is freed from the limits of the human meat-machine, and humans can change and improve their own hardware? It's hard to say. The changes would perhaps be goal-directed, but what goals would be chosen for self-directed evolution of the brain model? What does a brain representation become when freed from pain, hunger, lust, and pride? (If we knew the answer to this, we might be able to guess why we haven't detected any sign of other intelligences in the 100 billion stars of our galaxy!)

USER-INTERFACE OLYMPICS: USING COMPETITION TO DRIVE INNOVATION

Warren Robinett, Consultant

Has bicycle racing improved bicycles? Yes, it has. We humans like to win, and like Lance Armstrong pedaling through the Alps in the Tour de France, we demand the best tools that can be made. The competition, the prestige of being the world champion, the passion to win, publicity for the chosen tools of the winners — these forces squeeze the imaginations of bicycle engineers and the bank accounts of bicycle manufacturers to produce a stream of innovations: lighter and higher-strength materials, more efficient gearing, easier and more reliable gear-shifting, aerodynamic improvements such as farings and encased wheels... the list goes on and on.

Competition spawns rapid improvements. Sounds a bit like evolution, doesn't it? *Lack of competition* can lead to long periods of quiescence, where nothing much changes. (Did you know the QWERTY keyboard was designed 100 years ago?)

This principle that *competition spawns improvement* could be applied to drive innovations in user-interface design. We call the proposed competition the *User-Interface Olympics*. Here is a sketch of how it might work:

- It would be an annual competition sponsored by a prestigious organization — let's say, the U.S. National Science Foundation.

- The winners would get prestige and possibly prize money (like the Nobel Prize, Pulitzer Prize, Emmies, Academy Awards, Oscars, and so on).

- The competition would be composed of a certain number of events, analogous to Olympic events. Individual contestants, or teams of contestants, compete for the championship in each event. User-interface events would be such things as

  - a timed competition to enter English text into a computer as fast as possible. (Surely someone can do better than the QWERTY keyboard!)

  - a timed competition to select a specified series of items from lists. (Can we improve on the 40-year-old mouse?)

- Contestants would provide their own tools. This is analogous to the equipment used by athletes (special shoes, javelin, ice skates). However, for the User-Interface Olympics, the tools are the hardware and software used by each competitor.

- Since the goal is to stimulate innovation, contestants would have to fully disclose the working of their tools. A great new idea would get you one gold medal, not ten in a row. This is similar to the patent system, in which rewards during a limited period are bartered for disclosure and dissemination of ideas.

- An administrative authority would be needed, analogous in the Olympic Committee and its subordinate committees, to precisely define the rules for each event, for qualifying for events, and many other related matters. This Rules Committee would monitor the various events and make adjustments in the rules as needed.

- We would expect the rules of each event to co-evolve with the competitors and their tools. For example, the rule against goal tending in basketball was instituted in response to evolving player capabilities; in the 100-meter dash, precise rules for false starts must be continually monitored for effectiveness. Winning within the existing rules is not cheating, but some strategies that players may discover might not be really fair or might circumvent the intent of the competition. Of course, some competitors do cheat, and the rules must set reasonable penalties for each type of infraction. The Rules Committee would therefore have to evolve the rules of each event to keep the competition healthy.

- New events would be added from time to time.

These contests would be similar to multiplayer video games. The contestants would manipulate user-input devices such as the mouse, keyboard, joystick, and other input devices that might be invented. The usual classes of display devices (visual, aural, and haptic) would be available to the contestants, with innovations encouraged in this area, too. Most malleable, and therefore probably most fertile for spawning innovations, would be the software that defined the interaction techniques through which the contestant performed actions during the contest.

If we set things up right, perhaps we could tap some of the enormous energy that the youth of the nation currently pours into playing video games.

The rules for each contest, which would be published in advance, would be enforced by a computer program. Ideally, this referee program could handle all situations that come up in a contest; whether this actually worked, or whether a human referee would be needed, would have to determined in real contests. Making the referee completely automated would offer several advantages. Contests could be staged without hiring anyone. Computer referees would be, and would be perceived to be, unbiased. Early qualifying rounds could be held using the Internet, thus encouraging many contestants to participate. Figure B.14 shows a system diagram.

If this idea is to be attempted, it is critical to start with a well-chosen set of events. (Imagine that the Olympics had tried to start with synchronized swimming and sheep shearing!) A small, well-justified set of events might be best initially, just to keep it simple and try out the idea. One way to identify potential events for the UI Olympics is to look at input devices that currently are widely used:

- computer keyboard — suggests a text-entry event
- computer mouse — suggests an event based on selecting among alternatives

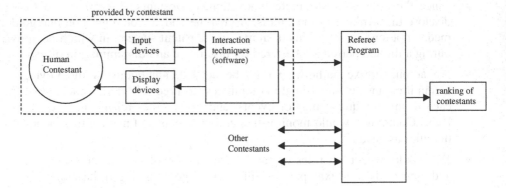

Figure B.14. System Diagram for a contest in the User-Interface Olympics, mediated by an automated referee program, with several contestants participating. The contestants provide their own hardware and software.

- joystick, car steering wheel — suggest one or more events about navigating through a 2-D or 3-D space

The real Olympics has events based both on raw power, speed, and stamina (weight lifting, races, and the marathon) and also events based on more complex skills (skiing, badminton, baseball). Similarly, the User-Interface Olympics could complement its events based on low-level skills (text entry, navigation) with some events requiring higher-level thinking. There are many kinds of "high-level thinking," of course. One class of well-developed intellectual contests is the mathematical competition. There are a number of well-known competitions or tests we can consider as examples: the MathCounts competitions run among middle schools and high schools; the Putnam Mathematical Competition run for undergraduates, and the math portion of the Scholastic Aptitude Test (or SAT, the college entrance test). Another similar competition is the annual student programming contest sponsored by the Association for Computing Machinery. One or more events based on solving well-defined categories of complex problems, using tools chosen by the contestant, would be desirable.

Strategy board games, such as chess and go, are another class of contests requiring complex skills. The rules for these games have already evolved to support interesting, healthy competitions and cultures. To focus on chess for a moment, by making chess an event in the User-Interface Olympics, we have an opportunity to reframe the false dichotomy between a human chess player and a chess-playing computer — we introduce a third possibility, a human contestant combined with her chess-analysis software. I personally believe that the combination of a good chess player, a good chess program, and a good user interface to integrate the two could probably beat both Deep Blue and Garry Kasparov. At any rate, this is a well-defined and testable hypothesis.

Therefore, the following events are proposed for the initial User-Interface Olympics:

- Text-entry speed competition
- Selection-among-alternatives race

- Navigation challenge: a race through a series of waypoints along a complex racecourse

- Timed math problems from the SAT (or equivalent problems)

- Timed chess matches

Each of these events would need precisely-formulated rules.

The strategy needed to achieve this vision of a thriving, well-known, self-perpetuating User-Interface Olympics that effectively drives innovation in user interface hardware and software is this:

- Fund the prizes for the first few years — let's say $100,000 for each of the four events

- Set up a governing committee and carefully choose its chairman and members. Give the committee itself an appropriate level of funding.

- Set an approximate date for the first User-Interface Olympics.

If the User-Interface Olympics were to become successful (meaning it had the participation of many contestants and user interface designers, it spawned good new ideas in user interface design, it had become prestigious, and it had become financially self-supporting), the benefits which could be expected might include the following:

- rapid innovation in user-interface hardware and software

- recognition for inventors and engineers — on a par with scientists (Nobel Prize), writers (Pulitzer Prize), and actors (Academy Award)

- improved performance on the tasks chosen as events

Sometimes prizes can have an inordinately large effect in relation to the amount of money put up. Witness the prize for the first computer to beat the (human) world chess champion (Hsu 1998; Loviglio 1997). Witness the prize for the first human-powered flying machine (Brown et al. 2001). A million dollars or so in prize money to jump-start the User-Interface Olympics might be one of the best investments ever made.

References

Brown, D.E., L.C. Thurow, and J. Burke. 2001. *Inventing modern America: From the microwave to the mouse.* MIT Press. Also at http://web.mit.edu/invent/www/ima/maccready_bio.html.

Hsu, F.H. 1998. Computer chess: The Deep Blue saga. http://www.stanford.edu/class/ee380/9798sum/lect03.html.

Loviglio, J. 1997. "Deep Blue Team Awarded $100,000 Fredkin Prize." New York Times CyberTimes, July 30. http://www.rci.rutgers.edu/~cfs/472_html/Intro/NYT_Intro/ChessMatch/DeepBlueTeamAwarded.html.

ACCELERATING CONVERGENCE OF NANOTECHNOLOGY, BIOTECHNOLOGY, AND INFORMATION TECHNOLOGY

Larry Todd Wilson, IEEE

My goal is to focus on a single NBIC-oriented idea that, if actualized, would unleash massive capabilities for improving all human performance. This single thing would have extreme interrelated, multiplicative effects. It's a bit like an explosion that starts consequential, far-reaching chain reactions. Furthermore, the one thing should accelerate and strengthen all other biotech ideas and fulfill a self-referential quality for advancing itself. It is difficult to negate the notion that some ideas, actions, or objects are more important than others. This perspective is characterized by statements like, "This is what should come first because if we had that ability or understanding, then we could (achieve these results)... and if we had those results, then we could actualize..."

The "One Thing"is, *Nullify the constraints associated with a human's inherent ability to assimilate information.*

Why should this receive favorable positioning? Advances in thinking performance are more important than advances in artifacts. This is due to the fact that the advances in artifacts are always a function of the human thinking system. The dynamics of innovation must be managed by human consciousness before it is "externally" managed at all. There are many naturally occurring phenomena that are not apparent to the senses or the imagination. However, a technology does not become a technology until it enters the realm of human consciousness.

Examples below deliver "as-is" versus "could be" explanations of the importance of enhancing how we assimilate information. From the examples, it is not difficult to imagine the transformations that may result due to the ripple effects. Overall, the focus on ways to enhance how humans assimilate information will result in significant increases in a human's ability to approach a complex need, achieve comprehension, and accomplish an intended result. Increased ability equates to gaining faster comprehension, better comprehension, comprehension in a situation that previously was unfathomable, faster solutions, and better solutions, and to finding solutions to problems that seemed unsolvable.

Assimilating information is a kind of human intellectual performance. There are three and only three types of human performance that could be the focus of improvement:

- intellectual performance (such as thinking, deciding, learning, and remembering)
- physical performance (such as moving, reaching, and lifting)
- emotional performance (feeling)

All human experiences are variations of one or more of these three.

Candidates of the "best thing" could be evaluated according to either criteria or questions like these:

- Is this idea/action/object fundamental to all dimensions and expressions of human performance (thinking, feeling, and moving)?

- Does this thing have a multiplicative nature in regards to all other biotech ideas, actions, and objects? Does this one thing produce fission-oriented and fusion-oriented results? Does its presence cause a reaction that in turn creates energy associated with pragmatic NBIC inventions and discoveries?

- *A priori*, does it have validity on its face? Does a listener agree that this one thing will indeed impact everything else?

- *A posteriori*, does it have perceptible, significant advances in several other areas? Did this one thing deliver a high return on investment? How do we know? What is measured? Does its presence actually increase the rate of all biotech inventions and discoveries?

Table B.1

AS IS	COULD BE
The span of judgment and the span of immediate memory impose severe limitations on the amount of information that we are able to receive, assimilate, and remember. In the mid-1950s, this was labeled as "seven, plus or minus two."	The innate limitations of human short-term memory are irrelevant due to the synergistic reliance upon "external" working memory, which is embedded in everything around us.
Short-term memory is working memory that works to retain sensory information presented by the mechanism of attention. No human being can hold many concepts in his head at one time. If he is dealing with more than a few, he must have some way to store and order these in an external medium, preferably a medium that can provide him with spatial patterns to associate the ordering, e.g., an ordered list of possible courses of action.	Increase the size and capability of working memory. Deliberate consideration of the items in external working memory can be called to mind upon demand. Manage how linguistic coding influences thought processes. Quantitatively measure stimulus (primarily in the form of linguistic-based prompts) and response (reactions in the form of decisions or feelings or movements).
Material is lost from short-term memory in two ways; it will not be committed to long-term memory if interference takes place or time decay occurs. One of the by-products related to the limitations of short-term memory is that there is great relief when information no longer needs to be retained. Short term memory is like a series of input and output buffers in which intermediate data can be stored during any thinking activity; this	Minimize the losses that naturally occur. Consciously add or delete items in working memory. Regulate the need for closure because the human is confident that it's "still there" (although I don't remember exactly what *it* is). Increase the number and rate of working memory instances. Engineer a seamless human mind/external memory interface, and

AS IS	COULD BE
memory has very limited capacity and can be easily overloaded. In order to alleviate the anguish of overload, there is a powerful desire to complete a task, reduce the memory load, and gain relief. This event is referred to as "closure," which is the completion of a task leading to relief.	thereby make human and machine intelligence coextensive. Basic analysis and evaluation of working memory contents are achieved in partnership or alone.
Bounded rationality refers to the limitations inherent in an individual's thought processes when there are more than a few alternatives being considered at the same time. Bounded rationality occurs because an individual has limited, imperfect knowledge and will seek satisfaction rather than strive for optimal decisions.	Effectively unbound "bounded rationality." The number and interrelationships of evaluations are dramatically expanded.
Individual thinking repertoires are limited (in their usefulness) and limiting (in their applicability).	Codify the elemental and compound thinking processes. Use the external working memory to manage the objects of the attention with novel ways of orchestrating the human's awareness of them. Increase the frequency, quantity (novel combinations), and throughput of these compounds. Gather more and more intelligence about the signals — the contextual nuances associated with variations of the compounds. Examples of compounds are Abstract Accept Accommodate Adopt Advise Agree Align Apply Appraise Approve Arrange Assign Assimilate Assume Authenticate Authorize Calculate Catalogue Categorize Change Check Choose Classify Close Compare Compile Compute Conclude Conduct Confirm Consider Consolidate Construct Contrast Contribute Coordinate Create Decide Decrease Deduce Define Delete Deliberate Deliver Deploy Derive Describe Determine Develop Differentiate Direct Disagree Disapprove Discern

AS IS	COULD BE
	Distinguish Elaborate Eliminate Emphasize Enable Enhance Enrich Establish Estimate Examine Exclude Execute Expand Explore Extrapolate Facilitate Find Focus Formulate Generalize Group Guess Guide Hypothesize Imagine Include Incorporate Increase Index Induce Infer Inform Initiate Insert Inspect Interpret Interview Invent Judge Locate Match Measure Memorize Merge Modify Monitor Observe Optimize Organize Originate Outline Pace Predict Prepare Presume Prevent Prioritize Probe Promote Provide Question Rank Rate Reason Receive Recognize Recommend Refine Reflect Regulate Reject Remove Report Resolve Respond Scan Schedule Scrutinize Search Seek Serve Settle Show Solicit Solve Sort Speculate Submit Support Suppose Survey Synthesize Translate Validate Verify Visualize.
Specialists often miss the point. The point is to swap advances among different disciplines. It's all about permutations and combinations. Discoveries from biology and chemistry are hooked up with synthesis and fabrication tools from engineering and physics. Each discipline has its own sets of problems, methods, social networks, and research practices. There are no effective ways in which the intellectual results of subdisciplines can be managed and thereby accelerate consilience and cross-disciplined performance breakthroughs.	Progress towards a new sense of the complex system. The most obvious change will be the benefits of working with many kinds of associations/relations. More people will be able to perceive loops and knots. Sense the complex system with a set of universal constructs for systematically managing the interrelationships among disciplines. Accurate visualization of many kinds of relations (not just parent-child relations) will shift the reliance of the satisficing mode of hierarchical interpretations to the closer-to-reality heterarchical structure. Continue to splinter the subdisciplines and achieve convergence when needed for important insights.
Today, many physicists spend time translating math into English. They hunt for metaphors that can serve as a basis for enhancing comprehension of	Integrate mathematics, verbal, and visual languages in order to allow individuals to traverse the explanation space.

AS IS	COULD BE
relatively imperceptible physical phenomena.	Aid the acceleration of new ways for more people to abandon their intuitive (perhaps innate) mode of sensory perception associated with the macro world. Achieve integration (and concise translation) between our symbol sets (math, verbal, and visual) and open up the chance to address more, apparently paradoxical, phenomena. The assumption is that many of these paradoxes are just illusions created when you look at an n-dimensional problem through a three-dimensional window.
Linguistic-based messages, which plod along the user's tolerance for listening, govern the rate of assimilation.	Establish the path more directly because all forms of intelligence, whether of sound or sight, have been reduced to the form of varying currents in an electric circuit.
Imaging modalities don't offer a concise way of observing the dynamics of how we assimilate information. PETs are more accurate in space, and EEGs are more accurate in time. EEGs can capture events on the scale of milliseconds, but they're only accurate to within centimeters. Scans are like slow motion — a thousand times slower — but they're accurate to the millionth of an inch.	Extend the visual languages to the actual visualization of localized neuronal activity. Understand the spatial-temporal nature of assimilation with a realtime movie stage where we watch thoughts as they gather and flow through the brain. Understand how the human perception of mind arises from the brain. Formalize in neural network models operating on traditional hardware. Thus, intelligences akin to humans will reside in the Internet. These intelligences, not being physically limited, will merge and transform themselves in novel ways. The notion of discrete intelligence will disappear.

C. IMPROVING HUMAN HEALTH AND PHYSICAL CAPABILITIES

THEME C SUMMARY

Panel: J. Bonadio, L. Cauller, B. Chance, P. Connolly, E. Garcia-Rill, R. Golledge, M. Heller, P.C. Johnson, K.A. Kang, A.P. Lee, R.R. Llinás, J.M. Loomis, V. Makarov, M.A.L. Nicolelis, L Parsons, A. Penz, A.T. Pope, J. Watson, G. Wolbring

The second NBIC theme is concerned with means to strengthen the physical or biological capabilities of individuals. The panel's work dovetailed with that of the first panel in the area of human cognition, especially the exciting and challenging field of brain performance. The *brain*, after all, is an organ of the human body and is the physical basis for that dynamic system of memory and cognition we call the *mind*. An extremely complex brain is the feature of human biology that distinguishes us from other animals, but all the other tissues and organs of the body are also essential to our existence and overall performance, and they thus deserve close scientific and technological attention.

The convergence of nano-bio-info-cogno technologies is bound to give us tremendous control over the well-being of the human body. In turn, it will change the way we think about health, disease, and how far we go to treat a patient. These new technologies will enable us to decipher the fundamental mechanisms of a living being, yet at the same time, they raise the fundamental questions of what life is and how human capability is defined. The panel gave highest priority to six technologies for the improvement of human health and capabilities in the next 10-20 years. In realizing these priorities, it will be essential to keep a "healthy" balance on human issues while seeking technological and social solutions.

1. Nano-Bio Processor

As the convergence of NBIC progresses, it will be imperative that the technology be focused on ways to help enhance human health and overall physical performance, be disseminated to a broad spectrum of the population, and be developed by a diverse group of scientists and engineers. One potential platform that will enable this would be a "bio-nano processor" for programming complex biological pathways on a chip that mimics responses of the human body and aids the development of corresponding treatments. An example would be the precise "decoration" of nanoparticles with a tailored dosage of biomolecules for the production of nanomedicines that target specific early biomarkers indicative of disease. The nanomedicine may be produced on one type of nano-bio processor and then tested on another that carries the relevant cellular mechanisms and resulting biomarker pathways. The nano-bio processor would parallel the microprocessor for electronics, such that the development of new processes, materials, and devices will not be limited to a handful of "nano specialists." With the advent of the nano-bio processor, knowledge from all fields (biology, chemistry, physics, engineering, mathematics) could be leveraged to enable advancements in a wide variety of applications that improve human health and enhance human capabilities.

2. Self-Monitoring of Physiological Well-Being and Dysfunction Using Nano Implant Devices

As the scales of nanofabrication and nanotransducers approach those of the critical biomolecular feature sizes, they give the technologist the toolset to probe and control biological functions at the most fundamental "life machinery" level. By the same token, this technology could profoundly affect the ways we manage our health.

One outcome of combining nanotechnology with biotechnology will be molecular prosthetics — nano components that can repair or replace defective cellular components such as ion channels or protein signaling receptors. Another result will be intracellular imaging, perhaps enabled by synthetic nano-materials that can act as contrast agents to highlight early disease markers in routine screening. Through self-delivered nano-medical intervention, patients in the future will be able in the comfort of their homes to perform noninvasive treatments autonomously or under remote supervision by physicians.

Metabolic and anatomical monitoring will be able to give humans the capability to track the energy balance of intake and consumption. Monitoring high-risk factors will be able to facilitate early diagnosis, when medical treatments can be most effective. Information systems designed to present medical data in ways that are intelligible to laypersons will allow anyone to monitor his or her health. As a result of NBIC-enabled "wonder medicines," there will be a need to develop technology and training modalities to make the patient an essential partner in the process of health monitoring and intervention.

As the population ages, more and more age-related diseases and deteriorating functions (e.g., hearing, memory, muscle strength, and sight) will be prevalent; an obvious example is Alzheimer's disease. Some of these dysfunctions are due to molecular changes over time, and some are due to the natural decay of bodily functions. NBIC will provide ways to slow down the aging process or even reverse it.

3. Nano-Medical Research and Intervention Monitoring and Robotics

The convergence of nano-bio-info-cogno technologies will enhance the toolset for medical research and allow medical intervention and monitoring through multifunctional nanorobots. For example, a nano brain surveillance camera could be developed. Imaging tools will be enhanced by nanomarkers as anchor points for hierarchical pinpointing in the brain. A range of nano-enabled unobtrusive tools will facilitate research on cognitive activities of the brain.

Nano-enabled unobtrusive tools will be invaluable for medical intervention, for example, nanorobots accomplishing entirely new kinds of surgery or carrying out traditional surgeries far less invasively than does a surgeon's scalpel. Technological convergence will also enhance post-surgery recovery. Although open surgical procedures will probably be reduced in numbers, the need for them will not be eliminated. Each procedure induces different side effects and risk factors. For instance, open-heart surgery increases the risk for stroke several days after the operation. NBIC technologies could enable devices that monitor such risk factors and immediately notify the physician at the first indication of a precursor to the onset of post-surgery traumas.

4. Multimodalities for Visual- and Hearing-Impaired

In the United States, there are eight million blind people and 80 million who are visually impaired. The current paradigm of electronic communication is visual and conducted through the use of monitors and keyboards. It will be important for NBIC technologists to address the need for multimodal platforms to communicate with, motivate, and utilize this population group. Examples of different modes of communication include talking environments and 3-D touch screens to enable access to the Internet.

While convergent technologies will benefit disabled persons, they in turn will contribute greatly to the development of the technology, thereby benefiting all people. In recognition of this fact, disabled scientists and engineers should be included in research and design teams. As NBIC blurs the boundaries of normal and abnormal, ethical and unethical, it will be important to include disabled members and advocates on advisory committees at all levels. This will include the private sector, academia, government, and international committees.

5. Brain-to-Brain and Brain-to-Machine Interfaces

The communication among people and between people and machines or tools has not been fully realized because of limitations in brain understanding and of the indirect interactions. The external tools need to be manipulated as an independent extension of one's body in order to achieve the desired goal. If machines and devices could be incorporated into the "neural space" as an extension of one's muscles or senses, they could lead to unprecedented augmentation in human sensory, motor, cognitive, and communication performance.

A major goal is to measure and simulate processes from the neuron level and then to develop interfaces to interact with the neural system. A visionary project by Llinás and Makarov proposes a nonintrusive retrievable cardiovascular approach to measure neuron and group-of-neuron activities, and on this basis, to develop two-way direct human communication and man-machine telepresence.

Another goal is to establish direct links between neuronal tissue and machines that would allow direct control of mechanical, electronic, and even virtual objects as if they were extensions of human bodies. Another visionary project by Nicolelis proposes electrophysiological methods to extract information about intentional brain processes and then translate the neural signals into models that are able to control external devices.

6. Virtual Environments

Nanotechnology will permit information technology to create realistic virtual environments and geographies. And biotechnology guided by cognitive science will produce interfaces that will allow humans to experience these environments intensely. Thus, the union of these technologies will transcend the biological limitations of human senses and create a new human relationship to the physical environment. It will be possible to simulate in humans the sensation of being at remote locations or at imaginary new buildings or facilities. This could be used for rapid design and testing of large projects, thereby saving the cost of errors. Other economically significant applications could be in the entertainment industry, and the tourist industry could use the technology to provide virtual samples of distant locations to prospective customers.

Applications of special relevance to improving health and enhancing human physical abilities include the use of virtual environments for education and interactive teaching. This will provide new ways for medical students to visualize, touch, enter, smell, and hear the human anatomy, physiological functions, and medical procedures, as if they were either the physician or a microscopic blood cell traveling through the body. Similarly, impaired users, ordinary people, athletic coaches, and a range of health-related professionals could train in these virtual environments.

Statements and Visions

Participants in the panel on human health and physical capabilities contributed statements and visions on a wide range of technological challenges and opportunities. Several contributors addressed life extension (P. Connolly); therapeutics at the cellular level (M.J. Heller, J. Bonadio), physiological level (A.T. Pope), and brain levels (B. Chance and K.A. Kang, E. Garcia-Rill, L. Cauller and A. Penz); as well as brain-machine interaction (R.R. Llinás and V. Makarov, M.A.L. Nicolelis) and improving the quality of life of disabled people (G. Wolbring and R. Golledge).

Reference

Gazzaniga, M.S., ed. 1995. *The cognitive neurosciences.* Cambridge, MA: MIT Press.

STATEMENTS

NANOBIOTECHNOLOGY AND LIFE EXTENSION

Patricia Connolly, University of Strathclyde

This paper concentrates on only one of the complex debates emerging due to the convergence of nano-bio-info-cogno (NBIC) and the ability to improve human performance: that is, how nanobiotechnology will affect life extension. To deal with this in a comprehensive manner, the concept of life extension will be discussed, along with a brief presentation of the major obstacles that can be defined from our current knowledge in bioscience and medicine. It is proposed that a successful strategy for the convergence of NBIC disciplines in human terms will require a holistic approach and consideration of the full pathway from the human, down through organ, cell, and molecule, analyzing where NBIC can successfully intervene in this complex cascade. Some examples are given of areas where nanobiotechnology has had, or could have, impact in the problem areas of human well-being and quality of life as they are understood today.

Life Extension and Nanobiotechnology: Some Key Criteria

Nanobiotechnology for the purposes of this discussion is defined as the application of nanotechnology or nanobiology to a biological environment that involves device or material interactions with biological or biomolecular systems. To consider nanobiotechnology and life extension, it is important to first consider which social groups might be targeted by this approach and then to examine their medical and social requirements, highlighting where the NBIC convergence will have an effect. For example, the problems of the developed and developing world are quite different in terms of life extension. The problem of environmental damage and rising world pollution threatens the quality and length of life span of both groups. Table C.1 summarizes some of the major problems that must be addressed in extending life in developed and developing countries (WHO 1998a; WHO 1998b; WHO 2000; WHO 2001).

Table C.1
The Challenges to Life Extension in Developed and Developing Countries

Target Groups	Quality of Life Problems	Major Causes of Death and Disability
Developed Countries: Aging Populations only	• Loss of strength and mobility • Loss of mental sharpness / neurological disease • Social isolation • Poverty	• Cardiovascular Disease • Diabetes and its complications • Inflammatory diseases including arthritis • Cancer • Neurological Disease or Impairment
Developing Countries: All age groups	• Environmental, lack of safe water & sanitation • Disease related loss of earnings • Poverty	• Malnutrition • Infectious diseases • Parasites • Cardiovascular disease

Governments in the developed world, including the United Kingdom (UK Foresight Consultation Document 1999), have started to develop an awareness of the needs of the increasingly aged populations that they have and will have in the first half of this century. Major disease groups or medical conditions that are the major causes of death or disability in the aging populations of the developed countries of the world have been identified. For example, according to the World Health Organization (WHO 2000), in 1999 around 30 percent of deaths worldwide were caused by cardiovascular disease and 12 percent by cancer.

The problems of the developing world are quite different, and it might be argued that unless life extension in this environment is addressed by those who have the technology and wealth to do so, then the stability of developed societies worldwide will be affected. The medical problems of developing countries are widespread:

many of these could be resolved by improvement in economic factors; however, some problems, such as parasitic infections, have eluded complete medical solutions. Toxoplasma infects 50 percent of the world population and leads to miscarriage, blindness, and mental retardation. The WHO (1998b) states that one child dies in the world every 30 seconds from malaria. There is much scope for improvement in the formulation of drugs, delivery modes, diagnostics, and effective vaccines for these and other diseases.

In addition, it is recognized that increasing levels of pollution with their consequent environmental changes drive aspects of both childhood and adult disease. The epidemiology of the disease patterns are being studied (WHO 2001), and nations are considering their role in reducing environmental emissions (EIA 1998). Nanobiotechnology may have a part to play here in land and water treatments through bioremediation strategies and in novel processes for industrial manufacture.

A Holistic Approach to Problem Definition

To effectively target emerging NBIC technologies, and in particular to make the most of the emerging field of nanobiotechnology, requires a strategic approach to identifying the problem areas in life extension. Biomedical problems currently exist on macro, micro, and nanoscales, and solutions to some apparently straightforward problems could enormously increase life expectancy and quality of life. A holistic approach would examine the key medical problems in the world's population that need to be solved to extend life, and at the same time, would consider the social environment in the aging population to ensure that quality of life and dignity are sustained after technological intervention.

A key element of this top-down approach is to consider the whole human being and not merely the immediate interface of nanobiotechnology with its target problem. The ability to view the needs in this area from a biomedical perspective that starts with the whole human and works down through organ and cellular levels to the molecular (nanoscale) level, will be an essential component of projects with successful outcomes in this field. There is little point in developing isolated, advanced technological systems or medical treatments to find that they solve one problem only to generate many others. For example, ingenious microdevices with nanoscale features that might patrol blood vessels or carry out tissue repairs have been suggested and designed (Moore 2001; Dario et al. 2000). However, there has been little detailed discussion or consideration at this stage regarding biocompatibility issues, particularly of the thrombogenicity (clot-forming potential) of these devices or of their immunogenicity (ability to stimulate an unwanted immune response). In this area, as in many others, there is a need for multidisciplinary teams to work together from the outset of projects to bring medicine and technology together. Ideally, these research teams would include clinicians, biomedical scientists, and engineers rather than being technologist-led projects that ignore much of the vast wealth of information we have already discovered about the human body through medical and biomedical research.

Accepting this need for biomedically informed project design also leads to the conclusion that understanding the cell-molecule interface, in other words the micro-nanoscale interactions, will be a factor in the extended application of nanobiotechnology. To create a holistic approach to widespread and successful

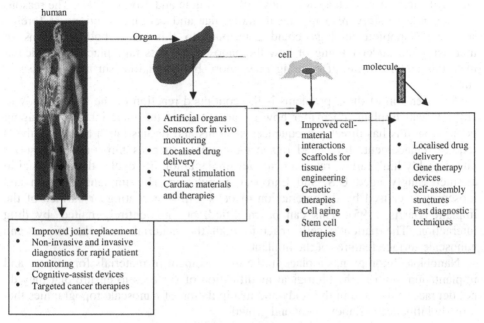

Figure C.1. Examples of levels for intervention of nanobiotechnology in human life extension.

introduction of nanobiotechnologies in life extension will require interdisciplinary teams and exchange of information. Figure C.1 illustrates the possible levels of intervention and some of the emerging solutions where nanobiotechnology will have a role in repair or replacement of damaged elements.

The Need for a Holistic Approach: Some Specific Problems

As previously stated, there are a number of identified medical challenges that might benefit from intervention with nanobiotechnology. Many of these are long-term problems that have not been resolved by current technological or medical solutions. The following section is intended to briefly introduce some of these problems.

The Human-Materials Interface

Many of the disease conditions in the human body, and deaths during surgical intervention, can be traced to the body's in-built ability to react to foreign materials or wound sites through its inflammatory response. In normal disease or wounds, this ensures the proper activation of the immune response or of a clotting response from coagulation factors in blood. In extreme conditions or at chronic wound sites, the cascade reaction triggers a full inflammatory response that is harmful to tissue. In cardiovascular surgery, for example, reaction to physical intervention and surgical materials can lead to Systemic Inflammatory Response Syndrome (SIRS), and in a small percentage of cases, this will in turn lead to multiple organ failure and death (Khan, Spychal, and Pooni 1997).

The appearance of an inflammatory response following blood contact with a biomaterial can be readily measured in the molecular markers that are generated during the response, such as cytokines. (Weerasinghe and Taylor 1998). The reasons for the inflammatory response lie in molecular and cellular reactions at foreign surfaces. Nanobiotechnology could contribute to this field, both in terms of increasing the understanding of how the nanoscale events take place on particular materials and in terms of creating new, more biocompatible surfaces for use in surgery.

An extension of these problems is the continued reaction of the human body to any artificial implant, no matter how apparently inert the material. For the aging population, this has direct consequences as joints and tissues (such as heart valves) require replacement. Implanted replacement joints such as hip joints still suffer unacceptably high failure rates and shorter implantation life cycles than are ideal in an increasingly aged U.S. and European population. Hip implant rejection and loosening is caused by the interaction of cells with the coating or surface of the implant (Harris 1995). This can be modified, but not entirely halted, by drug interaction. The patient's cells react to both the materials and the micro- and nanoscale surface features of the implant.

Nanobiotechnology has a place in the improvement of materials for surgery and implantation, both in the biological modification of surfaces to ensure that they do not degrade in use and in the study and manipulation of nanoscale topographies that directly influence cell movement and growth.

Neurological Disease

Both cellular decay and diseases such as Alzheimer's and Parkinson's contribute to loss of neural function, cognitive thought, and independence. In addition, events such as stroke leave many of the older population with impaired functions. It is here that implantable devices and cognitive science will have the greatest part to play in enhancing the quality of extended life.

Microdevices for cell-electrode interfacing for both cardiac and neural cells have been available for *in vitro* applications for many years. There are few examples of implanted systems. Some micro-array type devices have been implanted, for example, in rudimentary artificial vision systems (Greenberg 2000). On a slightly larger scale, electrode systems have been implanted in the brain to provide electrical signal patterns that alleviate some symptoms of Parkinson's disease (Activa®, Medtronic, Inc., USA).

Much remains to be done in neurological device development, including devising smaller systems capable of withstanding long-term implantation. Investigation of the submicron (synaptic) interface to devices from neurons may be an important area for consideration in this field. In the longer term, it may be that some conditions will be alleviated by local electrode and drug-release systems, but how to keep these devices in place for years so that they remain biologically or electrically viable remains a difficult problem. There will be a need to develop sub-micron arrays of electrodes and chemo-arrays in devices designed to replace diseased tissue. If nanoscale electrode-cell interactions are expected to be important, then a fuller understanding of the cell-nanoelectrode interface will be required both *in vitro* and *in vivo*.

Cell replacement technologies are also being developed to address neural decay, and success with this type of approach, such as stem cells (see discussion of artificial organs and tissue engineering below), may remove the need for extensive device development. Cell placement and growth techniques may still, however, require device intervention and nanobiotechnology know-how.

Artificial Organs and Tissue Engineering

In the field of tissue repair and replacement, advances are being made in the creation of artificial organs and replacement tissue. In the case of artificial organs, many of the components of the organ will not be linked to the body's own regulatory systems (e.g., artificial heart pumps). In engineered tissue for repair or replacement of damaged tissue, control of tissue growth and tissue integration are critical and will require monitoring.

To provide sensitive feedback control to artificial organs either within or external to the body (such as the artificial liver), biosensor systems will be required, perhaps coupled to drug or metabolite delivery systems. This is an ongoing problem, since no long-term implantation systems based on biosensors have become commercially available, even with the application of microtechnology (Moore 2001; Dario et al. 2000) — although improvements have been made for subcutaneous glucose sensors in recent years (Pickup 1999). There is opportunity here for the use of nanobiotechnology to both provide the sensors for monitoring and adjusting organ performance and to aid localized drug or metabolite delivery to artificial organs. It may be possible to create biosensors for long-term implantation by trapping "factory cells" in gels within the sensor system, which would, in turn, synthesize any required renewable nanocomponents in the sensors, thus avoiding the current problems of sensor degradation over time.

Significant amounts of time, money, and research effort are being directed to the field of tissue engineering for skin, cartilage, bone, and heart tissue regeneration or repair, as well as for other types of tissue. Biopolymer scaffolds are the material of choice for the seeding of cells to grow replacement tissue. At the macro or fiber level, much is known about these scaffolds, but little time has been devoted to the nanoscale effects of topography or surface molecular treatments that could be influenced by nanobiotechnology. Nanovesicles that could be incorporated into tissue scaffold structures for slow release of chemoattractants could greatly improve tissue uptake or repair. One group has recently successfully exploited the idea of self-assembly of molecules, in this case, peptide-amphiphile molecules, to create biopolymer scaffolds with nanoscale features for bone repair (Hartgerink, Beniah, and Stupp 2001). This group's experiments show that a key constituent of bone, hydroxyapatite, can be made to grow and align in the same manner as bone *in vivo* using these scaffolds.

Stem cell research promises to open up new possibilities for harvesting cells that can be transformed *in situ* into different tissue types for repair or regeneration of damaged tissue. This may require extensive technological intervention both for harvesting cells and in delivering cells for therapy.

Genetic Techniques

The explosion in the field of genetics has led to the availability of a range of diagnostic tests for predisposition to illnesses, including cancer, although final expression of many illnesses may have strong environmental factors that must be taken into account. Together with the possibility of gene therapy for specific diseases, this offers new hope of life extension to many people. For example, hereditary lung conditions such as cystic fibrosis are being targeted by gene therapy to replace missing or deficient genes (Douglas and Curiel 1998). Study of how cells age is being taken up by many research groups and, again, offers hope for many potential victims of cancer and degenerative disease. Nevertheless, any widespread genetic intervention in disease is still some way off. To quote one recent review paper, "Ideally, gene therapy should be efficient, cell-specific, and safe (Hu and Pathak 2000). One of the challenges of gene therapy is the efficient delivery of genes to target cells. Although the nucleic acids containing the genes can be generated in the laboratory with relative ease, the delivery of these materials into a specific set of cells in the body is far from simple." It is perhaps here in the design and development of efficient delivery devices and systems that nanobiotechnology will play its biggest role in gene therapy.

Drug Delivery

There are still many opportunities for nanobiotechnology in the field of drug delivery, particularly in delivery of those drugs unsuitable for the gastrointestinal system. Skin and lungs have become favorite alternative routes for drug delivery, with nanovesicles and microcrystals as popular drug carriers (Langer 1999). Cancer treatment has yet to fully benefit from the targeted delivery to tumors of drugs in microdevices with local nanoscale interactions. Likewise, cancer monitoring and surgery would benefit enormously from miniaturized sensor or other diagnostics systems that could be used in the pre-, peri-, and postoperative environment.

The Prospects for Life Extension

Any quantitative discussion on the prospects for life extension through nanobiotechnology intervention in disease must be purely hypothetical at this stage. However, speculating across the human-organ-cell-molecule model may give some idea of the possible times to application of some of the approaches under development. Table C.2 summarizes what is a very personal view of the likely outcome of convergence in NBIC.

Visions for the Future

Loss of mobility and therefore independence is critical in the onset of decay and isolation for many older people, and one area in the developed world where people are very dependent for mobility is in the use of a car. Confidence and cognizance decline for many people as they age; in the car of the future there is the possibility to see the true convergence of NBIC in extending independence and warding off part of the decline in the older person. Higher-speed, higher-density computers and effective sensors driven by nanotechnology may combine with on-board artificial intelligence in the car, helping the driver plan routes and avoid hazards and difficult traffic situations. Nanobiotechnology may also be present in on-board minimally invasive biosensors to monitor the driver's health, both in terms of physical stress and physiological condition, to be fed back to the car's computer. In a further interpretation, since the possibility of implanted devices to stimulate or improve cognizance are emerging, the driver may also benefit from neuronal stimulation designed to keep him or her alert and performing optimally during the trip.

Table C.2
Some Potential Gains in Life Extension from NBIC convergence

Level of Intervention	Key Advance	Timescale (years)	Life Extension
Human	Noninvasive diagnostics	5-10	Lifesaving for some conditions
	Cognitive assist devices	15-20	Higher quality of life for several years
	Targeted cancer therapies	5-10	Reduction in cancer deaths by up to 30%
Organ	Artificial heart	0-5	2-3 years awaiting transplant
	Neural stimulation or cell function replacement	5-20	10-20 years extra if successful for neurodegenerative patients
Cell	Improved cell-materials interactions	0-15	Lowering of death rates on invasive surgery by 10% and extending life of surgical implants to patient's lifetime
	Genetic therapies	30	Gains in the fight against cancer and hereditary diseases
	Stem cells	5-10	Tissue / brain repair Life extension of 10-20 years
Molecule	Localized drug delivery	0-10	Extending life through efficient drug targeting
	Genetic interventions	0-30	Life extension by targeting cell changes and aging in the fight against disease Likely to be a very complex environment to successfully manipulate

The convergence of NBIC in the field of life extension will lead to implanted devices such as sensors and drug delivery systems being developed to replace or monitor body function. Implanted devices, whether macro or micro in scale, present a problem today in terms of biocompatibility. Implantation of a heart valve in a patient means that a drug regime for anti-coagulation is mandatory — usually through administration of warfarin. Since inflammatory response and immunogenic response take place *in vivo,* many of the devices being discussed and designed today to improve human performance incorporating nanotechnology will not be implantable because of biocompatibility issues. A further complication will be how to keep a nanodevice biologically or electronically active (or both) during sustained periods of operation *in vivo.* Sustained exposure to physiological fluid, with its high salt and water content, destroys most electronic devices. Likewise, devices that emit biological molecules or are coated with biological molecules to ensure initial biocompatibility must have their biological components renewed or be destined to become nonfunctional some time after implantation. Little attention is being given to these problems, which may prove major stumbling blocks in the next 10-30 years to the successful application of nanotechnology in a range of medical conditions.

A "holistic human project" could bring together the best research clinicians, biomedical engineers, and biomedical scientists to discuss the main life-shortening diseases and conditions and current progress or problems in their treatment or eradication. Together with the nanotechnologists, areas where conventional medicine has not been successful could be identified as strategic targets for nanobiotechnology. Specific project calls could follow in these areas, with the condition that the applicants' teams must show sufficient interdisciplinary interaction to provide a comprehensive understanding of the nature of the problem. The opportunities are immense, but the resources available are not unlimited, and only strategic planning for project groups and project themes will realize the maximum benefit for biomedicine and society.

References

Dario, P., M.C. Carozza, A. Benvenuto, A. Menciassi. 2000. Micro-systems in biomedical applications. *J. Micromech. Microeng.* 10:235-244.

Douglas, J.T., and D.T. Curiel. 1998. Gene therapy for inherited, inflammatory and infectious diseases of the lung. *Medscape Pulmonary Medicine* 2, 3.

EIA (Energy Information Administration, U.S. Dept. of Energy). 1998. *Impacts of the Kyoto Protocol on U.S. energy markets and economic activity.* Report No. SR/OIAF/98-03.

Greenberg, R.J. 2000. Visual prostheses: A review. *Neuromodulation*, 3(3):161-165.

Harris, W.H. 1995. The problem is osteolysis. *Clinical Orthopaedics and Related Research* 311: 46-53.

Hartgerink, J.D., E. Beniah, and S.I. Stupp. 2001. Self-assembly and mineralization of peptide-amphiphile nanofibers. *Science* 294: 1684-1688 (November).

Hu, W-S., and V.K. Pathak. 2000. Design of retroviral vectors and helper cells for gene therapy. *Pharmacological Reviews* 52: 493-511.

Khan, Z.P., R.T. Spychal, and J.S. Pooni. 1997. The high-risk surgical patient. *Surgical Technology International* 9: 153-166 (Universal Medical Press).

Langer, R. 1999. Selected advances in drug delivery and tissue engineering. *J. of Controlled Release*, 62: 7-11.

Moore, A. 2001. *Brave small world.* EMBO Reports, 2(2): 86-89. (European Molecular Biology Organisation, Oxford University Press).

Pickup, J. 1999. Technological advances in diabetes care. Wellcome News Supplement Q3(S).

UK Foresight Consultation Document. 1999. *The aging population.* (http://www.dti.gov.uk).

Weerasinghe, A., and K.M. Taylor. 1998. The platelet in cardiopulmonary bypass. *Ann. Thorac. Surg*, 66:2145-52.

WHO (World Health Organization). 2000. *1997-1999 World Health Statistics Annual.* Geneva: World Health Organization.

WHO. 1998a. *The World Health Report 1998, Life in the Twenty-first Century: A Vision for All*
    1998, ISBN 92 4 156189 0

WHO. 1998b. *WHO Fact Sheet No. 94, Malaria.* Geneva: World Health Organization.

WHO. 2001. *Methodology for assessment of environmental burden of disease.* ISEE session on environmental burden of disease. Report of a WHO Consultation (WHO/SDE/WSH/00.7). Geneva: World Health Organization.

THE NANO-BIO CONNECTION AND ITS IMPLICATION FOR HUMAN PERFORMANCE

Michael J. Heller, University of California San Diego

Many aspects of nanotechnology will lead to significant improvements in human performance; however, the nano-bio area will be particularly important and relevant to such improvements. Technological advancements in the past decade have been nothing short of phenomenal. These advancements have led to an increasingly better understanding of human biology. We can expect that the new advancements in the nano-bio area will not just lead to a better understanding of human biology, but will also provide a new dimension and capability to affect human biology. The fact we are having this workshop and all know its true importance and underlying implications speaks for itself.

Individualized Treatment for Human Development

How nano-bio technologies will be applied in the most beneficial ways is dependent on the underlying basis for human performance. It is very likely that most of the underlying basis is genetic in origin (Wexler 1992; Ridley 2000). While this may still be widely debated and resisted for other reasons, it will (when proven) have profound implications, and it certainly needs to be considered in any planning on new technology application in human biology. The following is an example, which will hopefully not trivialize the issue.

Many individuals greatly enjoy a variety of sporting activities. However, a vast majority of individuals who do any of these sporting activities cannot approach the capabilities of a professional player, even with all the best new technology, instruction, and personal motivation. While some might feel this unfair, most people accept it and keep it in perspective. After all, people in general usually have something they do well, even if they never develop the desired trait. Not only is this true for athletic capabilities, but this is widely observed for other capabilities such as talent in art or music. Until recently, these perceptions were not based on any real scientific evidence. Now, with the first phase of the human genome project complete and a new geomics revolution occurring, good evidence is appearing that many human performance traits do indeed have a genetic basis.

This may also hold true for human behavior (Chorney et al. 1998; Dubnau and Tully 1998). Just a few years ago psychiatrists and psychologists would have doubted the genetic basis for many of the important mental illnesses. Today, there are few diseases left that are not known to be directly or indirectly genetically based (Kamboh 1995; Corder et al. 1994). Even infectious diseases are not really an exception to this premise, as there are always individuals who have a positive genetic component that provides varying degrees of resistance to the infection (Hill 1996).

A particularly relevant example of the importance of understanding the true basis of "cause and effect" in determining technological strategy now comes from the pharmaceutical industry. The new area of phamacogenomics is now proving for one drug after another that so-called drug toxicity is really based upon individual genetic polymorphisms. Usually, for any given drug, there are always a small number of individuals for whom that drug is toxic or less effective. As the genes and pathways for drug metabolism are better understood, this drug toxicity is usually found to correlate in some fashion with single nucleotide polymorphisms (point mutations) in the affected individuals. Not too long ago, most drug companies were investing huge amounts of money looking for "safe" drugs. Today, most accept or will soon accept the fact that patient stratification (via either genotyping or phenotyping) will be necessary to determine drug toxicity.

This represents a key example of how important it is to properly identify cause and effect in relation to technology development. The pharmaceutical industry spends enormous amounts of money developing new drugs, and many potentially useful drugs are being delayed or not used because they have serious toxicity for a small number of individuals. This also presents a view of how genetic determination is misunderstood. If we were to look at just a single drug, genetic testing of potential drug recipients would seem totally unfair and appear that genetic testing is being used to exclude some individuals from a potential benefit — even though some individuals truly don't benefit from that particular drug. However, at least in the area of therapeutics, we do not have to look at too many drugs until we find that, in general, the vast majority of humans will always have one or two important drugs that are not beneficial or are harmful to them. The lesson here is that it does not do a lot of good to pump enormous amounts of money into developing technology for new drug discovery without patient stratification — and this is genetics.

We should probably expect the same scenario to develop for human performance, and also, whether we like it or not, for human behavior.

Thus, now is really the time for scientists to put this issue into proper perspective. The misconception and fears about genetic determination are so misguided that we are delaying technology that can actually help improve existence for everyone. In medical diagnostic areas, we accept without any reservations tests and assays that try to determine whether we have a disease and the state of that disease. However, many people view with great concern genetic testing that is more direct and provides earlier detection. There are most certainly very important ethical issues relevant to the genetic determination. But even these are in some sense clouded by misconceptions, due to past behavior by groups who misunderstood the real meaning of genetic determination and/or intended to misuse it. It is time to correct this and gain the full benefits of our technology for everyone.

Tentative Plan for Understanding Genotype and Performance

We should start with the premise that almost every (physical) performance trait will be related to some distinct group of genotypes. (Genotypes from outside the group can also influence the trait, but this does not change the basic premise). This group of related genotypes will usually present itself in the general population as most individuals having average performance, some individuals having below-average performance, and another group of individuals having above-average performance. If we were to take *"running"* as an example, we can already begin to scientifically relate this trait to genetic polymorphisms in muscle tissue as well as other physiological characteristics. Even though we will ultimately identify the related group of genotypes that can accurately predict the performance level for any given physical trait, several problems do exist. The first problem is that there is considerable complexity in how different traits combine to affect "overall" performance. The second problem is to determine how these combinations of traits influence overall performance under different environmental challenges or stresses.

The goals for an initial plan to evaluate genotype and performance are listed below:

1. Begin to correlate physical (and related behavioral) performance characteristics with the genotypes and polymorphisms that are rapidly emerging from the human genome project. This would not be much different from what pharmaceutical companies are doing related to patient stratification for drug toxicity effects.

2. Begin to model how combinations of traits influence overall performance. Then separate the groups of directly related genotypes from those that indirectly influence the trait.

3. Begin to model and understand how a higher performance trait (or traits) that provide(s) an advantage under one set of environmental conditions and/or challenges, is not an advantage or is even a disadvantage under another set of environmental conditions and/or challenges.

This third point is probably the most difficult to deal with, because it leads to diversionary semantic and philosophical questions as to whether biology (genetics) or environment is in control, and what is cause and what is effect. These questions will be put into better perspective using examples of genetic disease in the human population (Jorder et al. 2000) and examples of how particular "types" of stress relate to heart disease (Ridley 2000; Marmot et al. 1991).

References

Hill, A.V.S. 1996. Genetics of infectious disease resistance. Opinion in *Genetics and Development* 6: 348-53.

Chorney, M.J., et al. 1998. A quantitative trait locus associated with cognitive ability in children. *Psychological Science* 9: 1-8.

Corder, E.H. et al. 1994. Protective effect of apolipoprotein E type 2 allele for late onset Alzheimer's disease. Nature Genetics 7: 180-84

Dubnau, J. and T. Tully. 1998. Gene discovery in drosophilia: New insights for learning and memory. Annual Review of Neuroscience 21: 407-44.

Jorde, L.B., J.C. Carey, M.J. Bamshed, and R.L. White. 2000. Medical genetics. 2nd ed. St. Louis, MO: Mosby.

Kamboh, M.I. 1995. Apolipoprotein E polymorphisms and susceptibility to Alzheimer's disease. Human Biology 67: 195-215.

Marmot, M.G., et al., 1991. Health inequalities among British civil servants: The Whitehall II study. Lancet 337: 1387-93.

Ridley, M. 2000. Genome: The autobiography of a species in 23 chapters. New York: Prennial/Harper Collins.

Wexler, N. 1992. Clairvoyance and caution: Repercussions from the Human Genome Project. In *The code of codes*. D. Kevles and L. Hood, eds. Cambridge, MA: Harvard University Press.

GENE THERAPY: REINVENTING THE WHEEL OR USEFUL ADJUNCT TO EXISTING PARADIGMS?

Jeffrey Bonadio, University of Washington

The availability of the human genome sequence should (a) improve our understanding of disease processes, (b) improve diagnostic testing for disease-susceptibility genes, and (c) allow for individually tailored treatments for common diseases. However, recent analyses suggest that the abundance of anticipated drug targets (yielded by the genome data) will acutely increase pharmaceutical R&D costs, straining the financial outlook of some companies. Therefore, to stay competitive, companies must couple a threshold infrastructure investment with more cost-effective validation/development technology. However, no such technology currently exists.

This paper discusses the potential advantages and disadvantages of gene therapy as a validation/delivery platform for the genomics era. Gene therapy is the use of recombinant DNA as a biologic substance for therapeutic purposes. Although significant technological hurdles exist, for certain drug targets the potential for gene therapy as a validation/delivery platform is enormous. Thus, one may see

- direct, efficient transitions from database query to preclinical validation to lead drug candidate development

- significant improvements in the patient care pathway of important common diseases such as cancer, diabetes, and osteoporosis; these improvements would be expected to yield improved compliance and significantly better control of disease manifestations

The vision is that in 10 to15 years, the U.S. private sector will have a drug discovery and development pathway that is significantly more cost-effective than what exists now and therefore is capable of taking full advantage of the promise of the human genome database. If this vision is realized, one can easily imagine that the process of transferring advances in drug development from the developed world to the undeveloped world will be significantly enhanced.

To traverse the technological hurdles associated with this vision, an interdisciplinary spirit will be required to advance our knowledge base in basic science and drug development, e.g., geneticists will (again) need to talk to physicists, physiologists to chemists, and cell biologists to engineers.

Drug Development Trends: Personalized Medicines

Human health is determined by the satisfaction of basic needs such as food and the avoidance of serious hazards such as trauma, environmental change, or economic disruption. In the world today, we find examples of almost all forms of social organization that have ever existed, including communities of hunter-gatherers, nomadic pastoralists, and primitive agriculturalists; unhygienic, large cities in the third world; and the modern, large cities of the developed world. This variation in living conditions is associated with differing patterns of human disease around the globe (McKeown 1988) as well as with patterns that shift in a dynamic manner, creating a rather large and varied number of therapeutic targets for the pharmaceutical industry to consider.

In contrast to the dynamic and varied patterns of human disease worldwide, the pharmaceutical industry has a long history of pursuing only those limited number of human proteins (G-protein coupled receptors, ion channels, nuclear hormone receptors, proteases, kinases, integrins, and DNA processing enzymes) that make the best drug targets (Wilson et al. 2001). Even so, a high percentage of drug candidates never reach the market because adverse reactions develop in a significant percentage of individuals, while many approved drugs are effective for only a fraction of the population in which they are prescribed. This variation in drug response depends on many factors, including gender, age, genetic background, lifestyle, living conditions, and co-morbidity.

Since the 1950s, pharmacogenetic studies have systematically identified allelic variants at genetic loci for relevant drug-metabolizing enzymes and drug targets (Evans and Relling 1999). These studies suggest that genetic tests may predict an individual's response to specific drugs and thereby allow medicines to be personalized to specific genetic backgrounds. For some drugs, the geographic distribution of allelic variants helps explain the differences in drug response across populations. The frequency of genetic polymorphisms in drug-metabolizing enzymes, which contribute significantly to phenotype, may vary among populations by as much as twelve-fold. For example, between 5 percent and 10 percent of Europeans, but only 1 percent of Japanese, have loss-of-function variants at *CYP2D6* (debrisoquine oxidation) that affect the metabolism of commonly used agents such as beta-blockers, codeine, and tricyclic antidepressants. Polymorphisms in drug-metabolizing enzymes can lead to acute toxic responses, unwanted drug–drug interactions, and therapeutic failure from augmented drug metabolism (Meyer and Zanger 1997). Therefore, one approach to drug development in the future may be to test candidate formulations in populations that are genetically homogenous for certain key genetic markers. Still, specific research challenges remain as to the most appropriate way to catalog human genetic variation and relate the inferred genetic structure to the drug response.

Impact of Genome Analysis Technology

The preceding 50 years have been a time of rapid and profound technological change. The elucidation of the genetic flow of biological information (i.e., information flow from DNA to RNA to protein) has provided a basis for the development of recombinant DNA technology; the rise of molecular cell biology; the advent of intellectual property in biology and medicine); the development of the biotechnology industry; the development of transgenic technologies (including human gene therapy); the elucidation of the modern definition of stem cells; and the advent of cloning technology. Arguably, the defining technological event of the last few years has been the development and large-scale implementation of tools for the global analysis of genomes. Less than a decade ago, it was relatively uncommon to have full-length cDNAs at hand for experimental purposes. Within a decade, it may be commonplace to freely access the atomic structure of proteins, often in the context of their molecular partners. We have entered a new era of life science discovery research in which structure-function relationships form the basis of our understanding of cellular physiology and pathology (Ideker, Galitski, and Hood 2001).

We have also entered a new era of pharmaceutical discovery in which structure-function relationships underlie the search for new therapies (Dry, McCarthy, and Harris 2001). Thus,

- *We still do not know how the transcription machinery regulates gene expression* (Strausberg and Riggins n.d.), despite the fact that the scientific literature richly describes the presence and functional significance of alternatively processed human transcripts — as derived from different transcription initiation sites, alternative exon splicing, and multiple polyadenylation sites. Therefore, genome sequences must be annotated and additional databases of information must be developed.

  Large-scale analysis of gene expression originates from the expressed sequence tag (EST) concept. In the EST approach, a unique identifier is assigned to each cDNA in a library. Sequence tags of more than 700 nucleotides are now common, and the EST approach has been aided by formation of the IMAGE consortium, an academic-industrial partnership designed to distribute clones. The Merck Gene Index and the Cancer Genome Anatomy Project have produced many of the human clones distributed through the IMAGE consortium (http://image.llnl.gov/).

  Imaginative new strategies complement the traditional EST approach. One of these, "serial analysis of gene expression" (Velculescu, Vogelstein, and Kinzler 2000), produces sequence tags (usually 14-nucleotides in length) located near defined restriction sites in cDNA. One advantage of this method is that each transcript has a unique tag, thereby facilitating transcript quantification. Tags are concatemerized, such that 30 or more gene tags can be read from a single sequencing lane, which also facilitates the effort to catalog genes. The Cancer Genome Anatomy Project, working together with the National Center for Biotechnology Information, has generated a SAGE database, SAGEmap, that includes over 4,000,000 gene tags. To proceed effectively with transcriptome efforts, there has been a significant shift in emphasis toward the sequencing of complete human transcripts.

  In this regard, in 1999 the National Institutes of Health announced the Mammalian Gene Collection Project (http://mgc.nci.nih.gov), which aims to identify and sequence human and mouse full-length cDNAs. To date, that project has produced over 5,000 human sequences (deposited in GenBank). The German Genome Project recently completed full-ORF human cDNA sequences derived from 1,500 human genes.

- *Functional genomics may provide a mechanism to understand how proteins collaborate in an integrated, regulated, adaptive manner.* Multiple technologies support the field of proteomics, including genomics, microarrays, new mass spectrometry approaches, global two-hybrid techniques, and innovative computational tools and methods (Fields 2001). Protein localization within cells is now feasible at a genomic level. For example, thousands of yeast strains were generated recently in which more than 2000 *S. cerevisiae* genes were marked by transposon tagging (Ross-Macdonald et al. 1999). Indirect immunofluorescence was used to determine the subcellular localization for over 1,300 of the tagged proteins.

Increasingly, proteomic strategies afford the opportunity for quantitative analysis of the cellular response to environmental change. Advances in direct analysis by mass spectrometry of peptide mixtures generated by the digestion of complex protein samples have led to an escalating number of protein identifications in one experiment. These and other advances suggest that human tissues one day may be evaluated this way to advance our understanding of disease etiology and pathogenesis.

Finally, protein expression and purification technologies will continue to improve, and procedures that make use of protein arrays will become commonplace. Potential applications include revealing interactions among proteins and between proteins and small molecules (drugs) or other ligands. The promise of this approach was suggested by the recent demonstration of proteins in nanoliter droplets immobilized by covalent attachment to glass slides: more than 10,000 samples could be spotted and assayed per slide with this technique (MacBeath and Schreiber 2001).

A shift from genomics to proteomics is likely to be complicated, because single genetic loci may yield multiple polypeptides; proteins may change conformation in order to carry out a particular function; protein levels often do not reflect mRNA levels; proteins may undergo post-translational modification and proteolysis; and the presence of an open reading frame does not guarantee the existence of a protein. Proteins may also adjust their stability, change locations in the cell, and swap binding partners.

Finally, protein function may depend on context, i.e., the function of an individual protein may be determined by the entire set of proteins operating in a microenvironment at a particular point in time — the concept of protein pleiotropism (Sporn 1999). When taken together, these considerations suggest that the proteome may be an order of magnitude more complex than the genome (Fields 2001; Hol 2000).

- *Structural genomics promises to capitalize upon numerous advances in cloning, protein expression, protein purification, characterization, crystallization, crystal drop inspection, crystal mounting, model building, and NMR spectra interpretation*, although high-throughput structure determination of drug candidates is not yet available (Russell and Eggleston 2000). With the potential to impact heavily on the design of new pharmaceuticals, structural genomics will take a place alongside high-throughput chemistry and screening as an integral platform approach underpinning modern drug discovery. Like the large-scale genomic sequencing projects that have been running for more than a decade, this will involve profound changes in thinking and approach. Instead of developing a specific biological justification in advance of working on a protein, crystallographers and NMR spectroscopists can now consider the determination of structures for all proteins in an organism.

    Bioinformatics will play several roles in structural genomics. Target selection involves database interrogation, sequence comparison, and fold recognition in order to aid selection of the best candidate proteins given a particular set of requirements, e.g., disease-associated genes, or those that are common to most organisms. Solved structures must be placed in an appropriate genomic context and annotated so that functional details may be predicted. Structural annotation may prove tricky, since large numbers of proteins of known structure but of unknown function have not previously been a major issue. Comparative modeling plays an essential role by providing structures for homologs of those determined experimentally, and efficient archiving of structural information is essential if the biological community is to make best use of all data. Given the biological and technological complexity associated with genome analysis technology, an interdisciplinary spirit will be essential to advance our knowledge base in basic science and drug development.

Drug Development in the Era of Genome Analysis: Applied Genomics

From SNP maps to individual drug response profiling, the human genome sequence should improve diagnostic testing for disease-susceptibility genes and lead to individually tailored treatment regimens for individuals with disease. Recent analyses (from both the public and private sector) suggest that the abundance of anticipated drug targets will dramatically increase pharmaceutical R&D costs. For example, it has been suggested that a threshold investment of $70-100 million will be required if companies are to profit from recent advances in bioinformatics. However, this investment may not yield a near-term return because current validation/development methods for drug targets are insufficiently robust to add value to R&D pipelines. Competitive considerations require companies to couple considerable infrastructure investment with cost-effective validation and/or development technology that has yet to be developed.

As described above, with advances in technology, the rational design and validation of new therapeutics increasingly will rely on the systematic interrogation of databases that contain genomic and proteomic information. One can imagine three pathways from database discovery to a validated product prototype, as shown in Figure C.2.

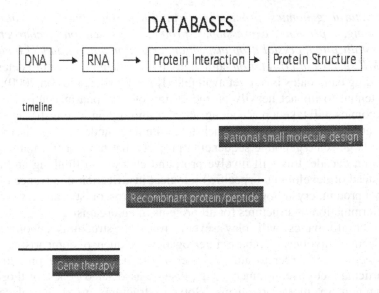

Figure C.2. Three pathways of drug discovery and development in the bioinformatics era.

For *Pathway 1, rational small-molecule design*, the methods for developing a small-molecule prototype are well established in the pharmaceutical industry, which reduces risk. However, it is not clear that small-molecule drugs can be designed, as shown above: the notion currently is without precedent (with perhaps the exception of inhibitors of HIV protease and influenza neuraminidase), and therefore is best considered as an unproven hypothesis.

A major advantage for *Pathway 2, recombinant protein/peptide design*, is that small-molecule prototypes need not be designed and validated at all, which may significantly accelerate product development. However, therapeutic peptides and recombinant proteins are generally ineffective when administered orally, and alternative routes of administration are generally associated with challenges in terms of formulation, compliance, efficacy, and safety.

A major advantage for *Pathway 3, gene therapy design*, is that one may proceed directly from database query to gene-based prototype — in theory, the shortest route to product validation and development. However, gene therapy is an early-stage technology, with known challenges in terms of efficacy and safety.

The Potential for Gene Therapy as a Validation / Delivery Platform

Gene therapy is the use of recombinant DNA as a biologic substance for therapeutic purposes (Bonadio 2000). Both viral and nonviral vectors have been employed. Nonviral vectors show many formulation and cost advantages, and they present a flexible chemistry. For example, the formulation of nonviral vectors with cationic agents results in nanometer-sized particles (synthetic polyplexes and lipoplexes) that show good efficiency (Felgner et al. 1997). Nonviral vectors have no theoretical sub-cloning limit, show a broad targeting specificity, transfect cells as episomes, and can be manufactured at scale relatively inexpensively. To enhance efficiency even further, one may use PEG to control surface properties of synthetic complexes, incorporate targeting moieties, use tissue-specific promoters, and incorporate fusogenic peptides and pH-responsive polymers.

On the other hand, the gain in gene-transfer efficiency associated with synthetic complexes must be balanced against the general lack of stability of polyplex and lipoplex vectors *in vivo* and the tendency of locally delivered cationic agents to cause tissue necrosis, which can be dramatic. Nonviral vectors are inefficient, and high doses may be required to achieve therapeutic effects. High-dose administration may be limited, however, by motifs in the vector backbone that stimulate the immune system (MacColl et al. 2001). While CpG-dependent immune stimulation is Th1-biased, SCID mice (Ballas, Rasmussen, and Krieg 1996) have shown increased levels of IFN- and IL-12 following plasmid-vector delivery (Klinman et al. 1996). Significantly, nonviral vector administration to animals has generated anti-DNA antibodies, leading to renal disease and premature death (Deng 1999). Relevant to the present application, Payette and colleagues (2001) recently showed that intramuscular delivery of a nonviral vector vaccine in mice led to destruction of antigen-expressing myocytes via a CTL-response.

Viruses are natural vectors for the transfer of recombinant DNA into cells. Recognition of this attribute has led to the design of engineered recombinant viral vectors for gene therapy. Viral vectors from retroviral, lentiviral, adenovirus, and herpes simplex species provide an important advantage in that they maximize gene transfer efficiency (Kay, Glorioso, and Naldini 2001). Viral genomes consist of genes and *cis*-acting gene regulatory sequences. Although overlap exists, most *cis*-acting sequences map outside viral coding sequences, and this spatial segregation is exploited in the design of recombinant viral vectors. Additionally, coding sequences work in *trans,* and viral genomes can be expressed by heterologous plasmids or be incorporated in the chromatin of producer cells to ensure stability and limit remobilization. Therefore, to generate vector particles, therapeutic genes and *cis*-acting sequences are first subcloned into separate plasmids, which are introduced into the same cell. Transfected cells produce replication-defective particles able to transduce target cells.

Viral vectors have inherent properties that affect suitability for specific gene therapy applications. A useful property of retroviral vectors, for example, is the ability to integrate efficiently into the chromatin of target cells. Disruption of the nuclear membrane is absolutely required for the pre-integration complex to gain access to chromatin (Roe et al. 1993), and productive transduction by retroviral vectors is strictly dependent on target cell mitosis (Miller, Adam, and Miller 1990). (Integration does not, however, guarantee stable expression of the transduced gene.) Because only a small fraction of muscle fibers pass through mitosis at any given time, this effectively prevents the use of regulated retroviral vectors (Rando and Blau 1994) in direct *in vivo* muscle gene therapy.

In contrast, replication-defective Ad vectors are attractive because they transduce post-mitotic cells very efficiently *in vivo* (Kozarsky and Wilson 1993). However, Ad vectors induce toxic immune responses that abrogate gene expression (Yang et al. 1995; Somia and Verma 2000). In a relevant example, Rivera et al. (1999) studied the feasibility of regulated Ad gene delivery after intramuscular injection in mice. The investigators employed an Ad vector cocktail encoding human growth hormone (hGH) under the control of transcriptional-switch technology. In initial experiments using immune-deficient mice, a single IP injection of rapamycin (5.0-mg/kg) resulted in a 100-fold increase in the plasma hGH level. Levels then diminished to baseline over the next 14 days. Similar induction profiles were noted after five subsequent injections (administered periodically over 6 months), and a direct relationship was observed between the peak hGH level and the amount of rapamycin administered (the i.v. dose range was 0.01 to 0.25 mg/kg). However, in immune-competent animals, peak levels of hGH were 50-fold lower, and no induction was observed after the first administration of rapamycin. These results were attributed to the destructive cellular and humoral immune responses to the Ad vector.

Experience with gene therapy suggests that this technology could serve as a broad validation and delivery platform for Pathway 3. To succeed, however, gene therapy must become a technology that more closely conforms to the current framework for drug development by pharmaceutical companies. Toward this end, gene therapy will need to be more easily managed by physician and patient; capable of producing therapeutic protein in a precise, dose-responsive, controllable manner; and formulated in a more simple, stable, and inexpensive manner. Ideally, a controllable gene-delivery system should feature low baseline transgene expression, a high induction ratio, and tight control by a small molecule drug. Indeed, it is difficult to imagine any gene therapy (for any indication) that does not involve regulated therapeutic gene expression as a way to avoid toxicity and still respond to the evolving nature of disease.

Among a multiplicity of DNA vector alternatives, recombinant adeno-associated viral (rAAV) vectors (Monahan and Samulski 2000) represent an attractive choice for a validation and delivery platform. rAAV vector particles efficiently transduce both dividing and nondividing cells, and the rAAV genome persists as integrated tandem repeats in chromosomal DNA. (Upon co-infection with a helper virus, AAV also transduces cells as an episome.) Elimination of AAV *rep* and *cap* coding sequences from rAAV prevents immune responses to viral gene products and the generation of wild-type helper virus (Hernandez et al. 1999; Xiao, Li, and Samulski 1996; Jooss et al. 1998). Transgene expression *in vivo* typically reaches a steady state after a gradual 2- to 10-week rise. Together, host chromosome integration and the absence of a cytotoxic T lymphocyte response provide a viable mechanism for long-term transgene expression, as demonstrated in the skeletal muscle (Herzog et al. 1999; Malik et al. 2000; Ye et al. 1999; Herzog et al. 1997) and brain (Davidson et al. 2000) of immunocompetent animals and in the skeletal muscle of human subjects (Kay et al. 2000). Importantly, the ability to conduct experiments is supported by the availability of small-scale procedures that allow the facile manufacture of sterile rAAV preparations at titers of 10^{11}-10^{12} vector genomes/mL (Auricchio et al. 2001). Even more importantly, rAAV gene therapy is controllable, as demonstrated below.

One promising technology (Figure C.3) employs a heterologous transcription factor that selectively binds the transgene promoter and activates transcription in response to a cell-permeant controller molecule (e.g., Rivera et al. 1996; Magari et al. 1997; Pollock et al. 2000). Activation is achieved by reconstitution of a transcription factor complex that couples independently expressed protein chimeras (Brown et al. 1994; Standaert et al. 1990). One protein consists of a unique DNA-binding domain called ZFHD1, genetically fused to FKBP. The other protein chimera consists of the activation domain of the p65 subunit of NFκB, fused with the rapamycin-binding domain of FRAP, which is termed FRB. Packaging limits of rAAV require that the three components of the system be incorporated into two vectors, one vector that expresses both transcription factors from a single transcriptional unit and a second vector containing the therapeutic gene driven by a

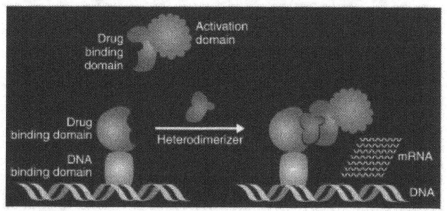

Figure C.3. Controlling gene expression using regulated transcription.

promoter recognized by the ZFHD1 DNA-binding domain. Infection of permissive human cells with equal quantities of the two AAV vectors at a high multiplicity of infection has resulted in full *in vitro* reconstitution of the regulated system with at least a 100-fold induction after exposure to rapamycin. (Effectiveness may be dramatically increased [Mateson et al. 1999] when chimeric transcriptional activators are expressed as noncovalent tetrameric bundles.)

The feasibility of reconstituting the regulated system *in vivo* has also been determined. Skeletal muscle has been selected for local delivery because muscle is permissive for rAAV transduction and because its component cells (muscle fibers) are long syncytia with extended nuclear domains that may be independently transduced with each vector. In one example (Ye et al. 1999), a controllable rAAV vector cocktail (2×10^8 infectious particles, with rAAV vectors at a 1:1 ratio) was injected into skeletal muscle of immune-competent mice. The administration of rapamycin resulted in 200-fold induction of erythropoietin in the plasma. Stable engraftment of this humanized system was achieved for six months, with similar results for at least three months in an immune-competent rhesus model.

The "transcriptional-switch" technology (described above) features an induction-decay response for the therapeutic protein that occurs on a time scale of days: transgene-encoded protein in blood typically peaks around 24 hours and then decreases to background over 4 to 14 days. This kinetic profile probably reflects the "early-point" of transgene regulation as well as the many potentially rate-limiting steps after therapeutic gene delivery. These steps involve the pharmacokinetics and pharmacodynamics of rapamycin (Mahalati and Kahan 2001) as well as the dynamic processes of transgene transcription, therapeutic protein translation and secretion, and therapeutic protein bioavailability. Such prolonged kinetics may be appropriate for certain proteins (e.g., erythropoietin) that govern relatively slow physiological

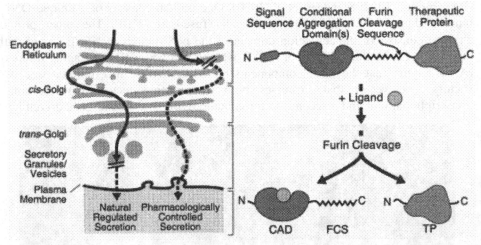

Figure C.4. Scheme for the pharmacologic control of protein secretion. (A) (*left*) Natural control of protein secretion (protein is stored in the secretory granules) is contrasted with the scheme for pharmacological control (protein is stored in the ER). (*right*) The therapeutic protein of interest (TP) is expressed as part of a fusion protein that contains, at its NH2-terminus, a signal sequence, a conditional aggregation domain (CAD), and a furin cleavage sequence (FCS). Processing and secretion of the TP is induced by ligand (Rivera et al. 2000).

processes. Prolonged kinetics may not be as appropriate, however, for proteins that regulate processes such as glucose homeostasis, which tend to be much faster.

To address this potential limitation of the transcriptional-switch system, Rivera et al. (2000), recently developed a second technology that allows protein secretion from the endoplasmic reticulum (ER) to be rapidly regulated Figure C.4). Therapeutic proteins are expressed as fusions with a conditional aggregation domain (CAD). CADs self-interact, and fusion proteins therefore form an aggregate in the ER that is far too large to be transported. Rivera and colleagues showed that the addition of cell-permeant ligand ("disaggregator") to transfected cells dissolves the aggregates and permits the rapid transport of therapeutic proteins from the ER via the constitutive secretory pathway.

To produce bioactive proteins, CAD moieties must be removed. Rivera et al. solved this problem by interposing a furin cleavage sequence between therapeutic protein and the CAD. In one example, Rivera et al. (2000) demonstrated that a natural version of hGH could be secreted in a controllable fashion using disaggregator technology. Thus, a single amino acid change (Phe^{36} to Met) converted monomeric FKBP12 into a CAD. Recombinant hGH was generated via a cDNA construct (Fig. C.3) consisting of a CMV promoter, signal sequence, four CAD motifs, a furin cleavage signal, and growth hormone (proinsulin was also used). Vectors were stably transfected into HT1080 cells, and fluorescence microscopy was used to demonstrate ER retention of both insulin and growth hormone in the absence of disaggregator. Cells expressing fusion proteins were then treated with increasing concentrations of disaggregator for two hours. The authors showed that accumulated protein was released by disaggregator administration, and the rate of release was controllable over an ~20-fold dose range. In the absence of ligand, fusion proteins were found only in cell lysate samples, whereas two hours after addition of ligand, fusion proteins were cleaved appropriately and secreted, as determined by Western analysis. Finally, myoblast transfer was used to demonstrate feasibility of the system in animal models. To this end, engineered cells were implanted into mice made diabetic by treatment with streptozotocin. Administration of vehicle failed to normalize serum glucose concentrations. However, after intravenous administration of ligand insulin was detected in serum within 15 minutes and peaked by 2 hours. Indeed, 2 hours after administration of a 10.0-mg/kg dose of ligand, the circulating insulin concentration increased to greater than 200.0-pM and serum glucose decreased concomitantly to normal. Lower doses of ligand were less effective.

Summary

Several trends have been identified:

a) Breakthroughs in controllable gene therapy technology have allowed therapeutic transgene expression to be regulated with precision over a period of months to years. The technology features low baseline transgene expression, a high induction ratio, and control via an orally available, cell permeant small molecule. Feasibility has been established in a series of elegant studies that employ recombinant adeno-associated viral (rAAV) vectors. *These breakthroughs are unique to gene therapy, i.e., similar levels of pro-drug stability and control simply do not exist for more traditional drug substances (small molecules, peptides, and proteins).*

b) *One may see enormous improvements in patient care pathways.* For diabetes and other endocrinopathies, the standard of care may change from "multiple daily injections" to a "single injection of gene therapy followed by ingestion of multiple tablets each day." Drug therapy could truly be personalized: once individual disease patterns are established (e.g., via sensor technology), the patient and physician could work together to develop a rational, personalized regimen of small molecule administration that would be expected to yield improved compliance and better control of disease; this in turn should lessen the cost of disease to U.S. society.

c) Given the availability of a panel of cell-permeant small molecules, gene therapy becomes a combined validation/development platform in which the therapy is a stable pro-drug that remains controllable for years following initial injection of tissues such as skeletal muscle. The small molecule panel would likely form an important core element of a company's intellectual property.

d) Given the biological and technological complexity associated with genome analysis technology, an interdisciplinary spirit will be required to advance our knowledge base in basic science and drug development. Although significant technological hurdles must be traversed, the potential advantages are enormous if controllable gene therapy can realize its potential as a validation and delivery platform. Drug discovery and development may one day be routine (a more-or-less turnkey process), characterized by direct, efficient transitions *from* database query *to* rational isolation of the relevant cDNA *to* preclinical validation, *to* validation in human clinical trials (Fig. C.1). Because the "drug substance" typically will consist of a recombinant gene and a small-molecule controller, many aspects of formulation, manufacturing, biodistribution, and toxicity would be well understood *prior to* initiation of a new development program. Obviously, companies would operate in an environment of significantly reduced risk relative to the current situation; this environment would allow companies to explore a much broader range of drug targets than typically is explored today.

e) Finally, we envision a pharmaceutical industry that possesses the technological tools and economic incentives to take full advantage of the power of genomics. Specifically, the vision proposed here is that in 10 to 15 years the U.S. private sector will have a drug discovery/drug development pathway that is significantly more cost effective (more turnkey and less risky) than what we now have and is capable of taking full advantage of the promise of the human genome sequence. (Pharmaceutical companies could actually take calculated risks!) If this vision is realized, one can easily imagine how the process of technology transfer from the developed to the undeveloped world would be incentivized for the first time.

References

Auricchio, A., Hildinger, M., O'Connor, E., Gao, G.P., and Wilson, J.M. 2001. Isolation of highly infectious and pure adeno-associated virus type 2 vectors with a single-step gravity-flow column. *Hum. Gene Ther.* 12:71.

Ballas, Z.K., Rasmussen, W.L., and Krieg, A.M. 1996. Induction of NK activity in murine and human cells by CpG motifs in oligodeoxynucleotides and bacterial DNA. *J. Immunol.* 157:1840.

Bonadio, J. 2000. Tissue engineering via local gene delivery: Update and future prospects for enhancing the technology. *Advanced Drug Delivery Reviews* 44:185.

Brown, E.J., Albers, M.W., Shin, T.B., Ichikawa, K., Keith, C.T., Lane, W.S., and Schreiber, S.L. 1994. A mammalian protein targeted by G1-arresting rapamycin-receptor complex. *Nature* 369:756.

Davidson, B.L., Stein, C.S., Heth, J.A., Martins, I., Kotin, R.M., Derksen, T.A., Zabner, J., Ghodsi, A., Chiorini J.A. 2000. Recombinant adeno-associated virus type 2,4, and 5 vectors: transduction of variant cell types and regions in the mammalian central nervous system. *Proc. Natl. Acad. Sci. U.S.A.* 97:3428.

Deng, G-M., Nilsson, I-M., Verdrengh, M., Collins, L.V., and Tarkowski, A. 1999. Intra-articularly localized bacterial DNA containing CpG motifs induces arthritis. *Nature Med.* 5:702.

Dry, S., McCarthy, S., and Harris, T. 2001. Structural genomics in the biotechnology sector. *Nat. Biotechnol.* 29:946.

Evans, W.E., and Relling, M.V. 1999. Pharmacogenomics: translation functional genomics into rational therapeutics. *Science* 286:487.

Felgner, P.L., Barenholz, Y., Behr, J.P., Cheng, S.H., Cullis, P., Huang, L., Jessee, J.A., Seymour, L., Szoka, F., Thierry, A.R., Wagner, E., and Wu, G. 1997. Nomenclature for synthetic gene delivery systems. *Hum. Gene Ther.* 8:511.

Fields, S. 2001. Proteomics in genomeland. *Science* 291:1221.

Hernandez, Y.J., et al. 1999. Latent adeno-associated virus infection elicits humoral but not cell-mediated immune responses in a nonhuman primate model. *J. Virol.* 73:8549.

Herzog, R.W., Hagstron, J.N., Kung, S-H., Tai, S.J., Wilson, J.M., Fisher, K.J., and High, K.A. 1997. Stable gene transfer and expression of human blood coagulation factor IX after intramuscular injection of recombinant adeno-associated virus. *Proc. Natl. Acad. Sci. U.S.A.* 94:5804.

Herzog, R.W., Yang, E.Y., Couto, L.B., Hagstron, J.N., Elwell, D., Fields, P.A., Burton, M., Bellinger, D.A., Read, M.S., Brinkhous, K.M., Podsakoff, G.M., Nichols, T.C., Kurtzman, G.J., and High, K.A. 1999. Long-term correction of canine hemophilia B by gene transfer of blood coagulation factor IX mediated by adeno-associated viral vector. *Nature Med.* 5:56.

Hol, W.G.J. 2000. Structural genomics for science and society. *Nat. Structural Biol.* 7 Suppl:964.

Ideker, T., Galitski, T., and Hood, L. 2001. A new approach to decoding life: systems biology. *Annu. Rev. Genomics Hum. Genet.* 2:343.

Jooss, K., Yang, Y., Fisher, K.J., and Wilson, J.M. 1998. Transduction of dendritic cells by DNA viral vectors directs the immune response to transgene products in muscle fibers. *J. Virol.* 72:4212.

Kay, M.A., Glorioso, J.C., and Naldini, L. 2001. Viral vectors for gene therapy: the art of turning infectious agents into vehicles of therapeutics. *Nature Med.* 7:33.

Kay, M.A., Manno, C.S., Ragni, M.V., Larson, P.J., Couto, L.B., McClelland, A., Glader, B., Chew, A.J., Tai, S.J., Herzog, R.W., Arruda, V., Johnson, F., Scallan, C., Skarsgard, E., Flake, A.W., and High, K.A. 2000. Evidence for gene transfer and expression of factor IX in haemophilia B patients treated with an AAV vector. *Nature Genet.* 24:257.

206   C. Improving Human Health and Physical Capabilities

Klinman, D.M., Yi, A.K., Beucage, S.L., Conover, J., and Krieg, A.M. 1996. CpG motifs present in bacterial DNA rapidly induce lymphocytes to secrete interleukin 6, interleukin 12, and interferon gamma. *Proc. Natl. Acad. Sci. U.S.A.* 93:2879.

Kozarsky, K.F., and Wilson, J.M. 1993. Gene therapy: adenovirus vectors. *Curr. Opin. Genet. Dev.* 3:499.

MacBeath, G., and Schreiber, S.L. 2000. Printing proteins as microarrays for high-throughput function determination. *Science* 289:1673.

MacColl, G., Bunn, C., Goldspink, G., Bouloux, P., and Gorecki, D.C. 2001. Intramuscular plasmid DNA injection can accelerate autoimmune responses. *Gene Ther.* 8:1354.

Magari, S.R., Rivera, V.M., Iuliucci, J.D., Gilman, M., and Cerasoli, F. 1997. Pharmacologic control of a humanized gene therapy system implanted into nude mice. *J. Clin. Invest.* 100:2865.

Mahalati, K., and Kahan, B.D. 2001. Clinical pharmacokinetics of sirolimus. *Clin. Pharmacokinet.* 40:573.

Malik, A.K., Monahan, P.E., Allen, D.L., Chen, B.G., Samulski, R.J., and Kurachi, K. 2000. Kinetics of recombinant adeno-associated virus-mediated gene transfer. J. Virol. 74:3555.

McKeown, T. 1988. *The origins of human disease*. Oxford, UK: Basil Blackwell Ltd., pp. 1-233.

Meyer, U.A., and Zanger, U.M. 1997. Molecular mechanisms of genetic polymorphisms of drug metabolism. *Annu. Rev. Pharmacol. Toxicol.* 37:269.

Miller, D.G., Adam, M.A., and Miller, A.D. 1990. Gene Transfer by retrovirus vectors occurs only in cells that are actively replicating at the time of infection. *Mol. Cell Biol.* 10:4239.

Monahan, P.E., and Samulski, R.J. 2000. AAV vectors: is clinical success on the horizon? *Gene Ther.* 7:24.

Nateson, S., Molinari, E., Rivera, V.M., Rickles, R.J., and Gilman, M. 1999. A general strategy to enhance the potency of chimeric transcriptional activators. *Proc. Natl. Acad. Sci. U.S.A.* 96:13898.

Payette, P.J., Weeratna, R.D., McCluskie, M.J., and Davis, H.L. 2001. Immune-mediated destruction of transfected myocytes following DNA vaccination occurs via multiple mechanisms. *Gene Ther.* 8:1395.

Pollock, R., Issner, R., Zoller, K., Natesan, S., Rivera, V.M., and Clackson, T. 2000. Delivery of a stringent dimerizer-regulated gene expression system in a single retroviral vector. *Proc. Natl. Acad. Sci. USA* 97: 13221.

Rando, T.A., and Blau, H.M. 1994. Primary mouse myoblast purification, characterization, and transplantation for cell-mediated gene therapy. *J. Cell Biol.* 125:1275.

Rivera, V.M., Clackson, T., Natesan, S., Pollock, R., Amara, J.F., Keenan, T., Magari, S.R., Phillips, T., Courage, N.L., Cerasoli, F. Jr., Holt, D.A., and Gilman, M. 1996. A humanized system for pharmacologic control of gene expression. *Nat. Med.* 2:1028.

Rivera, V.M., Wang, X., Wardwell, S., Courage, N.L., Volchuk, A., Keenan, T., Holt, D.A., Gilman, M., Orci, L., Cerasoli, F. Jr., Rothman, J.E., and Clackson, T. 2000. Regulation of protein secretion through controlled aggregation in the endoplasmic reticulum. *Science* 287:826.

Rivera, V.M., Ye, X., Courage, N.L., Sachar, J., Cerasoli, F. Jr., Wilson, J.M., and Gilman, M. 1999. Long-term regulated expression of growth hormone in mice after intramuscular gene transfer. *Proc. Natl. Acad. Sci. U.S.A.* 96:8657.

Roe, T., Reynolds, T.C., Yu, G., and Brown, P.O. 1993. Integration of murine leukemia virus DNA depends on mitosis. *EMBO J.* 12:2099.

Ross-Macdonald, P., Coelho, P.S., Roemer, T., Agarwal, S., Kumar, A., Jansen, R., Cheung, K.H., Sheehan, A., Symoniatis, D., Umansky, L., Heidtman, M., Nelson, F.K., Iwasaki, H., Hager, K., Gerstein, M., Miller, P., Roeder, G.S., and Snyder, M. 1999. Large-scale analysis of the yeast genome by transposon tagging and gene disruption. *Nature* 402:362.

Russell, R.B., and Eggleston, D.S. 2000. New roles for structure in biology and drug discovery. *Nat. Structural Biol.* 7 Suppl:928.

Somia, N., and Verma, I.M. 2000. Gene therapy: trials and tribulations. *Nature Rev. Genetics* 1:91.

Sporn, M.B. 1999. Microbes Infect. TGF-beta: 20 years and counting. 1:1251.

Standaert, R.F., Galat, A., Verdine, G.L., and Schreiber, S.L. 1990. Molecular cloning and overexpression of the human FK506-binding protein FKBP. *Nature* 346:671.

Strausberg, R.L., and Riggins, G.J. Navigating the human transcriptome. *Proc. Natl. Acad. Sci. U.S.A.* 98:11837.

Velculescu, V.E., Vogelstein, B., and Kinzler, K.W. 2000. Analyzing uncharted transcriptomes with SAGE. *Trends Genet.* 16:423.

Wilson, J.F., Weale, M.E., Smith, A.C., Gratrix, F., Fletcher, B., Thomas, M.G., Bradman, N., and Goldstein, D.B. 2001. Population genetic structure of variable drug response. *Nat. Biotechnol.* 29:265.

Xiao, X., Li, J., and Samulski, R.J. 1996. Efficient long-term gene transfer into muscle tissue of immunocompetent mice by adeno-associated virus vector. *J. Virol.* 70:8098.

Yang, Y., Li, Q., Ertl, H.C., and Wilson, J.M. 1995. Cellular and humoral immune responses to viral antigens create barriers to lung-directed gene therapy with recombinant adenoviruses. *J. Virol.* 69:2004.

Ye, X., Rivera, V.M., Zoltick, P., Cerasoli, F. Jr., Schnell, M.A., Gao, G., Hughes, J.V., Gilman, M., and Wilson J.M. 1999. Regulated delivery of therapeutic proteins after in vivo somatic cell gene transfer. *Science* 283:88.

IMPLICATIONS OF THE CONTINUUM OF BIOINFORMATICS

Peter C. Johnson, TissueInformatics, Inc.

The once impenetrable complexity of biology has come face to face with rapidly expanding microprocessing power and information management solutions, and this confluence is changing our world. The parallel development of tools needed to extract biological meaning from DNA, proteins, cells, tissues, organisms, and society as a whole has set the stage for improved understanding of biological mechanisms. This is being augmented by our ability to manage this information in uniform ways and to ask questions about relationships across broad levels of biological scale. This multiscalar description of biology from the molecular to the societal, with all of the tools needed to draw correlations across its landscape, is known as the continuum of bioinformatics (COB).

Though presently immature, the COB is growing in richness daily. Driven initially by the need to manage DNA and protein sequence data, it has grown with the inclusion of cellular imaging, tissue analysis, radiological imaging, and societal healthcare informatics inputs. It is presently virtual but, like the Internet before it, it is being tied together through the development of standard systems, query tools, and security measures. As it develops, the COB is changing our world through the enhancement of our understanding of biological process and the acceleration of development of products that can benefit man, animals, and plants. The unusual precision with which biological data is represented within the COB is making it possible to reduce the degrees of freedom normally accorded biological understanding — and therefore to enable the individualization of solutions that will protect life.

Nanotechnology will play a major role in the development of information gathering and processing systems for the COB.

Definition of Bioinformatics

The science of bioinformatics presents the rich complexity of biology in such a way that meaning can be extracted using digital tools. As a discipline having multiple parts, it can be defined overall in a number of ways. One definition of bioinformatics and its components is as follows (D'Trends n.d.):

(1) *Bioinformatics* - database-like activities involving persistent sets of data that are maintained in a consistent state over essentially indefinite periods of time

(2) *Computational biology* - the use of algorithmic tools to facilitate biological analyses

(3) *Bioinformation infrastructure* - the entire collective of information management systems, analysis tools and communication networks supporting biology

This composite definition points out the importance of three activities critical to the success of bioinformatics activities:

• The use of analytic methods to enable the presentation of biological information in digital fashion.

• The leveraging of massive digital storage systems and database technologies to manage the information obtained.

• The application of digital analytic tools to identify patterns in the data that clarify causes and effects in biological systems, augmented by visualization tools that enable the human mind to rapidly grasp these patterns.

Bioinformatics makes the complexity of biological systems tangible. Taken in the stepwise fashion described above, this complexity can often be reduced to terms that are understandable to scientists probing biological problems. Biological complexity is worthwhile to understand. A clear appreciation of cause and effect in biological systems can provide the knowledge needed to develop drugs and other medical therapies and also to provide a greater appreciation for what we are as humans beings. It is interesting to note that biological complexity is so extreme that it challenges the best that high-performance computing presently has to offer. Ironically, the fulfillment of the Socratic adage "Know Thyself" can now only be achieved through man's interaction with and dependence upon computing systems.

The recent accomplishment of sequencing the human genome (and now the genomes of several other species) focused attention on the information processing requirements at the molecular end of the biological spectrum. For a time, it seemed that "bioinformatics" was wholly concerned with the management and deciphering of genetic information. Soon, information descriptive of the patterns of expression of proteins and their interactions was added (proteomics). Since this information required stratification by disease type, cellular and tissue information became important to consider. Inevitably, it became apparent that information descriptive of the whole organism, such as radiological data, other morphometric data, chemistries, and other health record data should be included. Once this was done, aggregated societal data was the next logical addition.

The picture that has come into view is therefore one of a continuum of bioinformatics (Figure C.5). In the COB model, linked data at multiple scales of

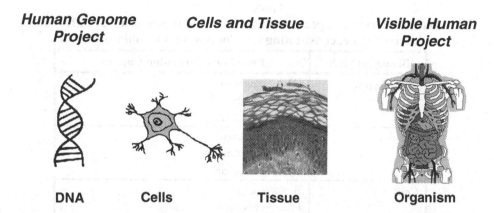

Human Genome Project   **Cells and Tissue**   **Visible Human Project**

DNA   **Cells**   **Tissue**   **Organism**

Figure C.5. Multiple scales of biological activity and form comprise the entire organism. The Continuum of Bioinformatics is an information system that includes and can correlate information from all of these scales of data. Though not shown in the figure, the aggregation of individual data into societal data (as in the form of healthcare statistics) is extremely valuable, since it places the individual's data within the context of the society as a whole.

biological complexity are considered together for both individuals and aggregates of individuals. The key to the value of the COB will be the ability to derive correlations between *causes* (such as gene expression, protein interactions, and the like) and *effects* (such as healthcare outcomes for individuals and societies). In this model, it may well be possible one day to determine the cost to society of the mutation of a single gene in a single individual! It will also be possible to predict with clarity which drugs will work for which individuals and why. By taking a reverse course through the COB from effects to causes, it will also be possible to sharply identify proteins that can serve as drug targets for specific disease states.

Information Capture

In order to benefit from the COB, information descriptive of biology at multiple scales must first be captured accurately and then managed such that different types of data can be interpreted with reference to one another. It is in this area that the convergence of nanotechnology and biotechnology will occur, since nanotechnology provides enabling mechanisms for the capture and management of complex biological information, particularly at the level of molecular expression data.

A simple way to look at this issue is to first note that to be useful in the COB context, all biological data must first be captured using techniques that enable its ultimate conversion to digital form. The mechanisms differ, depending upon the point in the COB under consideration. Table C.3 shows the levels of the COB and the tools needed to capture data digitally at the proper level of discretion to enable computerized correlation between data pools.

Table C.3
Tools required to capture human biological information at different levels of scale, constituting the Continuum Of Bioinformatics

Biological Scale	Tools For Information Capture
DNA, Genes	DNA Sequencers Electrophoresis Affinity Microarrays
Proteins	Electrophoresis Mass Spectrometry Affinity Microarrays
Cells	Bioassays Fluorescent probes Digital Imaging
Tissues	Digital Imaging Hyperquantitative Analysis
Organism	Digital Radiology (X-Ray, CT, MRI, PET) Chemistry Data Healthcare Record
Society	Aggregated Healthcare Records

Ideally, information at all levels of scale would be captured from the same individual and then aggregated into a societal record. Since this is impractical, aggregated information will most likely be used, and this will grow richer over time. Privacy concerns are often raised when highly discrete and potentially predictive personal information is gathered in this way. However, it is most likely that COB data (as this virtual network begins to merge together) will be anonymized sufficiently so that individuals will be protected. Indeed, one way to look at the COB is to envision it as a powerful reference database against which an individual's data can be compared in order to provide an individual with contextual information regarding his or her health at any point in time. This is the essence of what is known as "Systems Biology," as well.

Tissue Information as a Specific Instance

A specific example of the conversion of biological information to digital information occurs at the tissue level. Until recently, it was felt that only a pathologist could interpret the meaning of patterns of cells and other structural components of tissue. This meaning was summed up in the diagnosis that was applied to the tissue and used to guide healthcare decision-making. Over the past two decades, digital imaging of tissues on slides has created the basis for management of tissue information at the image level for ease of data sharing between pathologists and researchers. However, this did not convert the data completely into digital form, because human interpretation and diagnostic assignment of the overall image were still required. This limited the ability to correlate tissue data with other biological information to the level of resolution that diagnosis provided.

Recently, it has become possible to use automated machine vision analysis systems to measure all of the components that can be made visible within a tissue (both structural and functional) with reference to one another. This is known as *Hyperquantitative Analysis of Tissue* (Fig. C.6).

Preparation of tissue information in this way requires two steps:

a) automated imaging that enables location of tissue on a microscope slide and the capture of a composite image of the entire tissue — or tissues — on the slide

b) the application of image analytic software that has been designed to automatically segregate and co-localize in Cartesian space the visible components of tissue (including molecular probes, if applied)

Tissue information captured in this way enables very precise *mathematical comparison* of tissues to detect change (as in toxicology testing or, ultimately, clinical diagnostics). In each case, substantial work must first be done to collect normative reference data from tissue populations of interest.

More importantly, when tissue information is reduced to this level of scale, the data is made available for more precise *correlation* with other data sets in the continuum of bioinformatics in the following applications:

• *Backward correlation:* "Sorter" of genomic and proteomic data

Rationale: When gene or protein expression data are culled from a tissue that has undergone hyperquantitative analysis, tighter correlations are possible between molecular expression patterns and tissue features whose known biological roles help to explain the mechanisms of disease — and therefore may help to identify drug targets more sharply.

• *Forward correlation:* Stratifier of diagnosis with respect to prognosis

Rationale: When tissue information is collected along with highly detailed clinical descriptions and outcome data, subtle changes in tissue feature patterns within a diagnostic group may help to further stratify prognoses associated with

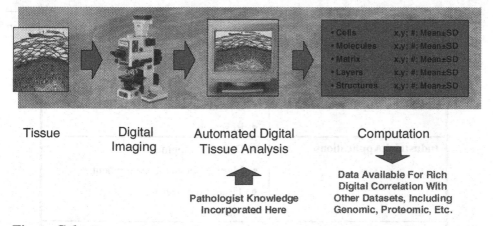

Figure C.6. Capture of tissue information in hyperquantitative fashion. All components of the tissue that can be made visible are located simultaneously after robotic capture of slide-based images. This step automates the analysis of tissue, putting it immediately into a form that enables sharing of images and derived data.

a diagnosis and may prompt more refined diagnostic classifications.
- *Pan Correlation:* Tighten linkage of prognosis with molecular diagnostics
  Rationale: Since tissue is the classical "site of diagnosis," the use of tissue information to correlate with molecular expression data and clinical outcome data validates those molecular expression patterns with reference to their associated diseases, enabling their confident application as molecular diagnostics.

Nanotechnology developments applicable to imaging and computational science will aid and abet these discoveries.

Information Management

The physical management of the large volumes of information needed to represent the COB is essentially an information storage and retrieval problem. Although only several years ago the amount of information that required management would have been a daunting problem, this is far less so today. Extremely large storage capacities in secure and fast computer systems are now commercially available. While excellent database systems are also available, none has yet been developed that completely meets the needs of the COB as envisioned. Database system development will continue to be required in order for the COB to be applied maximally. Several centers are now attempting the development of representative databases of this type.

Extracting Value From the Continuum of Bioinformatics

Once the COB is constructed and its anonymized data becomes available, it can be utilized by academia, industry, and government for multiple critical purposes. Table C.4 shows a short list of applications.

In order for COB data to be put to best use, considerable work will be needed to incorporate statistical methodology and robust graphical user interfaces into the COB. In some cases, the information gleaned will be so complex that new methods of visualization of data will need to be incorporated. The human mind is a powerful interpreter of graphical patterns. This may be the reason why tissue data — classically having its patterns interpreted visually by a pathologist — was the last in the continuum to be reduced to discrete digital form.

Table C.4
Applications of the COB in multiple sectors

Academic Applications	• Education
	• Research
Industrial Applications	• Drug Development
	• Medical Device Development
	• Tissue Engineering
	• Marketing
Government Applications	• Population Epidemiology
	• Disease Tracking
	• Healthcare Cost Management

As the COB develops, we are likely to see novel data visualization methods applied in ways that cannot be envisioned at all today. In each instance, the robustness of these tools will ultimately depend on the validity of the data that was entered into the COB and on the mode of application of statistical tools to the data being analyzed.

Impact on Human Health

The COB will significantly enhance our ability to put individual patterns of health and disease in context with that of the entire population. It will also enable us to better understand the mechanisms of disease, how disease extends throughout the population, and how it may be better treated. The availability of the COB will resect time and randomness from the process of scientific hypothesis testing, since data will be available in a preformed state to answer a limitless number of questions. Finally, the COB will enable the prediction of healthcare costs more accurately. All of these beneficial reesults will be accelerated through the application of nanotechnology principles and techniques to the creation and refinement of imaging, computational, and sensing technologies.

Reference

D'Trends, Inc. http://www.d-trends.com/Bioinformatics/bioinformatics.html.
West, J.L., and N.J. Halas. 2000. Applications of nanotechnology to biotechnology commentary, *Curr. Opin. Biotechnol.* 11(2):215-7 (Apr.).

SENSORY REPLACEMENT AND SENSORY SUBSTITUTION: OVERVIEW AND PROSPECTS FOR THE FUTURE

Jack M. Loomis, University of California, Santa Barbara

The traditional way of dealing with blindness and deafness has been some form of sensory substitution — allowing a remaining sense to take over the functions lost as the result of the sensory impairment. With visual loss, hearing and touch naturally take over as much as they can, vision and touch do the same for hearing, and in the rare cases where both vision and hearing are absent (e.g., Keller 1908), touch provides the primary contact with the external world. However, because unaided sensory substitution is only partially effective, humans have long improvised with artifices to facilitate the substitution of one sense with another. For blind people, braille has served in the place of visible print, and the long cane has supplemented spatial hearing in the sensing of obstacles and local features of the environment. For deaf people, lip reading and sign language have substituted for the loss of speech reception. Finally, for people who are both deaf and blind, fingerspelling by the sender in the palm of the receiver (Jaffe 1994; Reed et al. 1990) and the Tadoma method of speech reception (involving placement of the receiver's hand over the speaker's face) have provided a means by which they can receive messages from others (Reed et al. 1992).

Assistive Technology and Sensory Substitution

Over the last several decades, a number of new assistive technologies, many based on electronics and computers, have been adopted as more effective ways of promoting sensory substitution. This is especially true for ameliorating blindness. For example, access to print and other forms of text has been improved with these technologies: electronic braille displays, vibtrotactile display of optically sensed print (Bliss et al. 1970), and speech display of text sensed by video camera (Kurzweil 1989). For obstacle avoidance and sensing of the local environment, a number of ultrasonic sensors have been developed that use either auditory or tactile displays (Brabyn 1985; Collins 1985; Kay 1985). For help with large-scale wayfinding, assistive technologies now include electronic signage, like the system of Talking Signs (Crandall et al. 1993; Loughborough 1979; see also http://www.talkingsigns.com/), and navigation systems relying on the Global Positioning System (Loomis et al. 2001), both of which make use of auditory displays. For deaf people, improved access to spoken language has been made possible by automatic speech recognition coupled with visible display of text; in addition, research has been conducted on vibrotactile speech displays (Weisenberger et al. 1989) and synthetic visual displays of sign language (Pavel et al. 1987). Finally, for deaf-blind people, exploratory research has been conducted with electromechanical Tadoma displays (Tan et al. 1989) and finger spelling displays (Jaffe 1994).

Interdisciplinary Nature of Research on Sensory Replacement / Sensory Substitution

This paper is concerned with compensating for the loss of vision and hearing by way of sensory replacement and sensory substitution, with a primary focus on the latter. Figure C.7 shows the stages of processing from stimulus to perception for vision, hearing, and touch (which often plays a role in substitution) and indicates the

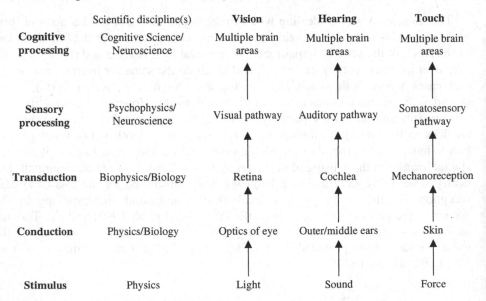

	Scientific discipline(s)	Vision	Hearing	Touch
Cognitive processing	Cognitive Science/ Neuroscience	Multiple brain areas	Multiple brain areas	Multiple brain areas
Sensory processing	Psychophysics/ Neuroscience	Visual pathway	Auditory pathway	Somatosensory pathway
Transduction	Biophysics/Biology	Retina	Cochlea	Mechanoreception
Conduction	Physics/Biology	Optics of eye	Outer/middle ears	Skin
Stimulus	Physics	Light	Sound	Force

Figure C.7. Sensory modalities and related disciplines.

associated basic sciences involved in understanding these stages of processing. (The sense of touch, or haptic sense, actually comprises two submodalities: kinesthesis and the cutaneous sense [Loomis and Lederman 1986]; here we focus on mechanical stimulation). What is clear is the extremely interdisciplinary nature of research to understand the human senses. Not surprisingly, the various attempts to use high technology to remedy visual and auditory impairments over the years have reflected the current scientific understanding of these senses at the time. Thus, there has been a general progression of technological solutions starting at the distal stages (front ends) of the two modalities, which were initially better understood, to solutions demanding an understanding of the brain and its functional characteristics, as provided by neuroscience and cognitive science.

Sensory Correction and Replacement

In certain cases of sensory loss, sensory correction and replacement are alternatives to sensory substitution. Sensory correction is a way to remedy sensory loss prior to transduction, the stage at which light or sound is converted into neural activity (Figure C.7). Optical correction, such as eyeglasses and contact lenses, and surgical correction, such as radial keratotomy (RK) and laser in situ keratomileusis (LASIK), have been employed over the years to correct for refractive errors in the optical media prior to the retina. For more serious deformations of the optical media, surgery has been used to restore vision (Valvo 1971). Likewise, hearing aids have long been used to correct for conductive inefficiencies prior to the cochlea. Because our interest is in more serious forms of sensory loss that cannot be overcome with such corrective measures, the remainder of this section will focus on sensory replacement using bionic devices.

In the case of deafness, tremendous progress has already been made with the cochlear implant, which involves replacing much of the function of the cochlea with direct electrical stimulation of the auditory nerve (Niparko 2000; Waltzman and Cohen 2000). In the case of blindness, there are two primary approaches to remedying blindness due to sensorineural loss: retinal and cortical prostheses. A retinal prosthesis involves electrically stimulating retinal neurons beyond the receptor layer with signals from a video camera (e.g., Humayun and de Juan 1998); it is feasible when the visual pathway beyond the receptors is intact. A cortical prosthesis involves direct stimulation of visual cortex with input driven by a video camera (e.g., Normann 1995). Both types of prosthesis present enormous technical challenges in terms of implanting the stimulator array, power delivery, avoidance of infection, and maintaining long-term effectiveness of the stimulator array.

There are two primary advantages of retinal implants over cortical implants. The first is that in retinal implants, the sensor array will move about within the mobile eye, thus maintaining the normal relationship between visual sensing and eye movements, as regulated by the eye muscle control system. The second is that in retinal implants, connectivity with the multiple projection centers of the brain, like primary visual cortex and superior colliculus, is maintained without the need for implants at multiple sites. Cortical implants, on the other hand, are technically more feasible (like the delivery of electrical power), and are the only form of treatment for blindness due to functional losses distal to visual cortex. For a discussion of other pros and cons of retinal and cortical prostheses, visit the Web site (http://insight.med.utah.edu/research/normann/normann.htm) of Professor Richard Normann of the University of Utah.

Interplay of Science and Technology

Besides benefiting the lives of blind and deaf people, information technology in the service of sensory replacement and sensory substitution will continue to play another very important role — contributing to our understanding of sensory and perceptual function. Because sensory replacement and sensory substitution involve modified delivery of visual and auditory information to the perceptual processes in the brain, the way in which perception is affected or unaffected by such modifications in delivery is informative about the sensory and brain processes involved in perception. For example, the success or lack thereof of using visual displays to convey the information in the acoustic speech signal provides important clues about which stages of processing are most critical to effective speech reception. Of course, the benefits flow in the opposite direction as well: as scientists learn more about the sensory and brain processes involved in perception, they can then use the knowledge gained to develop more effective forms of sensory replacement and substitution.

Sensory Replacement and the Need for Understanding Sensory Function

To the layperson, sensory replacement might seem conceptually straightforward — just take an electronic sensor (e.g., microphone or video camera) and then use its amplified signal to drive an array of neurons somewhere within the appropriate sensory pathway. This simplistic conception of "sensory organ replacement" fails to recognize the complexity of processing that takes place at the many stages of processing in the sensory pathway. Take the case of hearing. Replacing an inoperative cochlea involves a lot more than taking the amplified signal from a microphone and using it to stimulate a collection of auditory nerve fibers. The cochlea is a complex transducer that plays sound out in terms of frequency along the length of the cochlea. Thus, the electronic device that replaces the inoperative cochlea must duplicate its sensory function. In particular, the device needs to perform a running spectral analysis of the incoming acoustic signal and then use the intensity and phase in the various frequency channels to drive the appropriate auditory nerve fibers. This one example shows how designing an effective sensory replacement begs detailed knowledge about the underlying sensory processes. The same goes for cortical implants for blind people. Simply driving a large collection of neurons in primary visual cortex by signals from a video camera after a simple spatial sorting to preserve retinotopy overlooks the preprocessing of the photoreceptor signals being performed by the intervening synaptic levels in the visual pathway. The most effective cortical implant will be one that stimulates the visual cortex in ways that reflect the normal preprocessing performed up to that level, such as adaptation to the prevailing illumination level.

Sensory Substitution: An Analytic Approach

If sensory replacement seems conceptually daunting, it pales in comparison with sensory substitution. With sensory substitution, the goal is to substitute one sensory modality that is impaired or nonfunctioning with another intact modality (Bach-y-Rita 1972). It offers several advantages over sensory replacement: (1) Sensory substitution is suitable even for patients suffering sensory loss because of cortical damage; and (2) because the interface with the substituting modality involves normal sensory stimulation, there are no problems associated with implanting electrodes. However, because the three spatial modalities of vision, hearing, and touch differ greatly in terms of their processing characteristics, the hope that one modality, aided by some single device, can simply assume all of the functions of another is untenable. Instead, a more reasonable expectation is that one modality can only substitute for another in performance of certain limited functions (e.g., reading of print, obstacle avoidance, speech reception). Indeed, research and development in the field of sensory substitution has largely proceeded with the idea of restoring specific functions rather than attempting to achieve wholesale substitution. A partial listing follows of the functions performed by vision and hearing, which are potential goals for sensory substitution:

- **Some functions of vision = potential goals for sensory substitution**
  - access to text (e.g., books, recipes, assembly instructions, etc.)
  - access to static graphs/pictures
  - access to dynamic graphs/pictures (e.g., animations, scientific visualization)
  - access to environmental information (e.g., business establishments and their locations)
  - obstacle avoidance
  - navigation to remote locations
  - controlling dynamic events in 3-D (e.g., driving, sports)
  - access to social signals (e.g., facial expressions, eye gaze, body gestures)
  - visual aesthetics (e.g., sunset, beauty of a face, visual art)
- **Some functions of audition = potential goals for sensory substitution**
  - access to signals and alarms (e.g., ringing phone, fire alarm)
  - access to natural sounds of the environment
  - access to denotative content of speech
  - access to expressive content of speech
  - aesthetic response to music

An analytic approach to using one sensory modality (henceforth, the "receiving modality") to take over a function normally performed by another is to (1) identify what optical, acoustic, or other information (henceforth, the "source information") is most effective in enabling that function and (2) to determine how to transform the source information into sensory signals that are effectively coupled to the receiving modality.

The first step requires research to identify the source information necessary to perform a function or range of functions. Take, for example, the function of obstacle avoidance. A person walking through a cluttered environment is able to avoid bumping into obstacles, usually by using vision under sufficient lighting. Precisely what visual information or other form of information (e.g., ultrasonic, radar) best affords obstacle avoidance? Once one has identified the best information to use, one is then in a position to address the second step.

Sensory Substitution: Coupling the Required Information to the Receiving Modality

Coupling the source information to the receiving modality actually involves two different issues: sensory bandwidth and the specificity of higher-level representation. After research has determined the information needed to perform a task, it must be determined whether the sensory bandwidth of the receiving modality is adequate to receive this information. Consider the idea of using the tactile sense to substitute for vision in the control of locomotion, such as driving. Physiological and psychophysical research reveals that the sensory bandwidth of vision is much greater than the bandwidth of the tactile sense for any circumscribed region of the skin (Loomis and Lederman 1986). Thus, regardless of how optical information is transformed for display onto the skin, it seems unlikely that the bandwidth of tactile processing is adequate to allow touch to substitute for this particular function. In contrast, other simpler functions, such as detecting the presence of a bright flashing alarm signal, can be feasibly accomplished using tactile substitution of vision.

Even if the receiving modality has adequate sensory bandwidth to accommodate the source information, this is no guarantee that sensory substitution will be successful, because the higher-level processes of vision, hearing, and touch are highly specialized for the information that typically comes through those modalities. A nice example of this is the difficulty of using vision to substitute for hearing in deaf people. Even though vision has greater sensory bandwidth than hearing, there is yet no successful way of using vision to substitute for hearing in the reception of the raw acoustic signal (in contrast to sign language, which involves the production of visual symbols by the speaker). Evidence of this is the enormous challenge in deciphering an utterance represented by a speech spectrogram. There is the celebrated case of Victor Zue, an engineering professor who is able to translate visual speech spectrograms into their linguistic descriptions. Although his skill is an impressive accomplishment, the important point here is that enormous effort is required to learn this skill, and decoding a spectrogram of a short utterance is very time-consuming. Thus, the difficulty of visually interpreting the acoustic speech signal suggests that presenting an isomorphic representation of the acoustic speech signal does not engage the visual system in a way that facilitates speech processing.

Presumably there are specialized mechanisms in the brain for extracting the invariant aspects of the acoustic signal; these invariant aspects are probably articulatory features, which bear a closer correspondence with the intended message. Evidence for this view is the relative success of the Tadoma method of speech reception (Reed et al. 1992). Some deaf-blind individuals are able to receive spoken utterances at nearly normal speech rates by placing a hand on the speaker's face. This direct contact with articulatory features is presumably what allows the sense of touch to substitute more effectively than visual reception of an isomorphic representation of the speech signal, despite the fact that touch has less sensory bandwidth than vision (Reed et al. 1992).

Although we now understand a great deal about the sensory processing of visual, auditory, and haptic perception, we still have much to learn about the perceptual/cognitive representations of the external world created by each of these senses and the cortical mechanisms that underlie these representations. Research in cognitive science and neuroscience will produce major advances in the understanding of these topics in the near future. Even now, we can identify some important research themes that are relevant to the issue of coupling information normally sensed by the impaired modality with the processing characteristics of the receiving modality.

Achieving Sensory Substitution Through Abstract Meaning

Prior to the widespread availability of digital computers, the primary approach to sensory substitution using electronic devices was to use analog hardware to map optical or acoustic information into one or isomorphic dimensions of the receiving modality (e.g., using video to sense print or other high contrast 2-D images and then displaying isomorphic tactile images onto the skin surface). The advent of the digital computer has changed all this, for it allows a great deal of signal processing of the source information prior to its display to the receiving modality. There is no longer the requirement that the displayed information be isomorphic to the information being sensed. Taken to the extreme, the computer can use artificial intelligence algorithms to extract the "meaning" of the optical, acoustic, or other information needed for performance of the desired function and then display this meaning by way of speech or abstract symbols.

One of the great success stories in sensory substitution is the development of text-to-speech devices for the visually impaired (Kurzweil 1989). Here, printed text is converted by optical character recognition into electronic text, which is then displayed to the user as synthesized speech. In a similar vein, automatic speech recognition and the visual display of text may someday provide deaf people with immediate access to the speech of any desired interactant. One can also imagine that artificial intelligence may someday provide visually impaired people with detailed verbal descriptions of objects and their layout in the surrounding environment. However, because inculcating such intelligence into machines has proven far more challenging than was imagined several decades ago, exploiting the intelligence of human users in the interpretation of sensory information will continue to be an important approach to sensory substitution. The remaining research themes deal with this more common approach.

Amodal Representations

For 3-D space perception (e.g., perception of distance) and spatial cognition (e.g., large-scale navigation), it is quite likely that vision, hearing, and touch all feed into a common area of the brain, like the parietal cortex, with the result that the perceptual representations created by these three modalities give rise to amodal representations. Thus, seeing an object, hearing it, or feeling it with a stick, may all result in the same abstract spatial representation of its location, provided that its perceived location is the same for the three senses. Once an amodal representation has been created, it then might be used to guide action or cognition in a manner that is independent of the sensory modality that gave rise to it (Loomis et al. 2002). To the extent that two sensory modalities do result in shared amodal representations, there is immediate potential for one modality substituting for the other with respect to functions that rely on the amodal representations. Indeed, as mentioned at the outset of this chapter, natural sensory substitution (using touch to find objects when vision is impaired) exploits this very fact. Clearly, however, an amodal representation of spatial layout derived from hearing may lack the detail and precision of one derived from vision because the initial perceptual representations differ in the same way as they do in natural sensory substitution.

Intermodal Equivalence: Isomorphic Perceptual Representations

Another natural basis for sensory substitution is isomorphism of the perceptual representations created by two senses. Under a range of conditions, visual and haptic perception result in nearly isomorphic perceptual representations of 2-D and 3-D shapes (Klatzky et al. 1993; Lakatos and Marks 1999; Loomis 1990; Loomis et al. 1991). The similar perceptual representations are probably the basis both for cross-modal integration, where two senses cooperate in sensing spatial features of an object (Ernst et al. 2001; Ernst and Banks 2002; Heller et al. 1999), and for the ease with which subjects can perform cross-modal matching, that is, feeling an object and then recognizing it visually (Abravanel 1971; Davidson et al. 1974). However, there are interesting differences between the visual and haptic representations of objects (e.g., Newell et al. 2001), differences that probably limit the degree of cross-modal transfer and integration. Although the literature on cross-modal integration and transfer involving vision, hearing, and touch goes back years, this is a topic that is receiving renewed attention (some key references: Ernst and Banks 2002; Driver and Spence 1999; Heller et al. 1999; Martino and Marks 2000; Massaro and Cohen 2000; Welch and Warren 1980).

Synesthesia

For a few rare individuals, synesthesia is a strong correlation between perceptual dimensions or features in one sensory modality with perceptual dimensions or features in another (Harrison and Baron-Cohen 1997; Martino and Marks 2001). For example, such an individual may imagine certain colors when hearing certain pitches, may see different letters as different colors, or may associate tactile textures with voices. Strong synesthesia in a few rare individuals cannot be the basis for sensory substitution; however, much milder forms in the larger population, indicating reliable associations between intermodal dimensions that may be the basis for cross-modal transfer (Martino and Marks 2000), might be exploited to produce more compatible mappings between the impaired and substiting modalities. For example, Meijer (1992) has developed a device that uses hearing to substitute for vision. Because the natural correspondence between pitch and elevation is space (e.g., high-pitched tones are associated with higher elevation), the device uses the pitch of a pure tone to represent the vertical dimension of a graph or picture. The horizontal dimension of a graph or picture is represented by time. Thus, a graph portraying a 45° diagonal straight line is experienced as a tone of increasing pitch as a function of time. Apparently, this device is successful for conveying simple 2-D patterns and graphs. However, it would seem that images of complex natural scenes would result in a cacophony of sound that would be difficult to interpret.

Multimodal Sensory Substitution

The discussion of sensory substitution so far has assumed that the source information needed to perform a function or functions is displayed to a single receiving modality, but clearly there may be value in using multiple receiving modalities. A nice example is the idea of using speech and audible signals together with force feedback and vibrotactile stimulation from a haptic mouse to allow visually impaired people to access information about 2-D graphs, maps, and pictures (Golledge 2002, this volume). Another aid for visually impaired people is the "Talking Signs" system of electronic signage (Crandall et al. 1993), which includes transmitters located at points of interest in the environment that transmit infrared signals carrying speech information about the points of interest. The user holds a small receiver in the hand that receives the infrared signal when pointed in the direction of the transmitter; the receiver then displays the speech utterance by means of a speaker or earphone. In order to localize the transmitter, the user rotates the receiver in the hand until receiving the maximum signal strength; thus, haptic information is used to orient toward the transmitter, and speech information conveys the identity of the point of interest.

Rote Learning Through Extensive Exposure

Even when there is neither the possibility of extracting meaning using artificial intelligence algorithms nor the possibility of mapping the source information in a natural way onto the receiving modality, effective sensory substitution is not completely ruled out. Because human beings, especially when they are young, have a large capacity for learning complex skills, there is always the possibility that they can learn mappings between two sensory modalities that differ greatly in their higher-level interpretative mechanisms (e.g., use of vision to apprehend complex auditory signals or of hearing to apprehend complex 2-D spatial images). As mentioned earlier, Meijer (1992) has developed a device (The vOICe) that converts 2-D spatial images into time-varying auditory signals. While based on the natural correspondence between pitch and height in a 2-D figure, it seems unlikely that the higher-level interpretive mechanisms of hearing are suited to handling complex 2-D spatial images usually associated with vision. Still, it is possible that if such a device were used by a blind person from very early in life, the person might develop the equivalent of rudimentary vision. On the other hand, the previously discussed example of the difficulty of visually interpreting speech spectrograms is a good reason not to base one's hope too much on this capacity for learning.

Brain Mechanisms Underlying Sensory Substitution and Cross-Modal Transfer

In connection with his seminal work with the Tactile Vision Substitution System, which used a video camera to drive an electrotactile display, Bach-y-Rita (1967, 1972) speculated that the functional substitution of vision by touch actually involved a reorganization of the brain, whereby the incoming somatosensory input came to be linked to and analyzed by visual cortical areas. Though a radical idea at the time, it has recently received confirmation by a variety of studies involving brain imaging and transcranial magnetic stimulation (TMS). For example, research has shown that (1) the visual cortex of skilled blind readers of braille is activated when they are reading braille (Sadata et al. 1996), (2) TMS delivered to the visual cortex can interfere with the perception of braille in similar subjects (Cohen et al. 1997), and (3) that the visual signals of American Sign Language activate the speech areas of deaf subjects (Neville et al. 1998).

Future Prospects for Sensory Replacement and Sensory Substitution

With the enormous increases in computing power, the miniaturization of electronic devices (nanotechnology), the improvement of techniques for interfacing electronic devices with biological tissue, and increased understanding of the sensory pathways, the prospects are great for significant advances in sensory replacement in the coming years. Similarly, there is reason for great optimism in the area of sensory substitution. As we come to understand the higher level functioning of the brain through cognitive science and neuroscience research, we will know better how to map source information into the remaining intact senses. Perhaps even more important will be breakthroughs in technology and artificial intelligence. For example, the emergence of new sensing technologies, as yet unknown, just as the Global Positioning System was unknown several decades ago, will undoubtedly provide blind and deaf people with access to new types of information about the world around them. Also, the increasing power of computers and increasing sophistication of artificial intelligence software will mean that computers will be increasingly able to use this sensed information to build representations of the environment, which in turn can be used to inform and guide visually impaired people using synthesized speech and spatial displays. Similarly, improved speech recognition and speech understanding will eventually provide deaf people better communication with others who speak the same or even different languages. Ultimately, sensory replacement and sensory substitution may permit people with sensory impairments to perform many activities that are unimaginable today and to enjoy a wide range of experiences that they are currently denied.

References

Abravanel, E. 1971. Active detection of solid-shape information by touch and vision. *Perception & Psychophysics*, 10, 358-360.

Bach-y-Rita, P. 1967. Sensory plasticity: Applications to a vision substitution system. *Acta Neurologica Scandanavica*, 43, 417-426.

Bach-y-Rita, P. 1972. *Brain mechanisms in sensory substitution*. New York: Academic Press.

Bliss, J.C., M.H. Katcher, C.H. Rogers, and R.P. Shepard. 1970. Optical-to-tactile image conversion for the blind. *IEEE Transactions on Man-Machine Systems*, MMS-11, 58-65.

Brabyn, J.A. 1985. A review of mobility aids and means of assessment. In *Electronic spatial sensing for the blind*, D.H. Warren and E.R. Strelow, eds. Boston: Martinus Nijhoff.

Cohen, L.G., P. Celnik, A. Pascual-Leone, B. Corwell, L. Faiz, J. Dambrosia, M. Honda, N. Sadato, C. Gerloff, M.D. Catala, and M. Hallett. 1997. Functional relevance of cross-modal plasticity in blind humans. *Nature*, 389: 180-183.

Collins, C.C. 1985. On mobility aids for the blind. In *Electronic spatial sensing for the blind*, D.H. Warren and E.R. Strelow, eds. Boston: Martinus Nijhoff.

Crandall, W., W. Gerrey, and A. Alden. 1993. Remote signage and its implications to print-handicapped travelers. *Proceedings: Rehabilitation Engineering Society of North America RESNA Annual Conference*, Las Vegas, June 12-17, 1993, pp. 251-253.

Davidson, P.W., S. Abbott, and J. Gershenfeld. 1974. Influence of exploration time on haptic and visual matching of complex shape. *Perception and Psychophysics*, 15 : 539-543.

Driver, J., and C. Spence. 1999. Cross-modal links in spatial attention. In *Attention, space, and action: Studies in cognitive neuroscience*, G.W. Humphreys and J. Duncan, eds. New York: Oxford University Press.

Ernst, M.O. and M.S. Banks. 2002. Humans integrate visual and haptic information in a statistically optimal fashion. *Nature* 415: 429 - 433.

Ernst, M.O., M.S. Banks, and H.H. Buelthoff. 2000. Touch can change visual slant perception. *Nature Neuroscience* 3: 69-73.

Golledge, R.G. 2002. Spatial cognition and converging technologies. This volume.

Harrison, J., and S. Baron-Cohen. 1997. Synaesthesia: An introduction. In *Synaesthesia: Classic and contemporary readings,* S. Baron-Cohen and J.E. Harrison eds. Malden, MA: Blackwell Publishers.

Heller, M.A., J.A. Calcaterra, S.L. Green, and L. Brown. 1999. Intersensory conflict between vision and touch: The response modality dominates when precise, attention-riveting judgments are required. *Perception and Psychophysics* 61: 1384-1398.

Humayun, M.S., and E.T. de Juan, Jr. 1998. Artificial vision. *Eye* 12: 605-607.

Jaffe, D.L. 1994. Evolution of mechanical fingerspelling hands for people who are deaf-blind. *Journal of Rehabilitation Research and Development* 3: 236-244.

Kay, L. 1985. Sensory aids to spatial perception for blind persons: Their design and evaluation. In *Electronic spatial sensing for the blind,* D.H. Warren and E.R. Strelow, eds. Boston: Martinus Nijhoff.

Keller, H. 1908. *The world I live in.* New York: The Century Co.

Klatzky, R.L., J.M. Loomis, S.J. Lederman, H. Wake, and N. Fujita. 1993. Haptic perception of objects and their depictions. *Perception and Psychophysics* 54 : 170-178.

Kurzweil, R. 1989. Beyond pattern recognition. *Byte* 14: 277.

Lakatos, S., and L.E. Marks. 1999. Haptic form perception: Relative salience of local and global features. *Perception and Psychophysics* 61: 895-908.

Loomis, J.M. 1990. A model of character recognition and legibility. *Journal of Experimental Psychology: Human Perception and Performance* 16: 106-120.

Loomis, J.M., R.G. Golledge, and R.L. Klatzky. 2001. GPS-based navigation systems for the visually impaired. In *Fundamentals of wearable computers and augmented reality,* W. Barfield and T. Caudell, eds. Mahwah, NJ: Lawrence Erlbaum Associates.

Loomis, J.M., R.L. Klatzky, and S.J. Lederman. 1991. Similarity of tactual and visual picture perception with limited field of view. *Perception* 20: 167-177.

Loomis, J.M., and S.J. Lederman. 1986. Tactual perception. In K. Boff, L. Kaufman, and J. Thomas (Eds.), *Handbook of perception and human performance: Vol. 2. Cognitive processes and performance* (pp. 31.1-31.41). New York: Wiley.

Loomis, J.M., Y. Lippa, R.L. Klatzky, and R.G. Golledge. 2002. Spatial updating of locations specified by 3-D sound and spatial language. *J. of Experimental Psychology: Learning, Memory, and Cognition* 28: 335-345.

Loughborough, W. 1979. Talking lights. *Journal of Visual Impairment and Blindness* 73: 243.

Martino, G., and L.E. Marks. 2000. Cross-modal interaction between vision and touch: The role of synesthetic correspondence. *Perception* 29: 745-754.

_____. 2001. Synesthesia: Strong and weak. *Current Directions in Psychological Science* 10: 61-65.

Massaro, D.W., and M.M. Cohen. 2000. Tests of auditory-visual integration efficiency within the framework of the fuzzy logical model of perception. *Journal of the Acoustical Society of America* 108: 784-789.

Meijer, P.B.L. 1992. An experimental system for auditory image representations. *IEEE Transactions on Biomedical Engineering* 39: 112-121.

Neville, H.J., D. Bavelier, D. Corina, J. Rauschecker, A. Karni, A. Lalwani, A. Braun, V. Clark, P. Jezzard, and R. Turner. 1998. Cerebral organization for language in deaf and hearing subjects: Biological constraints and effects of experience. Neuroimaging of Human Brain Function, May 29-31, 1997, Irvine, CA. *Proceedings of the National Academy of Sciences* 95: 922-929.

Newell, F.N., M.O. Ernst, B.S. Tjan, and H.H. Buelthoff. 2001. Viewpoint dependence in visual and haptic object recognition. *Psychological Science* 12: 37-42.

Niparko, J.K. 2000. *Cochlear implants: Principles and practices.* Philadelphia: Lippincott Williams & Wilkins.

Normann, R.A. 1995. Visual neuroprosthetics: Functional vision for the blind. *IEEE Engineering in Medicine and Biology Magazine* 77-83.

Pavel, M., G. Sperling, T. Riedl, and A. Vanderbeek. 1987. Limits of visual communication: The effect of signal-to-noise ratio on the intelligibility of American Sign Language. *Journal of the Optical Society of America, A* 4: 2355-2365.

Reed, C.M., L.A. Delhorne, N.I. Durlach, and S.D. Fischer. 1990. A study of the tactual and visual reception of fingerspelling. *Journal of Speech and Hearing Research* 33: 786-797.

Reed, C.M., W.M. Rabinowitz, N.I. Durlach, L.A. Delhorne, L.D. Braida, J.C. Pemberton, B.D. Mulcahey, and D.L. Washington. 1992. Analytic study of the Tadoma method: Improving performance through the use of supplementary tactual displays. *Journal of Speech and Hearing Research* 35: 450-465.

Sadato, N., A. Pascual-Leone, J. Grafman, V. Ibanez, M-P Deiber, G. Dold, and M. Hallett. 1996. Activation of the primary visual cortex by Braille reading in blind subjects. *Nature* 380: 526-528.

Tan, H.Z., W.M. Rabinowitz, and N.I. Durlach. 1989. Analysis of a synthetic Tadoma system as a multidimensional tactile display. *Journal of the Acoustical Society of America* 86: 981-988.

Valvo, A. 1971. *Sight restoration after long-term blindness: the problems and behavior patterns of visual rehabilitation,* L L. Clark and Z.Z. Jastrzembska, eds.. New York, American Foundation for the Blind.

Waltzman, S.B., and N.L. Cohen. 2000. *Cochlear implants.* New York: Thieme.

Weisenberger, J.M., S.M. Broadstone, and F.A. Saunders. 1989. Evaluation of two multichannel tactile aids for the hearing impaired. *Journal of the Acoustical Society of America* 86: 1764-1775.

Welch, R.B., and D.H. Warren. 1980. Immediate perceptual response to intersensory discrepancy. *Psychological Bulletin* 88: 638-667.

VISION STATEMENT: INTERACTING BRAIN

Britton Chance, University of Pennsylvania, and Kyung A. Kang, University of Louisville

Brain functional studies are currently performed by several instruments, most having limitations at this time. PET and SPECT use have labeled glucose as an indicator of metabolic activity; however, they may not be used within a short time interval and also can be expensive. MRI is a versatile brain imaging technique, but is highly unlikely to be "wearable." MEG is an interesting technology to measure axon-derived currents with a high accuracy at a reasonable speed; this still requires minimal external magnetic fields, and a triply shielded micro-metal cage is required for the entire subject. While thermography has some advantages, the penetration is very small, and the presence of overlying tissues is a great problem. Many brain responses during cognitive activities may be recognized in terms of changes in blood volume and oxygen saturation at the brain part responsible. Since hemoglobin is a natural and strong optical absorber, changes in this molecule can be monitored by the near infrared (NIR) detection method very effectively without applying external contrast agents (Chance, Kang, and Sevick 1993). NIR can monitor not only the blood volume changes (the variable that most of the currently used methods are measuring) but also hemoglobin saturation (the variable that provides the actual energy usage) (Chance, Kang, and Sevick 1993;Hoshe et al. 1994; Chance et al

1998). Among the several brain imagers, the "NIR Cognoscope" (Figure C.8) is one of a few that have wearability (Chance et al. 1993; Luo, nioka, and Chance 1996; Chance et al 1998). Also, with fluorescent-labeled neuroreceptors or metabolites (such as glucose), the optical method will have a similar capability for such metabolic activities as PET and SPECT (Kang et al. 1998).

Nanotechnology and information technology (IT) can be invaluable for the development of future optical cognitive instruments. Nano-biomarkers targeted for cerebral function representing biomolecules will enable us to pinpoint the areas responsible for various cognitive activities as well as to diagnose various brain disorders. Nano-sized sources and detectors operated by very long lasting nano-sized batteries will be also very useful for unobstructed studies of brain function. It is important to acknowledge that in the process of taking cognitive function measurements, the instrument itself or the person who conducts the measurements should not (or should minimally) interfere with or distract the subject's cognitive activities. The ultimate optical system for cognitive studies, therefore, requires wireless instrumentation.

It is envisioned that once nanotech and IT are fully incorporated into the optical instrumentation, the sensing unit will be very lightweight, disposable Band-aid™ sensor/detector applicators or hats (or helmets) having no external connection. Stimuli triggering various cognitive activities can be given through a computer screen or visor with incorporating a virtual reality environment. Signal acquisition will be accomplished by telemetry and will be analyzed in real time. The needed feedback stimulus can also be created, depending on the nature of the analysis needed for further tests or treatments. Some of the important future applications of the kind of "cognoscope" described above are as follows:

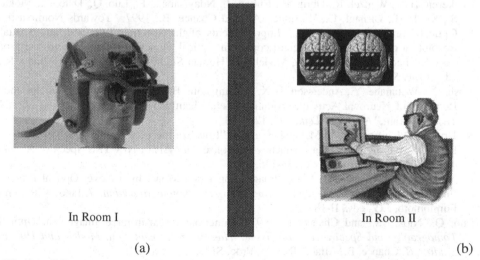

In Room I

In Room II

(a)   (b)

Figure C.8. A schematic diagram of the future NIR Cognosope. (a) A wireless, hat-like multiple source-detector system can be used for brain activities while the stimulus can be given though a visor-like interactive device. While a subject can be examined (or tested) in a room (room I) without any disturbance by examiners or other non-cognitive stimuli, the examiner can obtain the cognitive response through wireless transmission, can analyze the data in real-time, and also may be able to additional stimuli to the subjects for further tests, in another room (room II).

1. Medical diagnosis of brain diseases (Chance, Kang, and Sevick 1993)
2. Identification of children with learning disabilities (Chance et al. 1993; Hoshe et al. 1994; Chance et al. 1998)
3. Assessment of effectiveness in teaching techniques (Chance et al. 1993; Hoshe et al. 1994; Heekeren et al. 1997; Chance et al. 1998)
4. Applications for cognitive science — study of the thinking process (Chance et al. 1993; Hoshe et al. 1994; Chance et al. 1998)
5. Localization of brain sites responding to various stimuli (Gratton et al. 1995; Luo, Nioka, and Chance 19997; Heekeren et al. 1997; Villringer and Chance 1997)
6. Identification of the emotional state of a human being
7. Communicating with others without going through currently used sensory systems

References

Chance, B., Anday, E., Nioka, S., Zhou, S., Hong, L., Worden, K., Li, C., Overtsky, Y., Pidikiti, D., and Thomas, R., 1998. "A Novel Method for Fast Imaging of Brain Function, Noninvasively, with Light." Optical Express, 2(10): 411-423.

Chance, B., Kang, K.A., and Sevick, E., 1993. "Photon Diffusion in Breast and Brain: Spectroscopy and Imaging," *Optics and Photonics News*, 9-13.3.

Chance, B., Zhuang, Z., Chu, U., Alter, C., and Lipton, L., 1993. "Cognition Activated Low Frequency Modulation of Light Absorption in Human Brain," *PNAS*, 90: 2660-2774.

Gratton, G., Corballis, M., Cho, E., Gabiani, M., and Hood, D.C., 1995. "Shades of Gray Matter: Non-invasiveNoninvasive Optical Images of Human Brain Responses during Visual Stimulations," *Psychophysiology*, 32: 505-509.

Heekeren, H.R., Wenzel, R., Obrig, H., Ruben, J., Ndayisaba, J-P., Luo, Q., Dale, A., Nioka, S., Kohl, M., Dirnagl, U., Villringer, A., and Chance, B., 1997. "Towards Noninvasive Optical Human Brain Mapping - Improvements of the Spectral, Temporal, and Spatial Resolution of Near-infrared Spectroscopy," in Optical Tomography and Spectroscopy of Tissue: Theory, Instrumentation, Model, and Human Studies, II, Chance, B., Alfano, R., eds., *Proc. SPIE*, 2979: 847-857.

Hoshi, Y., Watanabe, Y., Andersson, J., X, X, Langstom, B., and X, 1994. "Non-synchronous Behavior of Neuronal Activity, Oxidative Metabolism and Blood Supply during Mental Tasks in Brain," *Neurosci. Lett.*, 197: 129-133.

X, Bruley, D.F., Londono, J.M., and X 1998. "Localization of a Fluorescent Object in a Highly Scattering Media via Frequency Response Analysis of NIR-TRS Spectra," *Annals of Biomedical Engineering*, 26:138-145.

X, Nioka, S., and Chance, B. 1996, "Imaging on Brain Model by a Novel Optical Probe - Fiber Hairbrush," in *Adv. Optical Imaging and Photon Migration*, Alfano, R.R., and Fumiomoto, J.G., eds., II-183-185.

Luo, Q., Nioka, S., and Chance, B. 1997. "Functional Near-infrared Image," in *Optical Tomography and Spectroscopy of Tissue: Theory, Instrumentation, Model, and Human Studies, II*, Chance, B., Alfano, R., eds., Proc. SPIE, 2979: 84-93.

Villringer, A., and Chance, B., 1997. "Noninvasive Optical Spectroscopy and Imaging of Human Brain Function," *Trends in Neuroscience*, 20: 435-442.

FOCUSING THE POSSIBILITIES OF NANOTECHNOLOGY FOR COGNITIVE EVOLUTION AND HUMAN PERFORMANCE

Edgar Garcia-Rill, University of Arkansas for Medical Sciences

Two statements are advanced in this paper:

1. Nanotechnology can help drive our cognitive evolution.
2. Nanotechnology applications can help us monitor distractibility and critical judgment, allowing unprecedented improvements in human performance.

The following will provide supporting arguments for these two positions, one general and one specific, regarding applications of nanotechnology for human performance. This vision and its transforming strategy will require the convergence of nanoscience, biotechnology, advanced computing, and principles in cognitive neuroscience.

Our Cognitive Evolution

How did the human brain acquire its incomparable power? Our species emerged less than 200,000 years ago, but it has no "new" modules compared to other primates. Our brains have retained vestiges from our evolutionary ancestors. The vertebrate (e.g., fish) nervous system is very old, and we have retained elements of the vertebrate brain, especially in the organization of spinal cord and brainstem systems. One radical change in evolution occurred in the transition from the aquatic to terrestrial environment. New "modules" arose to deal with the more complex needs of this environment in the form of the thalamic, basal ganglia, and cortical "modules" evident in the mammalian brain. The changes in brain structure between lower and higher mammals are related to size rather than to any novel structures. There was a dramatic growth in the size of the cerebral cortex between higher mammals and monkeys. But the difference between the monkey brain, the ape brain, and the human brain is again one of size. In comparing these three brains, we find that the size of the primary cortical areas (those dealing with sensory and motor functions) are similar in size, but in higher species, secondary and especially tertiary cortical areas (those dealing with higher-level processing of sensory and motor information) are the ones undergoing dramatic increases in size, especially in the human. That is, we have conserved a number of brain structures throughout evolution, but we seem to just have more of everything, especially cortex (Donald 1991).

In individuals, the factors that determine the anatomy of our cortex are genes, environment, and enculturation (Donald 1991). For instance, the structure of the basic computational unit of the cortex, the cortical column, is set genetically. However, the connectivity between cortical columns, which brings great computational power based on experience, is set by the environment, especially during critical stages in development. Moreover, the process of enculturation determines the plastic anatomical changes that allow entire circuits to be engaged in everyday human performance. This can be demonstrated experimentally. Genetic mutations lead to dramatic deficits in function; but if there is no genetic problem yet environmental exposure is prevented (such as covering the eyes during a critical period in development), lifelong deficits (blindness) result. If both genetic and environmental factors proceed normally but enculturation is withdrawn, symbolic skills and language fail to develop, with drastic effects.

The unprecedented growth of the cortex exposed to culture allowed us to develop more complex skills, language, and unmatched human performance. It is thought that it is our capacity to acquire symbolic skills that has led to our higher intelligence. Once we added symbols, alphabets, and mathematics, biological memory became inadequate for storing our collective knowledge. That is, the human mind became a "hybrid" structure built from vestiges of earlier biological stages, new evolutionarily-driven modules, and external (cultural "peripherals") symbolic memory devices (books, computers, etc.), which, in turn, have altered its organization, the way we "think" (Donald 1991). That is, just as we use our brain power to continue to develop technology, that technological enculturation has an impact on the way we process information, on the way our brain is shaped. This implies that we are more complex than any creatures before, and that we may not have yet reached our final evolutionary form. Since we are still evolving, the inescapable conclusion is that nanotechnology can help drive our evolution. This should be the charge to our nanoscientists: Develop nanoscale hybrid technology.

What kind of hybird structures should we develop? It is tempting to focus nanotechnology research on brain-machine integration, to develop *implantable* devices (rather than *peripheral* devices) to "optimize" detection, perception, and responsiveness, or to increase "computational power" or memory storage. If we can ever hope to do this, we need to know how the brain processes information. Recent progress in information processing in the brain sciences, in a sense, parallels that of advances in computation. According to Moore's Law, advances in hardware development enable a doubling of computing and storage power every 18 months, but this has not led to similar advances in software development, as faster computers seem to encourage less efficient software (Pollack 2002, this volume). Similarly, brain research has given us a wealth of information on the hardware of the brain, its anatomical connectivity and synaptic interactions, but this explosion of information has revealed little about the software the brain uses to process information and direct voluntary movement. Moreover, there is reason to believe that we tailor our software, developing more efficient "lines of code" as we grow and interact with the environment and culture. In neurobiological terms, the architecture of the brain is determined genetically, the connectivity pattern is set by experience, and we undergo plastic changes throughout our lives in the process of enculturation. Therefore, we need to hone our skills on the software of the brain.

What kind of software does the brain use? The brain does not work like a computer; it is not a digital device; it is an analog device. The majority of computations in the brain are performed in analog format, in the form of graded receptor and synaptic potentials, not all-or-none action potentials that, after all, end up inducing other grade potentials. Even groups of neurons, entire modules, and multi-module systems all generate waveforms of activity, from the 40 Hz rhythm thought to underlie binding of sensory events to slow potentials that may underlie long-term processes. Before we can ever hope to implant or drive machines at the macro, micro, or nano scale, the sciences of information technology and advanced computing need to sharpen our skills at analog computing. This should be the charge to our information technology colleagues: Develop analog computational software.

However, we do not have to wait until we make breakthroughs in that direction, because we can go ahead and develop nanoscale *peripherals* in the meantime.

Improving Human Performance

Sensory Gating

Human performance, being under direct control from the brain, is dependent on a pyramid of processes. Accurate human performance depends on practice gained from learning and memory, which in turn depend on selective attention to the performance of the task at hand, which in turn depends on "preattentional" arousal mechanisms that determine a level of attention (e.g., I need to be awake in order to pay attention). Human performance can be improved with training, which involves higher-level processes such as learning and memory. However, the most common factor leading to poor human performance is a lower-level process, lack of attention, or distractibility. Distractibility can result from fatigue, stress, and disease, to name a few. Is it possible to decrease the degree of distractibility, or at least to monitor the level of distractibility? Can nanotechnology provide a critical service in the crucial area of distractibility?

The National Research Council's Committee on Space Biology and Medicine (1998) has concluded,

> Cumulative stress has certain reliable effects, including psychophysiological changes related to alterations in the sympathetic-adrenal-medullary system and the hypothalamic-pituitary-adrenal axis (hormonal secretions, muscle tension, heart and respiration rate, gastrointestinal symptoms), subjective discomfort (anxiety; depression; changes in sleeping, eating and hygiene), interpersonal friction, and impairment of sustained cognitive functioning. *The person's appraisal of a feature of the environment as stressful and the extent to which he or she can cope with it are often more important than the objective characteristics of the threat.*

It is therefore critical to develop a *method for measuring our susceptibility under stress to respond inappropriately to features of the environment.* "Sensory gating" has been conceptualized as a critical function of the central nervous system to filter out extraneous background information and to focus attention on newer, more salient stimuli. By monitoring our sensory gating capability, our ability to appraise and filter out unwanted stimuli can be assessed, and the chances of successful *subsequent* task performance can be determined.

One proposed measure of sensory gating capability is the P50 potential. The P50 potential is a midlatency auditory evoked potential that is (a) rapidly habituating, (b) sleep state-dependent, and (c) generated in part by cholinergic elements of the Reticular Activating System (the RAS modulates sleep-wake states, arousal, and fight versus flight responses). Using a paired stimulus paradigm, sensory gating of the P50 potential has been found to be reduced in such disorders as anxiety disorder (especially post-traumatic stress disorder, PTSD), depression, and schizophrenia (Garcia-Rill 1997). Another "preattentional" measure, the startle response, could be used; however due to its marked habituation — measurement time is too prolonged (>20 min) — and because compliance using startling, loud stimuli could also be a problem, the use of the P50 potential is preferable.

Sensory gating deficits can be induced by stress and thus represent a serious impediment to proper performance under complex operational demands. *We propose the development of a nanoscale module designed for the use of the P50 potential as a measure of sensory gating (Figure C.9).*

A method to assess performance readiness could be used as a predictor of performance success, especially if it were noninvasive, reliable, and not time-consuming. If stress or other factors have produced decreased sensory gating, then remedial actions could be instituted to restore sensory gating to acceptable levels, e.g., coping strategies, relaxation techniques, pharmacotherapy. It should be noted that this technique also may be useful in detecting slowly developing (as a result of cumulative stress) chronic sensory gating deficits that could arise from clinical depression or anxiety disorder, in which case remedial actions may require psychopharmacological intervention with, for example, anxiolytics or antidepressants.

Implementation of this methodology would be limited to the ability to record midlatency auditory evoked responses in varied environments. The foreseen method of implementation would involve the use of an electronically shielded helmet (Figure C.9) containing the following: (1) P50 potential recording electrodes at the vertex, mastoids, and ground; (2) eye movement recording using a flip-down transparent screen to monitor the movements of one eye within acceptable limits that do not interfere with P50 potential acquisition; and (3) electrodes on the forehead to monitor muscle contractions that could interfere with P50 potential acquisition. The helmet would incorporate an audio stimulator for delivering click stimuli, operational amplifiers for the three measures, averaging software, wave detection software (not currently available), and simple computation and display on the flip-down screen of sensory gating as a percent. A high percentage compared to

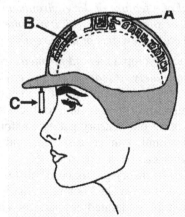

Figure C.9. Nanotechnology application: helmet incorporating P50 midlatency auditory evoked potential recording and near-infrared detection of frontal lobe blood flow to measure sensory gating and hypofrontality, respectively. A. Evoked potential module including audio stimulator (earphones), surface electrodes (vertex, mastoids, forehead), amplifiers, averager with wave recognition software, and data storage device for downloading. B. Near-infrared detection module for frontal lobe blood flow measurement. C. Flip-down screen for tracking eye movements and display of results from sensory gating and frontal blood flow measurements.

control conditions would be indicative of a lack of sensory gating (indicating increased distractibility, uncontrolled anxiety, etc.). An individual could don the helmet and obtain a measure of sensory gating *within 5-7 minutes.*

The applications for this nanotechnology would be considerable, including military uses for self-monitoring human performance in advance of and during critical maneuvers; for self-monitoring by astronauts on long-duration space missions; for pilots, drivers, and operators of sensitive and complex equipment, etc. It should be noted that this physiological measure can not be "faked" and is applicable across languages and cultures.

Hypofrontality

In general, the role of the frontal cortex is to control, through inhibition, those old parts of the brain we inherited from our early ancestors, the emotional brainstem (Damasio 1999). If the frontal cortex loses some of its inhibitory power, "primordial" behaviors are released. This can occur when the cortex suffers from decreased blood flow, known as "hypofrontality." Instinctive behaviors then can be released, including, in the extreme, exaggerated fight versus flight responses to misperceived threats, i.e., violent behavior in an attempt to attack or flee. "Hypofrontality" is evident in such disorders as schizophrenia, PTSD, and depression, as well as in neurodegenerative disorders like Alzheimer's and Huntington's diseases. Decreased frontal lobe blood flow can be induced by alcohol. Damage, decreased uptake of glucose, reduced blood flow, and reduced function have all been observed in the frontal cortex of violent individuals and murderers.

The proposed method described below could be used to detect preclinical dysfunction (i.e., could be used to *screen and select* crews for military or space travel operations); to determine individual performance under stress (i.e., could be used to *prospectively evaluate* individual performance in flight simulation/virtual emergency conditions); and to monitor the effects of chronic stressors (i.e., monitor sensory gating during long-duration missions). This nanomethodology would be virtually realtime; would not require invasive measures (such as sampling blood levels of cortisol, which are difficult to carry out accurately and are variable and delayed rather than predictive); and would be more reliable than, for example, urine cortisol levels (which would be delayed or could be compensated for during chronic stress). Training in individual and communal coping strategies is crucial for alleviating some of the sequelae of chronic stress, and the degree of effectiveness of these strategies could be *quantitatively assessed* using sensory gating of the P50 potential as well as frontal lobe blood flow. That is, these measures could be used to determine the efficacy of any therapeutic strategy, i.e., to measure outcome.

A detecting module located over frontal areas with a display on the flip-down screen could be incorporated in the helmet to provide a noninvasive measure of frontal lobe blood flow for self-monitoring in advance of critical maneuvers. The potential nanotechnology involved in such measures has already been addressed (Chance and Kang n.d.). Briefly, since hemoglobin is a strong absorber, changes in this molecule could be monitored using near-infrared detection. This promising field has the potential for monitoring changes in blood flow as well as hemoglobin saturation, a measure of energy usage.

Peripheral nanotechnology applications such as P50 potential recordings and frontal blood flow measures are likely to provide proximal, efficient, and useful improvements in human performance. Nanotechnology, by being transparently integrated into our executive functions, will become part of the enculturation process, modulating brain structure and driving our evolution.

References

Chance, B., Kang, K. 2002. Optical identification of cognitive state. *Converging technology (NBIC) for improving human performance* (this volume).
Damasio, A. 1999. *The Feeling of What Happens, Body and Emotion in the Making of Consciousness*, Harcourt Brace & Co., New York, NY.
Donald, M.W. 1991. *Origins of the Modern Mind, Three Stages in the Evolution of Culture and Cognition*, Harvard University Press, Cambridge, MA.
Garcia-Rill, E. 1997. Disorders of the Reticular Activating System. *Med. Hypoth.* 49, 379-387.
National Research Council Committee on Space Biology and Medicine. 1998. *Strategy for Research in Space Biology and Medicine into the Next Century*. National Academy Press: Washington, DC.
Pollack, J. 2002. The limits of design complexity. *Converging technology (NBIC) for improving human performance* (this volume).

SCIENCE AND TECHNOLOGY AND THE TRIPLE D (DISEASE, DISABILITY, DEFECT)

Gregor Wolbring, University of Calgary

Science and technology (S&T) have had throughout history — and will have in the future — positive and negative consequences for humankind. S&T is not developed and used in a value neutral environment. S&T activity is the result of human activity imbued with intention and purpose and embodying the perspectives, purposes, prejudice and particular objectives of any given society in which the research takes place. S&T is developed within the cultural, economical, ethical, and moral framework of the society in which the research takes place. Furthermore, the results of S&T are used in many different societies reflecting many different cultural, economical, ethical, moral frameworks. I will focus on the field of Bio/Gene/Nanomedicine. The development of Bio/Gene/Nanotechnology is — among other things — justified with the argument that it holds the promises to fix or help to fix perceived disabilities, impairments, diseases, and defects and to diminish suffering. But who decides what is a disability, disease, an impairment and a 'defect' in need of fixing? Who decides what the mode of fixing (medical or societal) should be, and who decides what is suffering? How will these developments affect societal structures?

Perception

The right answers to these questions will help ensure that these technologies will enhance human life creatively, rather than locking us into the prejudices and misconceptions of the past. Consider the following examples of blatant insensitivity:

> Fortunately the Air Dri-Goat features a patented goat-like outer sole
> for increased traction so you can taunt mortal injury without actually
> experiencing it. Right about now you're probably asking yourself

"How can a trail running shoe with an outer sole designed like a goat's hoof help me avoid compressing my spinal cord into a Slinky on the side of some unsuspecting conifer, thereby rendering me a drooling, misshapen non-extreme-trail-running husk of my former self, forced to roam the earth in a motorized wheelchair with my name embossed on one of those cute little license plates you get at carnivals or state fairs, fastened to the back?" (Nike advertisement, Backpacker Magazine, October 2000).

Is it more likely for such children to fall behind in society or will they through such afflictions develop the strengths of character and fortitude that lead to the head of their packs? Here I'm afraid that the word handicap cannot escape its true definition — being placed at a disadvantage. From this perspective seeing the bright side of being handicapped is like praising the virtues of extreme poverty. To be sure, there are many individuals who rise out of its inherently degrading states. But we perhaps most realistically should see it as the major origin of asocial behavior (Watson 1996).

American bioethicist Arthur Caplan said in regards to human genetic technology, "the understanding that our society or others have of the concept of health, disease, and normality will play a key role in shaping the application of emerging knowledge about human genetics" (Caplan 1992). I would add Nanomedicine/Nanotechnology into Caplan's quote because parts of nanotechnology development are inherently linked with bio/genetechnology as the following quote from a recent report on its societal implications illustrates:

Recent insights into the uses of nanofabricated devices and systems suggest that today's laborious process of genome sequencing and detecting the genes' expression can be made dramatically more efficient through use of nanofabricated surfaces and devices. Expanding our ability to characterize an individual's genetic makeup will revolutionize diagnostics and therapeutics (Roco and Bainbridge 2001).

In addition, nanomedicine and nanotechnologies must be added, to quote the report again, because they

...hold promise for contributing to a wide range of assistive solutions, from prosthetic limbs that adjust to the changes in the body, to more biocompatible implants, to artificial retinas or ears. Other opportunities lie in the area of neural prosthesis and the "spinal patch," a device envisioned to repair damage from spinal injuries (Roco and Bainbridge 2001).

Any of these solutions are linked to the normalcy concept, the ability concept, and to the perceptions of what needs to be assisted. Certainly, different responses will be made and different solutions will be sought depending on how the problem is defined; and how the problem will be defined depends on our concepts of and beliefs about such things as health, disease, disability, impairment, and defect. For example, whether being gay is seen as a disease and defect (medical model) or a variation of human diversity (social model) will lead to totally different intervention scenarios (medical cure versus social cure). In the same way, what if we would view women as a double X syndrome, or men as an XY syndrome?

In essence every biological reality can be shaped and seen as a defect, as a medical problem, or as a human rights and social problem. No one views nowadays — in western culture at least — the biological reality of being a women within a medical framework, although a women was still viewed at the end of last century in countries like the United Kingdom as too biologically fragile and emotional and thus too dependent, to bear the responsibility attached to voting, owning property, and retaining custody of their own children (Silvers et. al., 1998). Therefore, a societal cure of equal rights and respect is seen as the appropriate remedy for the existing disparity between women and men. Gays, lesbians, bisexuals, and other groups demand that their problems be seen within a social framework and not within a medical framework.

So what now about so-called disabled people? Are "disabled people" or differently said "people who do not fit society's expectation of normal ability" to be seen as a medical problem or as part of the diversity of humankind? Within the medical model, disability is viewed as a defect, a problem inherent in the person, directly caused by disease, trauma, or other health condition and a deviation from certain norms. Management of the disability of the disabled person or person-to-be is aimed at cure, prevention, or adaptation of the person (e.g. using assistive devices). Medical care and rehabilitation are viewed as the primary issues, and at the political level, the principal response is that of modifying or reforming health care policy.

The social model of disability, on the other hand, sees the issue mainly as a socially created problem and principally as a matter of the full integration of individuals into society. Disability is not an attribute of an individual, but rather a complex collection of conditions, many of which are created by the environment, particularly the social environment and socially mediated aspects of the physical environment. Hence, the management of the problem requires social action, and it is the collective responsibility of society at large to make the environmental modifications necessary for the full participation of people with disabilities in all areas of social life. The issue is therefore an attitudinal or ideological one requiring social change, which at the political level becomes a question of human rights to be seen in the same way as the issues of gender and sexual orientation. In essence able-ism is seen in the same light as racism, sexism, age-ism, homophobia, etc.

The social model of disability does not negate that a disabled person has a certain biological reality (like having no legs), which makes her/him different in her/his abilities, which make her/him not fit the norm. But it views the "need to fit a norm" as the disability and questions whether many deviations from the norm need a medical solution (adherence to the norm) or a social solution (change/elimination of norm).

Many bio/gene/nano technology applications (predictive testing, cures, adaptation) focus on the individual and his or her perceived shortcomings. They follow a medical, not a social evaluation of a characteristic (biological reality) and therefore offer only medical solutions (prevention or cure/adaptation) and no social solutions (acceptance, societal cures of equal rights and respect).

Furthermore the use and development focus of bio/gene/nanotechnology as it is perpetuates the medical, intrinsic, individualistic, defect view of disability. Not often discussed by clinicians, academics in general, or the general public is the view, commonly expressed by disabled people, that the demand for the technology is based too much on the medical model of disability and hardly acknowledges the social model of disability (Asch 1999, Miringoff 1991; Hubbard 1990: Lippman 1991; Field 1993; Fine & Asch 1982; Minden 1984; Finger 1987; Kaplan 1994; Asch 1989; Asch and Geller 1996).

The perception of disabled people as suffering entities with a poor quality of life, in need of cure and fixing, for the most part does not fit with the perceptions disabled people have of themselves. This fact is illustrated by Table C.5, which compares self esteem of people having spinal cord injury with the images many nondisabled people have of what this hypothetically would mean for themselves.

Table C.5: Self-esteem ratings following severe spinal cord injury (SCI)

Percent agreeing with each statement	Nondisabled Respondents	Nondisabled Respondents Imagining Self with SCI	SCI Survivors Comparison Group
I feel that I am a person of worth.	98%	55%	95%
I feel that I have a number of good qualities.	98%	81%	98%
I take a positive attitude.	96%	57%	91%
I am satisfied with myself on the whole.	95%	39%	72%
I am inclined to feel that I am a failure.	5%	27%	9%
I feel that I do not have much to be proud of.	6%	33%	12%
I feel useless at times.	50%	91%	73%
At times I feel I am no good at all.	26%	83%	39%

Clearly, most people with spinal cord injury have positive self-images, but nondisabled people have the false impression that a person with this injury would lack self-esteem. This table was adapted from Gerhart et al., 1994, but many other studies report similar findings (Cameron 1973; Woodrich and Patterson 1983; Ray and West 1984; Stensman 1985; Bach and Tilton 1994; Cushman and Dijkers 1990; Whiteneck et al. 1985; Eisenberg and Saltz 1991; Saigal et al. 1996 Tyson and Broyles 1996; Cooley et al. 1990).

The following passage provides an example of how many professionals view the effects of people with disabilities on their families.

> How did parents endure the shock [the birth of a thalidomide baby]?
> The few who made it through without enormous collateral damage to
> their lives had to summon up the same enormous reserves of courage
> and devotion that are necessary to all parents of children with special
> needs and disabilities; then, perhaps, they needed still more courage,
> because of the special, peculiar horror that the sight of their children

produced in even the most compassionate. Society does not reward
such courage... because [what] those parents experience represents
our own worst nightmare, ever since we first imagined becoming
parents ourselves. The impact upon the brothers and sisters of the
newborn was no less horrific. This was the defining ordeal of their
family life — leaving aside for now the crushing burden on their
financial resources from now on (Stephens and Brynner 2001).

While such negative views of the impact of children with disabilities on their
families have dominated clinical and research literature for decades, more recent
research has exposed these negative biases as empirically unsupportable and
clinically destructive (e.g., Helf and Glidden, 1998; Sobsey, 1990). Contemporary
research suggests that parents, like people with disabilities, do not view their
children with disabilities as their "worst nightmares," as sources of "peculiar horror"
or as "crushing burdens." In fact, most view them very much as they view children
without disabilities, as sources of significant demands but even greater rewards (e.g.,
Sobsey & Scorgie 2001). Yet, people with disabilities and their families are a part of
society and they can never be entirely free of the attitudes, beliefs, and biases held
by professionals and the general public.

Such attitudes and beliefs about disability contribute to the drive to fix people
with disabilities rather than accommodate them. For example, the quote from
Stephens and Brynner seems to suggest:

1. an implicit assumption of normalcy which requires two legs and two arms
2. an expectation that everyone has to be able to perform certain functions (e.g.,
    move from one place to another or eat)
3. an expectation that everyone has to perform this function in a the same way
    (e.g., walking upright on their own legs or eat with their hands)
4. an expectation that any variation in form, function, method will result in
    severe emotional distress for those involved in any way

These attitudes drive the development of artificial legs and arms and help to
explain why thalidomide kids and their parents were confronted with the single-
minded approach to outfit thalidomide kids with artificial limbs without exploring
different forms of functioning. Professionals typically persisted with this approach
in spite of the fact that artificial limbs were rather crude, not very functional, and
mostly cosmetic at the time and that they were being prescribed in great numbers.
The approach nearly completely excluded alternatives, such as crawling in the
absence of legs or eating with one's feet in the absence of arms. The sentiment
expressed by Stephens and Brynner also prevents adaptation by society to
alternative modes of function (e.g., moving and eating).

This kind of single-minded approach reflects an adherence to a certain norm,
which was more readily accepted by amputees who lost their arms or legs. They
were or are willing to accept this because in a large part due to the fact that they
were not allowed to adapt and get used to their new condition, a process that we all
know takes time. People take time to adapt to any change. Humankind is not known
for its ability to adapt easily to changes (e.g., divorce, career changes).
Thalidomiders did not have to readapt to a new body reality. That might explain
why most Thalidomiders threw away their artificial legs and arms as soon as they

were old enough to assert themselves against their parents and the medical profession. For them the reality was that they did not view their body as deficient and did not see artificial legs or arms as the most suitable mode of action. In light of the perception reflected in the Stephens and Brynner's quote, the question becomes whether the development of usable artificial legs and arms mean that someone without legs or arms will be even more stigmatized if he or she does not use them. If so, the presence of this option is not merely another free choice since existence of the option results in a coercive influence on those who might refuse it.

Choice

The question arises whether usable artificial limbs increase choice as an optional tool or establish a norm that restricts choice. Parents of Thalidomiders were not given much choice. Immense pressure was used to have the parents equip their kids with artificial limbs. Society already judges certain tools. A hierarchy regarding movement exists. Crawling is on the bottom of the acceptance list, below the wheelchair, which is seen as inferior to the artificial leg, particularly one that appears "natural." This hierarchy is not based on functionality for the person but rather on emotions, prejudice, and rigid adherence to a normative body movement. Tools like the wheelchair are frequently demonized in expressions such as "confined to the wheelchair." It is interesting that people do not say "confined to" artificial legs even though a wheelchair often leads to safer, easier, and more efficient mobility for an individual than artificial legs do. No one would use the phrase "confined to natural legs" for "normal" people, although in reality they are confined to their legs while many wheelchair users can leave their wheelchairs. Similarly, the negative concept of confinement is not used to describe driving a car, which is viewed as empowering rather than limiting, even though many of us are heavily dependent on this mode of transportation. In much the same way, most of us who live in the north would not survive a single winter without central heating, but we generally do not label ourselves as "technology dependent."

Cochlear implants provide another related example. Do we allow parents to say "No" to them if they feel there is nothing wrong with their kid using sign language, lip reading, or other modes of hearing? Will the refusal by the parents be viewed as child abuse (see Harris, 2000 for an ethical argument to view it as child abuse)? Might parents have been considered to commit child abuse if they had refused artificial limbs for their Thalidomide kids? Or in today's world, could a mother be considered to commit child abuse if she refused to terminate her pregnancy after ultrasound showed phocomelia (i.e., hands and feet attached close to the body without arms or legs) in the fetus. Of course, ultrasound wasn't an option when most of the Thalidomide cases occurred, but it is today. Furthermore, would the mother abuse society by not fixing (cure, adaptation, prevention) the "problem"?

A hint to the answer to these questions is given by the following results of a survey of genetic counselors in different countries (Wertz 1998):

> The majority in 24 countries believed it is unfair to the child to be
> born with a disability. 40% agreed in the USA, Canada, and Chile.
> 36% in Finland and the UK; 33% in Switzerland and the Netherlands;
> 29% in Argentina, 27% in Australia, 25% in Sweden, and 18% in
> Japan.

It is socially irresponsible knowingly to bring an infant with a serious [no legal document defines what is serious] genetic disorder into the world in an era of prenatal diagnosis." More than 50% agreed in South Africa, Belgium, Greece, Portugal, the Czech Republic, Hungary, Poland, Russia, Israel, Turkey, China, India, Thailand, Brazil, Columbia, Cuba, Mexico, Peru, and Venezuela. 26% of U.S. geneticists, 55% of U.S. primary care physicians, and 44% of U.S. patients agreed.

A high percentage of genetic counselors feels that societies will never provide enough support for people with disabilities. The percentage of agreement for the statement ranges from 18% as a lowest to 80% in the U.K. Germany is in the middle with 57%. The U.S.A. has a number of 65%.

These statements suggest that women don't have a free choice but are led to follow the path of medical intervention. In the absence of a possible social cure for disability, the only option left that may appear to be available is the medical cure in whatever shape and form, independent of its usefulness and need.

The treatment of Thalidomiders, the pressure to install cochlear implants, and prebirth counseling raise a more general question about whether advances in a wide range of assistive devices, partly due to advances in micro- and nanotechnologies, will lead to increased or restricted choices. We can hope that technological convergence offers humanity so many choices that false stereotypes about the disabled are discredited once and for all. But this can happen only if we recognize the alternatives as real choices that must be considered with sensitivity, imagination, and — most importantly — the judgment of disabled people themselves.

Consequences

The history of the debate around bio/gene/nano-technology as it relates to disability shows a strong bias towards a medical, individualistic, intrinsic defect view of disability focusing on medical/technological cures without addressing societal components. People who promote the use of bio/genetechnology often denounce the social model of disability (Harris 2000; Singer 2001).

The medical model of disability can also show itself in court rulings, such as some recent U.S. Supreme Court rulings. The Supreme Court ruled on the "definition of disability" in Sutton v. United Airlines (130 F.3d 893, 119 S. Ct. 2139), Albertsons Inc. v. Kirkingburg (143 F.3d 1228, 119 S. Ct. 2162), and Murphy v. United Parcel (141 F.3d 1185, 119 S. Ct. 1331), stating that the Americans with Disabilities Act does not cover those persons with correctable impairments.[2]   In other words, as soon as adaptations are available, all problems must be fixed and no protections through civil rights laws, such as the ADA, are allowed anymore. Not only that, the ruling implies that disability is something that can be fixed through medical technological means. A social view of disability does not fit with the above ruling.

---

[2] National Council on Disability USA, 2000; Civil Rights, Sutton v. United Airlines, Albertsons Inc. v. Kirkingburg, and Murphy v. United Parcel (http://www.ncd.gov/newsroom/publications/policy98-99.html#1).

We see a disenfranchisement of disabled people from the equality/human rights movement. (Wolbring 1999, 2000, and 2001). So far, bio/genetechnology has led to an increase in discrimination against characteristics labeled as disabilities, as the following three examples illustrate.

First, we see a proliferation of legal cases involving wrongful life or wrongful birth suits (Wolbring, 2001,2002a). Wrongful life suits are only accepted if the child is disabled. And wrongful birth suits are specific by now for disability with special rulings whereas cases based on non-disability are called wrongful pregnancy. The remedies in the case of wrongful birth/pregnancy cases are quite different. The following quotations illustrate the logic of such cases.

> Two other justices based their agreement of wrongful life suits on the view that the physician's wrongful life liability towards the disabled infant rests on the right to life without a handicap. Thus the damage is measured by comparing the actual impaired life of the plaintiff to a hypothetical unimpaired life (CA 518, 540, 82 Zeitzoff versus Katz (1986) 40 (2) PD 85 Supreme Court of Israel (482); Shapiro 1998).

> ...in essence ... that [defendants] through their negligence, [have] forced upon [the child] the worse of ... two alternatives, ... that nonexistence — never being born — would have been preferable to existence in the diseased state (Soeck v. Finegold, 408 A.2d 496(Pa. 1970)).

> "Thus the legislature has recognized," the judge said, "as do most reasonable people, that cases exist where it is in the interest of the parents, family, and possibly society that it is better not to allow a fetus to develop into a seriously defective person causing serious financial and emotional problems to those who are responsible for such person's maintenance and well-being (Strauss 1996).

Second, anti-genetic discrimination laws cover discrimination on genetic characteristics which might lead in the future to 'disabilities' in a medical sense but are for the time being asymptomatic. In essence, the feature of genetic discrimination is the use of genetic information about an asymptomatic disabled person. The vogue for the establishment of an Anti-Genetic Discrimination law for asymptomatic disabled people highlights one other reality, namely that symptomatic disabled people are excluded from exactly the benefits the Anti-Genetic Discrimination laws try to address. With the new laws these symptomatic disabled people will still be discriminated against whereas the asymptomatic ones will be safe. Not only that, ability becomes a measure to justify these new laws, as the following statement from the American Civil Liberties Union illustrates.

> The ACLU believes that Congress should take immediate steps to protect genetic privacy for three reasons. First, it is inherently unfair to discriminate against someone based on immutable characteristics that do not limit their abilities... (ACLU 2000)

> In sum, the ACLU believes that Americans should be judged on their actual abilities, not their potential disabilities. No American should lose a job or an insurance policy based on his or her genetic predisposition. (ACLU 2000)

A third consequence of the current mindset is differential use of predictive genetic testing. We see an Animal Farm Philosophy in regards to what to test for. Testing to eliminate any so-called disability, disease, defect is acceptable; but testing to determine and select on the basis of a characteristic like sex is not (Wolbring 2000, 2001).

Where should we go from here? To prevent further stigmatization, recommendations such as those quoted below from the UNESCO World Conference on Sciences 1999 conference should be implemented.

> 25. ...that there are barriers which have precluded the full participation of other groups, of both sexes, including disabled people, indigenous peoples and ethnic minorities, hereafter referred to as "disadvantaged groups..."

> 42. Equality in access to science is not only a social and ethical requirement for human development, but also a necessity for realizing the full potential of scientific communities worldwide and for orienting scientific progress towards meeting the needs of humankind. The difficulties encountered by women, constituting over half of the population in the world, in entering, pursuing, and advancing in a career in the sciences and in participating in decision-making in science and technology should be addressed urgently. There is an equally urgent need to address the difficulties faced by disadvantaged groups, which preclude their full and effective participation.

Thus, it is essential that the greatest possible diversity of people participate in the development of convergent technologies and contribute to the associated sciences:

> 17. Scientists, research institutions, and learned scientific societies and other relevant non-governmental organizations should commit themselves to increased international collaboration including exchange of knowledge and expertise. Initiatives to facilitate access to scientific information sources by scientists and institutions in the developing countries should be especially encouraged and supported. Initiatives to fully incorporate women scientists and other disadvantaged groups from the South and North into scientific networks should be implemented. In this context efforts should be made to ensure that results of publicly funded research will be made accessible.

> 79. The full participation of disadvantaged groups in all aspects of research activities, including the development of policy, also needs to be ensured.

> 81. Governments and educational institutions should identify and eliminate, from the early learning stages, educational practices that have a discriminatory effect, so as to increase the successful participation in science of individuals from all sectors of society, including disadvantaged groups.

91. Special efforts also need to be made to ensure the full
participation of disadvantaged groups in science and technology, such
efforts to include:

> removing barriers in the education system;
>
> removing barriers in the research system;
>
> raising awareness of the contribution of these groups to
> science and technology in order to
>
> overcome existing stereotypes;
>
> undertaking research, supported by the collection of data,
>
> documenting constraints;
>
> monitoring implementation and documenting best
> practices;
>
> ensuring representation in policy-making bodies and
> forums (UNESCO 2000)

We should strive to eliminate able-ism and promote the acceptance of diversity in abilities for the sake of humankind as the best defense against gene-ism, which might affect 60 percent of society according to a New Zealand study. This acceptance of diverse abilities is actually also needed for the thriving of assistive technologies. For example, if an assistive technology leads to better vision than humankind has normally, should we discard the now majority of people who are less able? Or should we force all to use the new adaptive devices? Or should we demonize the ones who are more able?

The labeling of people and groups within a medical disease defect model against their will is unacceptable. In essence every scientist whose work has societal consequences has to become a societal activist to prevent these consequences.

Conclusion

The views expressed here are not opposed to progress in science and technology. As a lab bench biochemist, it would be strange for me to oppose S&T in general. Rather, this essay emphasizes the importance of openness to different perspectives on what qualifies as progress (Wolbring, 2002b). Science and technology can be extremely useful, but certain perceptions, stereotypes, and societal dynamics can lead scientists and engineers to focus on certain types of S&T, quite apart from their objective utility to potential users.

This is not merely an issue of fairness to diverse groups of people, including the disabled. It is also an issue of imagination and insight. Convergent technologies will accomplish much more for humanity, and unification of science will lead to much greater knowledge, if they are free of the ignorant prejudices of the past. Specifically, science and engineering will benefit from the varied perspectives that the disabled may have about what it means to improve human performance. One essential tool to achieve this is to make sure that the teams of researchers, designers, and policy makers include many talented people who happen to be disabled.

References

ACLU. 2000. http://www.aclu.org/congress/rightgenetics.html and July 19, 2000, Letter to The Honorable Edward M. Kennedy. Ranking Member Committee on Health, Education, Labor & Pensions. 428 Dirksen Senate Office Building. Washington, D.C. 20510. http://www.aclu.org/congress/l071900a.html.

Asch, A. 1999. Prenatal diagnosis and selective abortion: A challenge to practice and policy. *American Journal of Public Health.* 89, 11, 1649.

Asch A. 1989. Reproductive technology and disability. In Cohen S, Taub N. *Reproductive laws for the 1990s.* Clifton, NJ: Humana Press. 69_124.

Asch A, Geller G. 1996. Feminism, bioethics and genetics. In: Wolf S, ed. *Feminism and bioethics: Beyond reproduction.* New York, NY. Oxford University Press. 318-350.

Bach, J.R. and Tilton, M.C. 1994. Life satisfaction and well-being measures in ventilator assisted individuals with traumatic tetraplegia. *Archives of Physical Medicine and Rehabilitation.* vol. 75, 626-632.

Cameron, P. et al. 1973. The life satisfaction of nonnormal persons, *Journal of Consulting and Clinical Psychology.* vol. 41, 207-214.

Caplan, Arthur L. 1992. If gene therapy is the cure, what is the disease? in Annas, George J.; Elias, Sherman, eds. *Gene mapping: Using law and ethics as guides.* New York. Oxford University Press. page 128-14.

Cooley, W.C. et al. 1990, Reactions of mothers and medical profession to a film about Down Syndrome *Am. J. Dis. Child.* 144 1112.

Cushman, L.A. and Dijkers, M.P. 1990. Depressed mood in spinal cord injured patients: staff perceptions and patient realities. *Archives of Physical Medicine and Rehabilitation.* vol. 71. 191-196.

Eisenberg M.G. and Saltz. C.C. 1991. Quality of life among aging spinal cord injured persons: long term rehabilitation outcomes. *Paraplegia.* vol. 29, 514-520.

Field, N.I.A. 1993. Killing "the handicapped" before and after birth. *Harvard Women's Law J* 16:79-138.

Fine M., Asch A. 1982. The question of disability: no easy answers for the women's movement. *Reproductive Rights Newsletter.* 4(3). 19-20.

Finger, A. 1987. *Past due: Disability, pregnancy and birth.* Seattle. Washington Seal Press.

Kaplan D. 1994. Prenatal screening and diagnosis: the impact on persons with disabilities. In Rosenberg KHZ, Thompson JED, eds. *Women and prenatal testing: Facing the challenges of genetic technology.* Columbus. Ohio State University Press. 49-61.

Gerhart, K.A. et al. 1994. Quality of life following spinal cord injury. Knowledge and attitudes of emergency care providers. *Annals of Emergency Medicine.* vol. 23. 807-812.

Harris, J. 2000. Is there a coherent social conception of disability? *J. of Medical Ethics* 26. pp. 95-100.

Helf, C.M., and Glidden, L.M. 1998. More positive or less negative? Trends in research on adjustment of families rearing children with developmental disabilities. *Mental Retardation,* 36. 457-464.

Hubbard, R. 1990. *The politics of women in biology.* New Brunswick, NJ. Rutgers University Press. chap 12-14.

Lippman, A. 1991. Prenatal genetic testing and screening: constructing needs and reinforcing inequities. *Am J Law Med.* 17(1_2): 15_50;

Minden, S. 1984. Born and unborn: the implications of reproductive technologies for people with disabilities. In: Aridity R, Duello-Klein R, Minding S, eds. *Test-tube women: What future for motherhood?* Boston, Mass. Pandora Press. 298-312.

Miringoff, M.L. 1991. *The social costs of genetic welfare.* New Brunswick, NJ: Rutgers University Press.

Ray, C., West, J. 1984. Social, sexual and personal implications of paraplegia. *Paraplegia.* 22:75-86.

Roco, M.C., and W.S. Bainbridge (eds). 2001. *Societal Implications of Nanoscience and Nanotechnology*. Dordrecht, Netherlands: Kluwer (also available at http://nano.gov).

Saigal, S., et al.; 1996. Self-perceived health status and health-related quality of life of extremely low-birth-weight infants at adolescence. *JAMA*. 276: 453-459.

Silver, A. et al. 1998. Disability, Difference, Discrimination: Perspective on justice in bioethics and public policy. Rowman & Littlefield Publishers, INC. Landham, Bolder, New York, Oxford.

Scorgie, K., and Sobsey, D. 2000. Transformational outcomes associated with parenting children with disabilities. *Mental Retardation*. 38(3), 195-206

Shapiro, A. 1998 'Wrongful life' lawsuits for faulty genetic counselling: Should the impaired newborn be entitled to sue? *J. of Medical Ethics* 24. 369-375.

Sobsey, D. 1990. Too much stress on stress? Abuse and the family stress factor. *Quarterly Newsletter of the American Association on Mental Retardation*. 3, 2, 8.

Singer. 2001. Response to Mark Kuczewski. *American Journal of Bioethics*. Volume 1. Number 3. p. 55–56

Stensman, S., 1985. Severely mobility-disabled people assess the quality of their lives. *Scandinavian Journal of Rehabilitation Medicine*. vol. 17. 87-99.

Stephens, T. and Brynner, R. 2001. *Dark Remedy; the impact of thalidomide*. Perseus Publishing. Cambridge Massachusetts. USA page 65/66.

Strauss, S.A. 1996. 'Wrongful conception', 'wrongful birth' and 'wrongful life': the first South African cases. *Med. Law*. 15: 161-173.

Tyson J.E., Broyles RS. 1996. Progress in assessing the long-term outcome of extremely low-birth-weight infants. *JAMA*. 276: 492-493.

UNESCO, 2000. http://unesdoc.unesco.org/images/0012/001229/122938eo.pdf#xml=http://unesdoc.unesco. org/ulis/cgi-bin/ulis.pl?database=ged&set=3BE443E4_0_108&hits_rec=3&hits_lng=eng

Watson, J.D. 1996. President's essay: genes and politics. *Annual Report Cold Springs Harbor*. 1996:1-20." http://www.cshl.org/96AnReport/essay1.html. Exact page http://www.cshl.org/96AnReport/essay14.html.

Wertz, D.C. 1998. Eugenics is alive and well. *Science in Context* 11. 3-4. pp 493-510 (p501).

Whiteneck, G.C. et al. 1985. *Rocky mountain spinal cord injury system*. Report to the National Institute of Handicapped Research. 29-33.

Wolbring, G. 1999. *Gene Watch* June 1999 Vol.12 No.3; http://www.bioethicsanddisability.org/Eugenics,%20Euthanics,%20Euphenics.html

Wolbring, G. 2000. Science and the disadvantaged; http://www.edmonds-institute.org/ wolbring.html

Wolbring, G. 2001. Surviving in a technological world. In *Disability and the life course: Global perspectives*. Edited by Mark Priestley. Cambridge University Press.

Wolbring, G. 2001. 150 page single spaced expert opinion for the Study Commission on the Law and Ethics of Modern Medicine of the German Bundestag with the title "Folgen der Anwendung genetischer Diagnostik fuer behinderte Menschen" (Consequences of the application of genetic diagnostics for disabled people) http://www.bundestag.de/gremien/ medi/medi_gut_wol.pdf

Wolbring, G. 2002a. International Center for Bioethics, Culture and Disability. http://www.bioethicsanddisability.org/wrongfulbirth.html

Wolbring, G. 2002b. Wrongful birth/life suits. http://www.bioethicsanddisability.org

Woodrich, W. and Patterson, J.B. 1983. Variables related to acceptance of disability in persons with spinal cord injuries. *Journal of Rehabilitation*. July-Sept. 26-30.

VISIONARY PROJECTS

BRAIN-MACHINE INTERFACE VIA A NEUROVASCULAR APPROACH

Rodolfo R. Llinás and Valeri A. Makarov, NYU Medical School

The issue of brain-machine (computer) interface is, without doubt, one of the central problems to be addressed in the next two decades when considering the role of neuroscience in modern society. Indeed, our ability to design and build new information analysis and storage systems that are sufficiently light to be easily carried by a human, will serve as a strong impetus to develop such peripherals. Ultimately, the brain-machine interface will then become the major bottleneck and stumbling block to robust and rapid communication with those devices.

So far, the interface improvements have not been as impressive as the progress in miniaturization or computational power expansion. Indeed, the limiting factor with most modern devices relates to the human interface. Buttons must be large enough to manipulate, screens wide enough to allow symbol recognition, and so on. Clearly, the only way to proceed is to establish a more direct relation between the brain and such devices, and so, the problem of the future brain-machine interface will indeed become one of the central issues of modern society. As this is being considered, another quite different revolution is being enacted by the very rapid and exciting developments of nanotechnology (n-technology). Such development deals with manufactured objects with characteristic dimensions of less than one micrometer. This issue is brought to bear here, because it is through n-technology that the brain-machine bottleneck may ultimately be resolved. Obviously, what is required is a robust and noninvasive way to both tap and address brain activity optimized for future brain-machine interaction.

Needless to say, in addition to serving as a brain-machine interface, such an approach would be extraordinarily valuable in the diagnosis and treatment of many neurological and psychiatric conditions. Here, the technology to be described will be vital in the diagnosis and treatment of abnormal brain function. Such technology would allow constant monitoring and functional imaging, as well as direct modulation of brain activity. For instance, an advanced variation of present-day deep brain stimulation will be of excellent therapeutic value. Besides, interface with "intelligent" devices would significantly improve the quality of life of disabled individuals, allowing them to be more involved in everyday activity.

The problem we consider has two main parts to be resolved: (1) hardware and (2) software. To approach these issues, we propose to develop a new technology that would allow *direct* interaction of a machine with the human brain and that would be secure and minimally invasive.

The Neurovascular Approach

One of the most attractive possibilities that come to mind in trying to solve the hardware problem concerns the development of a vascular approach. The fact that the nervous system parenchyma is totally permeated by a very rich vascular bed that supplies blood gas exchange and nurturing to the brain mass makes this space a very

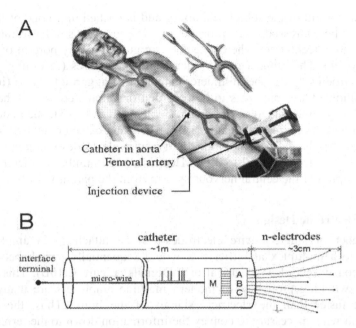

Figure C.10. The neurovascular approach. A. Present day procedure utilized to guide catheters to the brain via the vascular system. Catheters are introduced into femoral, subclavial, or carotid artery. B. The general electronic design includes n-electrodes (diameter of 0.5 micron and length not more than 3 cm) to record/stimulate neuronal activity; Amplifier-Binary Converter (ABC) block that converts acquired analog signals into binary form; Multiplex (M) unit that transforms analog input into serial form by fast switching between all signals; and microwire (approx. 1 m long) that conveys information to the terminal. (Only one logic set is shown.)

attractive candidate for our interface. The capillary bed consists of 25,000 meters of arterio-venous capillary connections with a gauge of approximately 10 microns. At distances more proximal to the heart, the vessels increase rapidly in diameter, with a final dimension of over 20 millimeters. Concerning the acquisition of brain activity through the vascular system, the use of n-wire technology coupled with n-technology electronics seems very attractive. It would allow the nervous system to be addressed by an extraordinarily large number of isolated n-probes via the vascular bed, utilizing the catheter-type technology used extensively in medicine and in particular in interventional neuro-radiology.

The basic idea consists of a set of n-wires tethered to electronics in the main catheter such that they will spread out in a "bouquet" arrangement into a particular portion of the brain's vascular system. Such arrangement could support a very large number of probes (in the millions). Each n-wire would be used to record, very securely, electrical activity of a single or small group of neurons without invading the brain parenchyma. Obviously, the advantage of such system is that it would not interfere with either the blood flow exchange of gases or produce any type of disruption of brain activity, due to the tiny space occupied in the vascular bed.

In order to give a more precise description of the proposed interface, an illustration of the procedure is shown in Figure C.10. A catheter is introduced into the femoral carotid or the sub-clavial artery and is pushed up to one of the vascular territories to be addressed. Such procedure is, in principle, similar to interventional neuro-radiology techniques where catheters are guided to any portion of the central nervous system. The number of 0.5 micron diameter wires (recording points) that could be introduced in a one-millimeter catheter is staggeringly large (in the range of few million). Once the areas to be recorded or stimulated are reached, a set of leads held inside the catheter head would be allowed to be extended and randomly distributed into the brain's circulatory system. Since a catheter can be placed in any major brain vessels, the maximum length of n-wire electrodes required to reach any capillary bed is of the order 2 to 3 cm. Hence, a large number of electrodes would cover any region of the central nervous system from the parent vessels harboring the stem catheters.

General Electronic Design

A number of single n-wire electrodes can be attached via amplifier-binary converter to a multiplex amplifier that would sequentially switch between large, "simultaneously recorded" electrical brain signals (Figure C.10B). This is possible since the switching properties of modern multiple amplifiers are many orders of magnitude faster than the electrical signals of the brain. Thus, the number of independent wires necessary to convey the information down to the terminals of the interface would be a small fraction of the total number of n-wires, and thus, inexpensive and robust microwires can be used along the catheter length.

Many technical issues concerning hardware problems, such as n-amplifiers and multiplex units, can in fact be solved by present technology. The actual size of the expected extracellular recording wiring is given inFigure C.11 by comparing the size of one-micrometer wire with the size of a capillary in the brain parenchyma. In this case, an individual Purkinje cell is drawn to show where the capillary spaces reside within the dendritic tree of such neurons. Note that the number of capillaries

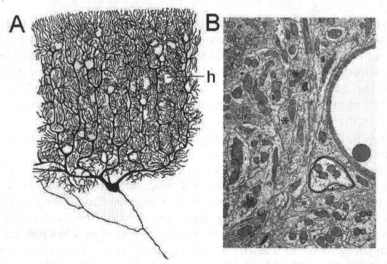

Figure C.11. Illustration of comparative size scales for a neuron, a capillary, and an n-wire. A. Purkinje cell with dendritic tree penetrated by many capillaries foramen. h. B. Elentronmicrograph of a corresponding site in the dendritic as shown in h with a 1μ electrode (spot) drawn inside a capillary.

traversing each cell is numerous (in this particular case, more than 20). On the right inset is an electron micrograph of the same area that gives an accurate representation of the size relation between one such n-wire (in this case 0.9 micron) and the diameter of the smallest of capillaries in that portion of brain parenchyma.

Thus, at this point, the materials and methodology required to implement a mega electrode system are basically within our technology over the next decade.

Software Requirements

The second significant issue is that of the computational requirements that would allow the reading, storing, and contextualizing of the enormous amount of neuronal information that would become available with the vascular approach described above. While this may prove to be more challenging than the hardware component of this interface, it would also be most valuable, as the proper understanding of such activity would give us an significant window into brain function, further defining the relations between electrophysiology and cognitive/motor properties of the brain.

Attempting to investigate this problem, the second step in this proposal, would be the development of mathematical algorithms able to classify brain states based on neuronal unit activity and field potential analysis. Initially, we plan to correlate, in real time, the moment-to-moment electrical activity of neurons with large functional brain states. It is assumed that the electrical properties of neurons define all possible brain states and that such states co-vary systematically with the global state dynamics. However, this does not imply that there exists one-to-one correspondence between purely local patterns of brain activity and a particular set of functional states. The generation of a new functional state in the brain, for instance, transition "sleep-wakefulness," is known to correspond to activity reorganization over many groups of neurons. Needless to say, there is a large number of possible patterns that differs minimally from one other. The approach is to map the small variance patterns into relatively small sets of different functional states. For example, in the simplest case only three global functional states may be considered: (1) sleep, (2) wakefulness, and (3) "none of the above" or uncertain state, e.g., drowsiness. The last state is an absolutely necessary form to be included, for two reasons: (a) mathematically, the output domain of the algorithm must be closed in order to address correctly "any possible input pattern," including those that have unavoidable noise impact or belong to intermediate, non-pure states without a reliable answer within statistical significance level; and (b) from the conceptual viewpoint, the third state is vital, as for instance, seeing can only occur during wakefulness, and during sleep, this state is uncertain.

The design of the hardware part of the interface (see Figure C.10B) has not been dictated by electronic purposes only but also pursues the goal of preliminary signal processing. Here, we use the commonly accepted hypothesis that neurons interact with each other mostly via action potentials and related synaptic interactions. Thus, it seems to be natural to convert electrical signals taken from n-electrodes into binary form. This approach has many advantages. In particular, if the threshold level for digitalization is appropriately chosen, we would be able to overcome the following problems:

- Not all electrodes would be placed at "right" positions (some of them may be far enough from any neuron to produce reliable data), or just damaged.

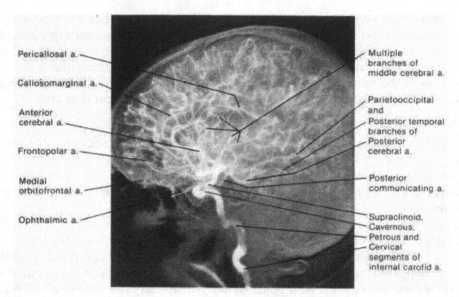

Figure C.12. Lateral view of brain arteries.

- Two electrodes placed in vicinity of a single neuron but at diverse distances from it will produce output voltage traces of different amplitude.
- The signal-to-noise ratio may not be optimal if an electrode records from more than one neuron, as one of them may be selected and others suppressed by the threshold system.

Moreover, binary form is computer friendly and supports efficient operation. Also additional processing logic can be easily included between a computer and the terminals of microwires that would significantly speed up data acquisition, storage, and contextualization.

Memory Requirements

A rough estimate of memory requirements to support resident information and input bandwidth (informational flow rate) will be $10^6 \times 10^3 = 10^9$ bits/s, assuming input signals from 10^6 independent binary variables with a sampling rate of 1 kHz. That is 100 MB per second for the total output, which is attainable with present day technologies. Utilization of additional intermediate logic would even afford a greater performance increase.

Classification Algorithms

As mentioned above, the computational algorithm must be designed to spot alterations in the brain activity that relate to a global change of states. This activity is represented by the set of binary time series taken from many neurons, i.e., by spatiotemporal patterns. Thus, we have the pattern classification problem mentioned above. For an algorithm to be useful, it must be optimized to (1) determine the minimal number of hypotheses (possible functional states) concerning the data set; (2) economize on data storage and subsequent data manipulation/calculation; (3)

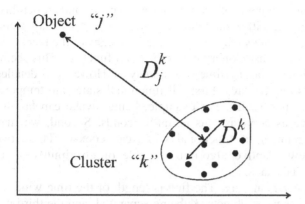

Object "j"

D_j^k

D^k

Cluster "k"

Figure C.13. Qualitative illustration of dissimilarity of object "j" to cluster "k" and mean dissimilarity within the cluster.

scale for increasing data sets and for the number of functional states; and (4) be robust. The approach to the problem we propose below is based on *cluster analysis* (Kaufman 1990) and measures of dissimilarity between time series (see, for example, Kantz 1994; Schreiber 1997; Schreiber and Schmitz 1997).

In the first step, the data set will be split into J short time intervals by shifting a time window of length T. The time scale T can be varied for different purposes, and its choice is a compromise between speed and reliability in data analysis. Each window will be referred to as "an object" or entity, assuming that a window encompasses an unchanged functional state. Assuming a correct set of hypotheses concerning the number of clusters, K, (e.g., for three global functional states: wakefulness, sleep, and uncertain state, $K=3$), the J different objects must be related to K functional states.

The algorithm starts with K random clusters and then moves objects between those clusters in order to split objects into clusters such that variance in each cluster would be minimal, while variance between clusters would be maximal. This can be realized by minimization of the so-called cost function (Schreiber and Schmitz 1997). To implement this function, a measure of dissimilarity between objects must be obtained. This can be, for instance, determined by calculating Euclidean distances between objects in a multidimensional space. Figure C.13 shows a sketch of average dissimilarity of object j to cluster k (distance between j and k) and average dissimilarity within cluster k. The optimization strategy to determine the absolute minimum of the cost function will employ simulated annealing (Kirkpatrick, Gelatt, and Vecchi 1983; Press et al. n.d.), which follows local gradients, but can move against the gradient in order to escape "local minima" shadowing an absolute minima.

The algorithm described above works well under the assumption that the correct dissimilarity has been determined. For time series objects, in the simplest case, neuronal firing rates can be used as coordinates in a multidimensional space. However, application of this measure is rigid (although it has its own advantages), as it takes into account only local oscillatory properties. Another useful procedure will be the dissimilarity matrix calculation introduced (Schreiber and Schmitz 1997) based on the Grassberger-Procaccia cross-correlation sum (Grassberger and Procaccia 1983).

The classification algorithm given here may be referred to as unsupervised. It is based on the hypothesis of a "good" dissimilarity measure and does not include any optimization. This approach can be upgraded to a supervised training data set, where the correct results of classification are known *a priori* for a part of data and may be used as a feedback for improvement of computational speed and reliability. However, even after tuning, the algorithm may fail, since brain plasticity may occur. Thus, the possibility of sudden mistakes may be corrected by means of the feedback.

The basic problem here is the nonstationary nature of brain function. This seems at first glance to be a big obstacle for any time series analysis. However, a detailed study of the problem indicates two features: First, all functional states are temporal and have essentially different time scales. For example, being awake can last for hours, while cognition can be as short as tens of milliseconds. Second, we may assume that only a limited number of functional states can coexist. These two features allow building a new adaptive algorithm capable of discriminating, in principle, any possible functional states.

There are three main parameters at play. The first is length of the time window, T; the next is the number of clusters of objects, K, being separated; and the third is a dissimilarity measurement. We can start the process of classification with relatively long T and small K. Thus, fast processes (functional states) would be eliminated due to averaging over a protracted time. Moreover, functional states with intermediate time scales and with a strong influence onto others would be left out due to very rough classification, since we have split patterns into a few clusters only. Then, when a first approximation of cluster boundaries is determined and it can reliably detect functional states of the top level, a step down can be taken by decreasing window size T and by including finer functional states (increasing K). Moreover, it is possible to work "within" a functional state of the upper level and reject all non-fitting. Such modification of the algorithm allows scalability and a method of exploration of all possible functional states. One problem here is that the deeper we go into the functional state hierarchy, the heavier the computation needed. However, the main parts of the algorithm can be easily paralleled and hence effectively performed by parallel computers or even by specially designed electronics.

Conclusions

We proposed that a novel brain-machine interface is realizable that would allow a robust solution to this important problem. This hardware/software approach allows a direct brain interface and the classification of its functional states using a benign invasive approach. We propose that this approach would be very helpful in human capacity augmentation and will yield significant new information regarding normal and abnormal brain function. Because its development and utilization is inevitable given the extraordinarily attractive feature of being retrievable, in the sense that the recording/stimulating filaments are small enough that the device can be removed without violating the integrity of the brain parenchyma.

Because such interfaces will probably be streamlined over the coming years in efforts such as "hypervision" (Llinás and Vorontsov in preparation), two-way direct human communication, and man-machine telepresence (which would allow actuator-based distant manipulation), this approach should be fully examined. Finally, the development of new nanotechnology instrumentation may ultimately be an important tool in preventive medicine and in diagnostic/therapeutic outcome monitoring of physiological parameters.

References

Grassberger, P. and I. Procaccia. 1983. *Physica (Amstredam)* D 9, 189 (1983).

Kantz, H. 1994. Quantifying the Closeness of Fractal Measures, *Phys. Rev. E* 49, 5091.

Kaufman, L. 1990. *Finding Groups in Data: An Introduction to Cluster Analysis* (Wiley, New York).

Kirkpatrick, S., C.D. Gelatt Jr., and M.P. Vecchi. 1983. Optimization by Simulated Annealing. *Science* 220, N. 4598, 671.

Press, W.H., B.P. Flannery, S.A. Teukolsky and W.T. Vetterling. ND. *Numerical Recipes, The Art of Scientific Computing* (Book series: Cambridge Univ. Press).

Schreiber, T. 1997. Detecting and Analyzing Nonstationarity in a Time Series Using Nonlinear Cross Predictions, *Phys. Rev. Lett.* 78, 843.

Schreiber, T. and A. Schmitz. 1997. Classification of Time Series Data with Nonlinear Similarity Measures, *Phys. Rev. Lett.* 79, 1475.

HUMAN-MACHINE INTERACTION: POTENTIAL IMPACT OF NANOTECHOLOGY IN THE DESIGN OF NEUROPROSTHETIC DEVICES AIMED AT RESTORING OR AUGMENTING HUMAN PERFORMANCE

Miguel A.L. Nicolelis, Duke University Medical Center
and Mandayam A. Srinivasan, MIT

Throughout history, the introduction of new technologies has significantly impacted human life in many different ways. Until now, however, each new artificial device or tool designed to enhance human motor, sensory, or cognitive capabilities has relied on explicit human motor behaviors (e.g., hand, finger, foot movements), often augmented by automation, in order to translate the subject's intent into concrete goals or final products. The increasing use of computers in our daily lives provides a clear example of such a trend. In less than three decades, digital computers have permeated almost every aspect of our daily routine and, as a result, have considerably increased human capabilities. Yet, realization of the full potential of the "digital revolution" has been hindered by its reliance on low-bandwidth and relatively slow user-machine interfaces (e.g., keyboard, mice, etc.). Indeed, because these user-machine interfaces are far removed from the way one's brain normally interacts with the surrounding environment, the classical Von Neuman design of digital computers is destined to be perceived by the operator just as another external tool, one that needs to be manipulated as an independent extension of one's body in order to achieve the desired goal. In other words, the reach of such a tool is limited by its inherent inability to be assimilated by the brain's multiple internal representations as a continuous extension of our body appendices or sensory organs. This is a significant point, because in theory, if such devices could be incorporated into "neural space" as extensions of our muscles or senses, they could lead to unprecedented (and currently unattainable) augmentation in human sensory, motor, and cognitive performance

It is clear that recent advances in nanotechnology could significantly impact the development of brain-machine interfaces and neuroprosthetic devices. By establishing direct links between neuronal tissue and machines, these devices could significantly enhance our ability to use voluntary neuronal activity to directly control mechanical, electronic, and even virtual objects as if they were extensions of our own bodies.

Main Goals

For the past few years, we and others have proposed that a new generation of tools can be developed in the next few decades in which direct brain-machine interfaces (BMIs) will be used to allow subjects to interact seamlessly with a variety of actuators and sensory devices through the expression of their voluntary brain activity. In fact, recent animal research on BMIs has supported the contention that we are at the brink of a technological revolution, where artificial devices may be "integrated" in the multiple sensory, motor, and cognitive representations that exist in the primate brain. Such a demonstration would lead to the introduction of a new generation of actuators/sensors that can be manipulated and controlled through direct brain processes in virtually the same way that we see, walk, or grab an object.

At the core of this new technology is our growing ability to use electrophysiological methods to extract information about intentional brain processes (e.g., moving an arm) from the raw electrical activity of large populations of single neurons, and then translate these neural signals into models that control external devices. Moreover, by providing ways to deliver sensory (e.g., visual, tactile, auditory, etc.) feedback from these devices to the brain, it would be possible to establish a reciprocal (and more biologically plausible) interaction between large neural circuits and machines and hence fulfill the requirements for artificial actuators of significantly augmenting human motor performance to be recognized as simple extensions of our bodies. Using this premise and taking advantage of recent developments in the field of nanotechnology, one can envision the construction of a set of closed-loop control BMIs capable of restoring or augmenting motor performance in macro, micron, and even nano environments (Fig. C.14).

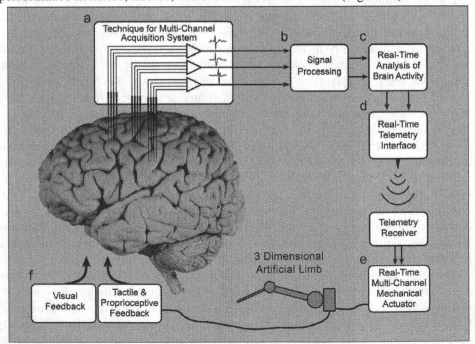

Figure C.14. General architecture of a closed-loop control brain-machine interface: Neuroprosthesis for restoring motor function of damaged brain areas.

Envisioned Utility of BMIs

The full extent to which BMIs would impact human behavior is vastly unknown. Yet, short-term possibilities are innumerable. For example, there is a growing consensus that BMIs could provide the only viable short-term therapeutic alternative to restore motor functions in patients suffering from extensive body paralysis (including lack of communication skills) resulting from devastating neurological disorders.

Assuming that noninvasive techniques to extract large-scale brain activity with enough spatial and temporal resolution can be implemented, BMIs could also lead to a major paradigm shift in the way normal healthy subjects can interact with their environment. Indeed, one can envision a series of applications that may lead to unprecedented ability to augment perception and performance in almost all human activities. These applications would involve interactions with either real or virtual environments. According to this view, real environments can also include local or remote control relative to the human subject, while virtual environments can be realistic or intentionally unrealistic. Here are some examples.

1. **Local, real environment:** Restoration of the motor function in a quadriplegic patient. Using a neurochip implanted in the subject's brain, neural signals from healthy motor brain areas can be used to control an exoskeletal or prosthetic robotic arm used to restore fundamental motor functions such as reaching, grabbing, and walking.

2. **Remote, real environment:** Superhuman performance, such as clearing heavy debris by a robot controlled by the brain signals of a human operator located far away from the danger zone. Recent results by the P.I. and his collaborators have demonstrated that such remote control could be achieved even across the Internet.

3. **Realistic virtual environment:** Training to learn a complex sequence of repair operations by the trainee's brain directly interacting with a virtual reality program, with or without the involvement of the trainee's peripheral sensorimotor system.

4. **Unrealistic virtual environment:** Experiencing unrealistic physics through a virtual reality system for a "what if" scenario, in order to understand deeply the consequences of terrestrial physics.

Given the significant degree of plasticity documented even in the adult brain, repeated use of BMIs will likely transform the brain itself, perhaps more rapidly and extensively than what is currently possible with traditional forms of learning. For example, if a robot located locally or remotely is repeatedly activated via a BMI, it is likely that cortical areas specifically devoted to representing the robot will emerge, causing the robot to effectively become an extra limb of the user.

What real advantages might we obtain from future BMI based devices, compared to more conventional interfaces such as joysticks, mice, keyboards, voice recognition systems, and so forth? Three possible application domains emerge:

1. *Scaling of position and motion,* so that a "slave" actuator, being controlled directly by the subject's voluntary brain activity, can operate within workspaces that are either far smaller (e.g., nanoscale) or far bigger (e.g., space robots; industrial robots, cranes, etc.) than our normal reach

2. *Scaling of forces and power,* so that extremely delicate (e.g., microsurgery) or high-force tasks (e.g., lifting and displacing a tank) can be accomplished

3. *Scaling of time,* so that tasks can be accomplished much more rapidly than normal human reaction time, and normally impossible tasks become possible (e.g., braking a vehicle to a stop after seeing brake lights ahead; catching a fly in your hand; catching something you have dropped; responding in hand-to-hand combat at a rate far exceeding that of an opponent)

To some extent, all these tasks, with the exception of time scaling, can, in principle, be accomplished though conventional teleoperator systems in which the human using his limbs operates a master device, which, in turn, controls a local or remote slave device. There is a history of five decades of research in this area of robotics, with moderate success, such as recent commercial development of teleoperated surgical systems. Major difficulties have been the design of appropriate master devices that the human can interact with naturally and the destabilizing effects of long time delay between the master and the slave. BMIs offer unique advantages in two ways:

1. They eliminate the need for master devices that interact with the human

2. Since the human is directly operating through his brain, the time delays associated with the signal transmission from the peripheral sensors to the CNS (~ 10–30 msec) and from CNS to the muscles (~10-30 msec), and then the time required of a limb to complete the needed action (~100-900 msec), can be reduced by an order of magnitude.

Elimination of the need for a master device is a radical departure from conventional teleoperation. Furthermore, the reduction of time delays leads to the exciting possibility of superhuman performance. For example, moving an arm from point A to point B can take ~500 msec from the time muscles are commanded by the brain, because of the force generation limitations of the muscles, the inertia of the arm, and the need to accelerate from A and to decelerate to B. But if a slave robot that is much better than the human arm in terms of power/mass ratio is directly controlled though a BMI, all three types of time delays (peripheral sensory, motor signal transmission, and limb motion) can be minimized or eliminated, possibly leading to faster and more stable operation of the slave robot. For instance, it is possible for an impaired or unimpaired person to wear an arm exoskeleton that directly interacts with the brain much faster than the natural arms.

In recent years, work developed by our laboratories has demonstrated the feasibility of building BMIs dedicated to the task of utilizing brain-derived signals to control the 1-D and 3-D movements of artificial devices. In a series of studies, we have provided the first demonstrations in animals that such BMIs can be built, that animals can learn to operate these devices in order to obtain a reward, and that motor control signals derived from the extracellular activity of relatively small populations of cortical neurons (50-100 cells) can be used to reproduce complex 3-D arm movements in a robotic device in real time.

Recent advances in nanotechnology could help significantly the advance of this area of research. First, this technology could provide new ways to extract large-scale brain activity by reducing the degree of invasiveness of current electrophysiological methods. Investment in research aimed at designing a new generation of VLSI aimed at both conditioning and analyzing large-scale electrical brain activity will also be required. Finally, a complete new generation of actuators, designed to operate in micro- or nanospaces needs to be built, since there are many new applications that can be envisioned if brain-derived signals can be employed to directly control nanomachines.

NANOTECHNOLOGY: THE MERGING OF DIAGNOSTICS AND TREATMENT

Abraham Phillip Lee, University of California at Irvine

The key to advancing from the discovery stage of nanoscience to commercially feasible nanotechnology is the ability to reliably manufacture nanoscale features and control nanoscale functions. The application of nanotechnology to biology further requires the functional nano-interface between artificial and biological components. From a systems perspective, this requires signal transduction at matching impedances so that sensitivity and specificity are adequate to decipher the biological events. The maturation of these capabilities will enable the probing and manipulating of the fundamental building blocks of biology, namely biomolecules such as carbohydrates, lipids, nucleic acids, and proteins.

The biological cell has proven to be the most intricate functional system of its scale. Unique functionalities include its ability to regulate and adapt, hierarchical self-assembly, repair and maintenance, parallel processing, just-in-time processes, asynchronous control and signaling, and scalability from nano to macro. However, these features and functions are hard to quantify, model, engineer, and reprogram. On the other hand, microfabrication and nanofabrication techniques have given us integrated nanoscale electronics, microfluidics, microelectromechanical systems (MEMS), and microphotonics. These top-down fabrication techniques allow addressability of large-scale component platforms. On the other hand, bottom-up nanofabrication techniques (such as self-assembly) mimic how biology builds very complex systems out of simple molecules. As the scale of these two fields overlaps, devices can be developed with high sensitivity and selectivity for detecting and interfacing to biomolecules.

Projects exemplifying the field of nanobiotechnology include single molecule detection studies, functional imaging of cells and biomolecules by scanning probe microscopy, nanoparticles for targeted therapy, nanomechanical devices to measure biomolecular force interactions, etc. These research efforts represent the start towards interfacing with biological functions at the most fundamental level. However, biology is the intertwined combination of many single molecular events, each being coupled with one another either synchronously or asynchronously. To truly unveil biological events such as cell signaling pathways, genetic mutation processes, or the immune responses to pathogens, one must have a method to generate large-scale, multifunctional nano-bio interfaces with readout and control at the single biomolecule level.

I provide three visions for features of the nanobiotechnology roadmap:

1. The development of a "biological microprocessor" for synthesizing and analyzing biomolecules on nano platforms (liposomes, nanoparticles, self-assembled monolayers, and membranes) in fluids. These "biomolecular

nanotransducers" will be able to function (1) as multiplexed nanomedicines capable of long duration, *in vivo* targeted detection, diagnosis, and treatment of molecular diseases; (2) as key ingredients of smart coatings for versatile environmental monitoring of toxins/pathogens; and (3) as engineered biomolecular nanosystems that mimic cellular functions for fundamental biology experiments.

2. The coupling of biomolecular units — whether they be DNA, receptors, antibodies, or enzymes — with MEMS for reassembling cell components and reprogramming cell functions. This will enable the rewiring of biological cell pathways in artificially controlled platforms such that it will be possible to carry out preclinical experiments without the use of animals or humans.

3. The coupling of "nano guards for health" (e.g., nanoparticles) with microfluidic controllers for long-term control of certain health parameters. For instance, the feedback loop of a glucose sensor and delivery of nano artificial islets can enable the merging of detection, diagnosis, and treatment into one MEMS device.

ARTIFICIAL BRAINS AND NATURAL INTELLIGENCE

Larry Cauller and Andy Penz, University of Texas at Dallas

It is widely accepted that nanotechnology will help push Moore's Law to, or past, its prediction that the next few decades will witness a truly amazing advance in affordable personal computing power. Several visionary techno-futurists have attempted to estimate the equivalent power of the human brain to predict when our handheld personal computers may be able to convince us that they occasionally feel, well, unappreciated, at least. With the advent of nano-neuro-techniques, neuroscience is also about to gain unfathomable insight into the dynamical mechanisms of higher brain functions. But many neuroscientists who have dared to map the future path to an artificial brain with human intelligence do not see this problem in simple terms of "computing power" or calculations per second. We agree that the near future of nano-neuro-technology will open paths to the development of artificial brains with natural intelligence. But we see this future more in terms of a coming nano-neuro-cogno-symbiosis that will enhance human potential in two fundamental ways: (1) by creating brilliant, autonomous artificial partners to join us in our struggle to improve our world and (2) by opening direct channels of natural communication between human and artificial nervous systems for the seamless fusion of technology and mind.

Human brain function emerges from a complex network of many billion cooperating neurons whose activity is generated by nanoscale circuit elements. In other words, the brain is a massively parallel nanocomputer. And, for the first time, nanotechnology reveals approaches toward the design and construction of computational systems based more precisely upon the natural principles of nervous systems. These natural principles include (1) enormous numbers of elementary nonlinear computational components, (2) extensive and interwoven networks of modifiable connectivity patterns, (3) neurointeractive sensory/motor behavior, and (4) a long period of nurtured development (real or virtual). We believe human-like functions will likewise emerge from artificial brains based upon these natural principles.

A simple nanoelectronic component, the resonant tunneling diode, possesses nonlinear characteristics similar to the channel proteins that are responsible for much of our neurons' complex behavior. In many ways, nanoscale electronics may be more suitable for the design of nonlinear neural networks than as simple switching elements in digital circuits. At this NBIC meeting, Phil Kuekes from Hewlett-Packard described a nanoscale cross-link connection scheme that may provide an approach to solving the truly difficult problem of how to interconnect enormous networks of these nanocomponents. But as a beginning, these initial steps to realization of a nano-neuro-computer permit consideration of the much greater density that is possible using nanoelectronic neurons than has so far been possible with microelectronic solutions, where equivalent chip architectures would need to be millions of times larger. If the size of the artificial brain were small enough to mount on a human-size organism, then it might be simpler to design nurturing environments to promote the emergence of human-like higher functions.

Decades of neuroscience progress have shed much light upon the complexity of our brain's functional neuro-architecture (e.g., Felleman and Van Essen 1991). Despite its extreme complexity (>100,000 miles of neuron fibers), fundamental principles of organization have been established that permit a comprehensive, although highly simplified sketch of the structure responsible for natural intelligence. In addition, neuroscience has characterized many of the principles by which the network's connections are constantly changing and self-organizing throughout a lifetime of experience (e.g., Abbott and Nelson 2001). While some futurists have included the possibility that it will be possible to exactly replicate the cellular structure of the human brain (Kurzweil 1999), it seems impossible from a neuroscience point of view, even with nanotechnology. But it is not necessary to be too precise. Genetics is not that precise. We know many of the principles of neuro-competition and plasticity that are the basis for the continuous refinement of neural functions in the midst of precise wiring and environmental complexity. But the only test of these far-reaching principles is to construct a working model and learn to use it.

Constrained by the limits of microtechnology, previous attempts to mimic human brain functions have dealt with the brain's extreme complexity using mathematical simplifications (i.e. neural networks) or by careful analysis of intelligent behavior (i.e. artificial intelligence). By opening doors to the design and construction of realistic brain-scale architectures, nanotechnology is allowing us to rethink approaches to human-like brain function without eliminating the very complexity that makes it possible in the first place. The tools of nonlinear dynamical mechanics provide the most suitable framework to describe and manage this extreme complexity (e.g. Kelso 1995; Freeman 2000). But the first step is to recognize and accept the natural reality that the collective dynamics of the neural process responsible for the highest human functions are not mathematically tractable.

Instead, higher functions of the brain are emergent properties of its neuro-interactivity between neurons, between collections of neurons, and between the brain and the environment. While purely deterministic, it is no more possible to track the cause-effect path from neuron activity to higher functions such as language and discovery than it is to track the path from an H_2O molecule to the curl of a beach wave. Unfortunately, appeals to emergence always leave an unsatisfying gap in any attempt to provide a complete explanation, but nature is full of examples, and classical descriptions of human intelligence have depended strongly upon the concept of emergence (i.e. Jean Piaget, see Elman et al. 1997). But modern emergent doctrine is gaining legitimacy from the powerful new tools of nonlinear dynamical mathematics for the analysis of fractals and deterministic chaos. Instead of tracking cause-effect sequence, the new paradigm helps to identify the dynamical mechanisms responsible for the phase shifts from water to ice, or from exploring to understanding.

From the perspective of neuro-interactive emergence, brain function is entirely self-organized so it may only be interpreted with respect to the interactive behavior of the organism within meaningful contexts. For instance, speech communication develops by first listening to one's own speech sounds, learning to predict the sensory consequence of vocalization, and then extending those predictions to include the response of other speakers to one's own speech. This natural process of self-growth is radically different from the approaches taken by artificial intelligence and "neural net" technologies. The kernel of this natural process is a proactive hypothesis-testing cycle spanning the scales of the nervous system that acts first and learns to predict the resulting consequences of each action within its context (Cauller, in press; see also Edelman and Tonomi 2001). Higher functions of children emerge as a result of mentored development within nurturing environments. And emergence of higher functions in artificial brains will probably require the same kinds of care and nurturing infrastructure we must give our children.

So the future of the most extreme forms of machine intelligence from this neuroscience perspective differs in many respects from popular visions: (1) "artificial people" will be very human-like given that their natural intelligence will develop within the human environment over a long course of close relationships with humans; (2) artificial people will not be like computers any more than humans are. In other words, they will not be programmable or especially good at computing. And (3) artificial people will need social systems to develop their ethics and aesthetics.

An optimal solution to the problem of creating a seamless fusion of brain and machine also needs to be based upon these neurointeractive principles. Again, nanotechnology, such as minimally invasive nano-neuro transceivers, is providing potential solutions to bridge the communication gap between brain and machine. But the nature of that communication should be based upon the same neural fundamentals that would go into the design of an artificial brain.

For instance, sensory systems cannot be enhanced by simply mapping inputs into the brain (e.g., stimulating the visual cortex with outputs from an infrared camera won't work). The system must be fused with the reciprocating neurointeractivity that is responsible for ongoing conscious awareness. This means that brain control over the sensory input device is essential for the system to interpret the input in the form of natural awareness (e.g., there must be direct brain control over the position of the video source). In other words, brain enhancements will involve the externalization of the neurointeractive process into peripheral systems that will respond directly to brain signals. These systems will become an extension of the

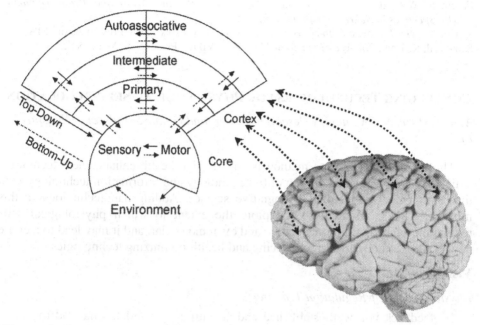

Figure C.15. Neurointeractive artificial brain/human brain interface for neuroprosthesis or enhancement.

human mind/body over a course of accommodation that resembles the struggle of physical therapy following cerebral stroke.

Fusion of artificial brains into larger brains that share experience is a direct extension of this line of reasoning. This also would not be an immediate effect of interconnection, and the fusion would involve give and take on both sides of the connection over an extended course of active accommodation. But the result should surpass the sum of its parts with respect to its ability to cope with increasing environmental complexity.

Speculation leads to the next level of interconnection, between human and artificial brains. On the face of it, this appears to be a potential path to cognitive enhancement. However, the give and take that makes neurointeractive processes work may be too risky when humans are asked to participate.

References

Abbott, L.F., and Nelson S.B. 2000. Synaptic plasticity: taming the beast. *Nat Neurosci 3:*1178-83

Cauller, L.J. (*in press*). The neurointeractive paradigm: dynamical mechanics and the emergence of higher cortical function. In: *Theories of Cerebral Cortex*, Hecht-Neilsen R and McKenna T (eds).

Edelman G.M. and G. Tonomi. 2001. *A Universe of Consciousness: How Matter Becomes Imagination*, Basic Books.

Elman, J.L., D. Parisi, E.A. Bates, M.H. Johnson, A. Karmiloff-Smith. 1997. *Rethinking Innateness: A Connectionist Perspective on Development*, MIT Press, Boston.

Felleman, D.J., and Van Essen D.C. 1991. Distributed hierarchical processing in the primate cerebral cortex. *Cereb Cortex 1(1):*1-47.

Freeman, W.J. 2000. *Neurodynamics: An Exploration in Mesoscopic Brain Dynamics (Perspectives in Neural Computing)*. Springer Verlag.

Kelso, S. 1995. *Dynamic Patterns (Complex Adaptive Systems)*. MIT Press, Boston, MA.

Kurzweil, R. 1999. *The Age of Spiritual Machines*. Viking Press, New York, NY.

CONVERGING TECHNOLOGIES FOR PHYSIOLOGICAL SELF-REGULATION

Alan T. Pope, NASA Langley Research Center, and Olafur S. Palsson, Mindspire, LLC

The biofeedback training method is an effective health-enhancement technique, which exemplifies the integration of biotechnology and information technology with the reinforcement principles of cognitive science. Adding nanotechnology to this mix will enable researchers to explore the extent to which physiological self-regulation can be made more specific and even molecular, and it may lead to a entire new class of effective health-enhancing and health-optimizing technologies.

Vision

Physiological Self-Regulation Training

Biofeedback is a well-established and scientifically validated method to treat a variety of health problems and normalize or enhance human physiological functioning. It consists of placing sensors on the body to measure biological activity and enabling patients to self-correct their physiological activity by showing them on a computer screen (typically in the form of dynamic graphs) what is going on inside their bodies.

> Biofeedback means "the feeding back of information to the individual about change in a physiological system." It implies that the subject is continuously, or discontinuously, informed about change in a particular physiological system under study. The information is believed to act as a reinforcer for further changes in either the same or the opposite direction. As a result of instrumental learning, a physiological response may come under "instructional" or "volitional" control as a function of the feedback of information. (Hugdahl 1995, 39)

When patients are able to observe the moment-to-moment changes in their physiological activity in this way, they can learn over time to control various body functions that are usually beyond conscious control, such as heart rate, muscle tension, or blood flow in the skin:

> According to a basic premise in biofeedback applications, if an individual is given information about biological processes, and changes in their level, then the person can learn to regulate this activity. Therefore, with appropriate conditioning and training techniques, an individual can presumably learn to control body processes that were long considered to be automatic and not subject to voluntary regulation. (Andreassi 2000, 365)

Biofeedback has been used for 40 years with considerable success in the treatment of various health problems, such as migraine headaches, hypertension, and muscle aches and pains. More recently, biofeedback training has been used to enhance performance in a number of occupations and sports activities (Norris and Currieri 1999). At NASA Langley Research Center, work in physiological self-regulation is directed at reducing human error in aviation:

> Our work has focused on a number of areas with the goal of
> improving cognitive resource management, including that of
> physiological self-regulation reported here. Other areas include
> adaptive task allocation, adaptive interfaces, hazardous unawareness
> modeling, cognitive awareness training, and stress-counter-response
> training. (Prinzel, Pope, and Freeman 2002, p. 196)

Intrasomatic Biofeedback: A New Frontier

The exclusive reliance upon sensing physiological functions from the surface of the body has limited biofeedback's specificity in targeting the physiological processes that underlie human performance and the physiological dysregulation implicated in several disorders. Biofeedback technology has yet to incorporate recent advances in biotechnology, including nanoscale biosensors, perhaps because biofeedback research and practice is dominated by a focus on traditional and proven training protocols rather than on biotechnology.

> As a result of the development of new analytical tools capable of
> probing the world of the nanometer, it is becoming increasingly
> possible to characterize the chemical and mechanical properties of
> cells (including processes such as cell division and locomotion) and to
> measure properties of single molecules. These capabilities
> complement (and largely supplant) the ensemble average techniques
> presently used in the life sciences. (Roco and Bainbridge 2001, 7)

Current biofeedback technology still mostly detects, processes, and feeds back to trainees broad signals from sensors on the skin. Such surface sensors are only suited for providing summary information about broad functional characteristics of the organism, like overall autonomic functioning, summative brain activity in a large portion of the cortex, or activity levels of large masses of striated muscle.

Nanoscale technologies, specifically nanoscale biosensor technology, hold the potential for realtime sensing and feedback of internal bodily processes that are the origins or precursors of the physiological signals sensed on the skin surface by current biofeedback technology. Intrasomatic signals, closer to the physiological source of the body activity of interest than surface-detectable signals, could be used for more targeted and precise feedback conditioning of physiological functions and physiological dysregulation. They could also be used to dynamically feed back to patients the consequences and benefits of exercises and practices, or warnings of hazardous alterations in physiology, in order to provide education as well as motivation for adhering to prescribed behavioral treatment regimens. Furthermore, the presence of such small intrasomatic sensors could enable physicians or surveillance computers to titrate or fine-tune the treatment of a patient's disorder (such as medication flow-rate) in ways otherwise not possible.

Early work by Hefferline, Keenan, and Harford (1959) demonstrated that covert physiological responses could be conditioned by attaching consequences, in a traditional psychological reinforcement paradigm, to the production of the responses without the trainee's conscious, deliberate effort to control the responses. Most biofeedback training successes do indeed operate without the necessity for the trainee to be able to articulate the exact nature of the efforts they employ in the learning process, and sometimes without their even trying to consciously control the process. Nevertheless, an additional application of feedback of nanoscale biosensed parameters may be to inform the trainee of the results of his/her overt efforts to facilitate management of a physiological function. An example would be the moment-to-moment feedback of blood oxygenation level or oxygen/CO_2 balance in respiration training for hyperventilation in panic disorder (Ley 1987).

Roles of Converging Technologies

The roles of NBIC technologies in the intrasomatic biofeedback vision are illustrated schematically in Figure C.16.

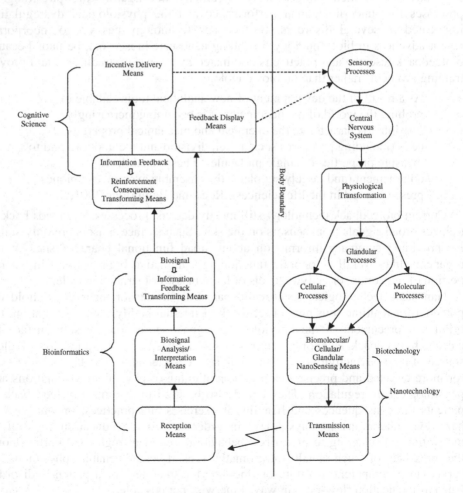

Figure C.16. Intrasomatic biofeedback

Cognitive Science

Mainly used in psychophysiology as an applied technique, the
principle of biofeedback goes back to the idea that nonvolitional,
autonomic behavior can be instrumentally conditioned in a stimulus-
reinforcement paradigm.

Traditional learning theory at the time of the discovery of the
biofeedback principle held that an autonomic, involuntary response
could be conditioned only through the principles of classical, or
Pavlovian, conditioning. Instrumental, operant learning could be
applied only to voluntary behavior and responses. However, in a
series of experiments, Miller (1969) showed that autonomic behavior,
like changes in blood pressure, could be operantly conditioned in rats
(Hugdahl 1995, 40).

In the beginning of the biofeedback field, researchers, working with animals,
experimented with more precisely accessing internal physiological phenomena to
provide the signals and information representing the functions to be conditioned:

The experimental work on animals has developed a powerful
technique for using instrumental learning to modify glandular and
visceral responses. The improved training technique consists of
moment-to-moment recording of the visceral function and immediate
reward, at first, of very small changes in the desired direction and then
of progressively larger ones. The success of this technique suggests
that it should be able to produce therapeutic changes (Miller 1969,
443-444).

Miller identified critical characteristics that make a symptom (or physiological
function) amenable to instrumental conditioning through biofeedback:

Such a procedure should be well worth trying on any symptom,
functional or organic, that is under neural control, that can be
continuously monitored by modern instrumentation, and for which a
given direction of change is clearly indicated medically — for
example, cardiac arrhythmias, spastic colitis, asthma, and those cases
of high blood pressure that are not essential compensation for kidney
damage (Miller 1969, 443-444).

The mechanism of neural control that would enable instrumental conditioning of
basic molecular physiological processes has yet to be identified. Current
understanding is limited to the notion that it generally involves a "bucket brigade"
effect where willful cognitive influences in the cortex are handed down through the
limbic system and on down into the hypothalamus, which disseminates the effect
throughout the body via various neural and endocrine avenues.

Similarly, researchers in the field of psychoneuroimmunology have yet to find
the exact biological mechanisms linking the brain and the immune system.
Nevertheless, Robert Ader, one of the first to present evidence that immune
responses could be modified by classical conditioning (Ader and Cohen 1975),
states:

There are many psychological phenomena, and medical phenomena for that matter, for which we have not yet defined the precise mechanisms. It doesn't mean it's not a real phenomenon (Azar 1999).

Nanobiotechnology

Miller's (1969, 443-444) requirement that the physiological function be "continuously monitored by modern instrumentation" is now made possible by nanoscale biosensors, enabling the investigation of the instrumental conditioning of biomolecular phenomena.

> Implantable sensors or "smart" patches will be developed that can monitor patients who are at risk for specific conditions. Such sensors might monitor, for example, blood chemistry, local electric signals, or pressures. The sensors would communicate with devices outside the body to report results, such as early signals that a tumor, heart damage, or infection is developing. Or these sensors could be incorporated into "closed loop" systems that would dispense a drug or other agent that would counteract the detected anomaly. For chronic conditions like diabetes, this would constitute a great leap forward. Nanotechnology will contribute critical technologies needed to make possible the development of these sensors and dispensers (NSTC 2000, 54, 55).

Another "closed loop system" that would "counteract the detected anomaly" is intrasomatic biofeedback training. In this case, remediation of a physiological anomaly or suboptimal condition would be achieved by self-regulation learned through instrumental conditioning, rather than by an external agent such as a drug or nanodevice.

Freitas (1999, section 4.1) describes "nanosensors that allow for medical nanodevices to monitor environmental states at three different operational levels," including "local and global somatic states (inside the human body)," and cellular bioscanning:

> The goal of cellular bioscanning is the noninvasive and non-destructive *in vivo* examination of interior biological structures. One of the most common nanomedical sensor tasks is the scanning of cellular and subcellular structures. Such tasks may include localization and examination of cytoplasmic and nuclear membranes, as well as the identification and diagnostic measurement of cellular contents including organelles and other natural molecular devices, cytoskeletal structures, biochemical composition, and the kinetics of the cytoplasm (Freitas 1999, section 4.8).

The function of "communicating outside the body to report results" (NSTC 2000, 54, 55) is essential for an intrasomatic biofeedback application. Freitas (1999) describes a similar function for nanorobots:

> In many applications, *in vivo* medical nanodevices may need to communicate information directly to the user or patient. This capability is crucial in providing feedback to establish stable and reliable autogenous command and control systems (Chapter 12).

Outmessaging from nanorobot to the patient or user requires the nanodevice to manipulate a sensory channel that is consciously available to human perception, which manipulation can then be properly interpreted by the patient as a message.

Sensory channels available for such communication include sight, audition, gustation and olfaction, kinesthesia, and somesthetic sensory channels such as pressure, pain, and temperature (Freitas 1999, section 7.4.6).

In this application, "outmessaging" is described as enabling user control of a nanorobot; for intrasomatic biofeedback, this function would provide the information that acts as a reinforcer for conditioning changes in cellular and molecular processes (Figure C.16).

Transforming Strategy

A Technical Challenge

Early on, Kamiya (1971) specified the requirements for the biofeedback training technique, and these have not changed substantially:

- The targeted physiological function must be monitored in real time.

- Information about the function must be presented to the trainee so that the trainee perceives changes in the parameter immediately.

- The feedback information should also serve to motivate the trainee to attend to the training task.

The challenges for the fields of nanotechnology, biotechnology, information technology, and cognitive science (NBIC) in creating the technology to enable internally targeted physiological self-regulation technology can be differentiated according to the disparities between (1) the time response of existing physiometric technology, (2) the time course of the targeted physiological processes, and (3) the requirements for feedback immediacy in the biofeedback paradigm. Realtime sensing is essential to make the processes available for display and attaching sensory feedback consequences to detected changes.

The physiological processes most readily amenable to biofeedback self-regulation are those where the internal training targets are available in real time with current or emerging technologies, such as electrical (e.g. brainwave) and hydraulic (e.g. blood flow) physiological signals.

Instruments using microdialysis, microflow, and biosensor technologies to deliver blood chemistry data such as glucose and lactate in real time (European Commission 2001) will need to reduce test cycle time from minutes to seconds to meet the feedback immediacy criterion required for biofeedback training. Even then, it may be discovered that time delays between the initiation of the production of these chemicals and their appearance in the bloodstream require that signals from upstream stages in the formation process are more appropriate targets for feedback in the self-regulation training loop.

Flow cytometry is an example of an offline, non-realtime technology, in this case for measuring certain physical and chemical characteristics, such as size, shape, and internal complexity, of cells or particles as they travel in suspension one by one past a sensing point. For the blood cell formation process that controls these characteristics of cells, hematopoiesis, to become a candidate for physiological self-regulation training will require advances in molecular-scale technology. These advances will probably need to occur in the upstream monitoring of molecular or biosignal (hormonal, antibody, etc.) precursors of the blood cell formation process, bringing tracking of the process into the realtime scale required for feedback immediacy.

Internal nanosensors will similarly solve the time-response problem that has prevented the utilization of brain functional monitoring and imaging in biofeedback.

> Thus, current functional imaging methods are not in real time with brain activity; they are too slow by a factor of 100 or more. The big advance will be to develop functional imaging techniques that show us — as it is happening — how various areas of the brain interact. ... Do not ask me what the basis of this new imaging will be. A combination of electrical recording and changes in some other brain properties perhaps? (McKhann 2001, 90)

> The precision and speed of medical nanodevices is so great that they can provide a surfeit of detailed diagnostic information well beyond that which is normally needed in classical medicine for a complete analysis of somatic status (Freitas 1999, section 4.8).

Enabling Collaborations

The collaboration of key institutions will be necessary to expedite the development of the intrasomatic biofeedback vision. Potentially enabling joint efforts are already in place (National Aeronautics and Space Administration [NASA] and the National Cancer Institute [NCI] 2002):

> NASA and the National Cancer Institute (NCI) cosponsor a new joint research program entitled "Fundamental Technologies for the Development of Biomolecular Sensors." The goal of this program is to develop biomolecular sensors that will revolutionize the practice of medicine on Earth and in space.

> The Biomolecular Systems Research Program (BSRP) administrates the NASA element of the new program, while the Unconventional Innovations Program (UIP) does so for NCI.

> NASA and NCI are jointly seeking innovations in fundamental technologies that will support the development of minimally invasive biomolecular sensor systems that can measure, analyze, and manipulate molecular processes in the living body. (National Aeronautics and Space Administration [NASA] 2002)

One of the purposes that this program is designed to serve is NASA's requirement "for diagnosis and treatment of injury, illness, and emerging pathologies in astronauts during long duration space missions ... Breakthrough technology is needed to move clinical care from the ground to the venue of long duration space flight ... Thus, the space flight clinical care system must be autonomous ..." (NASA/NCI 2001). Intrasomatic biofeedback's potential for self-remediation of physiological changes that threaten health or performance would be useful in many remote settings.

The nanotechnology, biotechnology, and information technology (NBI) components of the NASA/NCI joint project are specified in a NASA News Release:

> The ability to identify changes such as protein expression or gene expression that will develop into cancer at a later date may enable scientists to develop therapies to attack these cells before the disease spreads. "With molecular technologies, we may be able to understand the molecular signatures within a cell using the fusion of biotechnology, nanotechnology, and information technology," [John] Hines [NASA Biomolecular Physics and Chemistry Program Manager] said.
>
> [NASA] Ames [Research Center] will focus on six key areas in molecular and cellular biology and associated technologies. Biomolecular sensors may some day be able to kill tumor cells or provide targeted delivery of medication. Molecular imaging may help scientists understand how genes are expressed and how they control cells. Developments in signal amplification could make monitoring and measurement of target molecules easier. Biosignatures — identification of signatures of life — offer the possibility of distinguishing cancerous cells from healthy cells. Information processing (bioinformatics) will use pattern recognition and modeling of biological behavior and processes to assess physiological conditions. Finally, molecular-based sensors and instrumentation systems will provide an invaluable aid to meeting NASA and NCI objectives (Hutchison 2001).

The NASA/NCI project is designed to "develop and study nanoscale (one-billionth of a meter) biomedical sensors that can detect changes at the cellular and molecular level and communicate irregularities to a device outside the body" (Brown 2001). This communication aspect of the technology will make possible the external sensory display of internal functioning that is essential to the intrasomatic biofeedback vision (Figure C.16).

Collaborations such as this NASA/NCI project provide the NBI components of the intrasomatic biofeedback vision. The participation of organizations devoted to the development and application of cognitive science (C), such as those specified in Figure C.17, would complete the set of disciplines necessary to realize the vision.

Estimated Implications: The Promise of Intrasomatic Biofeedback

It has not been widely appreciated outside the highly insular field of psychophysiology that humans, given sufficiently informative feedback about their own physiological processes, have both the capacity and inherent inclination to learn to regulate those processes. This phenomenon has, however, been established conclusively in numerous biofeedback applications across a range of different biological functions, including the training of brain electrical activity and of autonomic responses. The integration of NBIC technologies will enable the health- and performance-enhancing benefits of this powerful methodology to be extended to

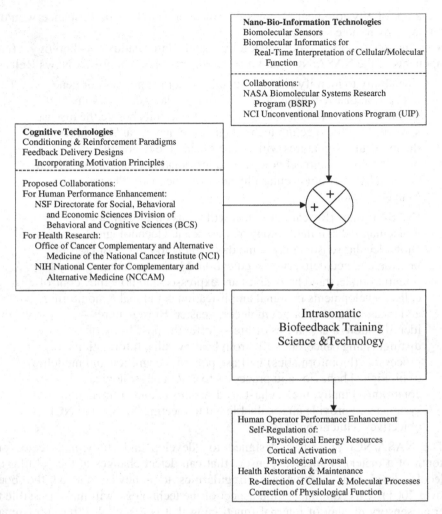

Figure C.17. Enabling collaborations.

other critical physiological processes not previously considered amenable to change by training.

While self-regulation of basic molecular physiological processes may seem fantastical at the present time, it is worth keeping in mind that therapeutic conditioning of autonomic and brainwave signals, now well established, was similarly considered in the fantasy realm no more than four decades ago. The discovery of the human capacity for physiological self-regulation awaited the inventiveness of pioneers, who, in a bold empowering stroke, displayed physiological signals, previously scrutinized only by the researcher, to the subjects whose signals they were, with the aim of giving the subjects control of these processes. This innovation began the discovery process that has demonstrated that, given the right information about their bodily processes in the right form, people can exert impressive control over those responses. The integration of NBIC with the biofeedback method opens an entirely new frontier, inviting the pioneers of a new era in psychophysiology to explore the extent to which this physiological self-regulation can be made more precise, perhaps even to the point of reliably modifying specific molecular events. These developments will enable human beings to willfully induce inside their own bodies small and highly specific biological changes with large health- and performance-enhancing consequences.

References

Ader, R. and Cohen, N. 1975. Behaviorally conditioned immunosuppression. *Psychosomatic Medicine*, 37(4): 333-340.

Andreassi, J.L. 2000. Psychophysiology: human behavior and physiological response, 4th edition, New Jersey: Lawrence Erlbaum Associates.

Azar, B. (1999). Father of PNI reflects on the field's growth. *APA Monitor*, 30(6). Retrieved April 7, 2002 from http://www.apa.org/monitor/jun99/pni.html.

Brown, D. 2001. "Joint NASA/NCI Research to Develop Sensors for Health Monitoring Inside the Human Body". News Release 01-229, Nov. 21, 2001, NASA Headquarters, Washington, DC. Retrieved May 3, 2002 from http://spaceresearch.nasa.gov/general_info/OBPR-01-230.html.

European Commission. 2001. Blood Chemistry in Real Time. Innovation in Europe: Research and Results: Medicine and Health. Retrieved November 5, 2001, from http://europa.eu.int/comm/research/success/en/med/0011e.html.

Freitas, R.A., Jr. 1999. Nanomedicine, Volume I: Basic Capabilities. Landes Bioscience. Retrieved April 7, 2002 from http://www.landesbioscience.com/nanomedicine/.

Hefferline, R.F., Keenan, B., and Harford, R.A. 1959. Escape and Avoidance Conditioning in Human Subjects without Their Observation of the Response. *Science*, 130, 1338-1339.

Hugdahl, K. 1995. *Psychophysiology: The Mind-Body Perspective*. Cambridge, MA: Harvard University Press.

Hutchison, A. 2001. "NASA Biotechnology Project May Advance Cancer Research." News Release 01-96AR, Dec. 5, 2001, NASA Ames Research Center, Moffett Field, Calif. Retrieved May 3, 2002 from http://amesnews.arc.nasa.gov/releases/2001/01_96AR.html.

Kamiya, J. 1971. Biofeedback and Self-Control: Preface. Chicago: Aldine-Atherton. ix-xvi.

Ley, R. 1987. Panic Disorder: A Hyperventilation Interpretation. In Michelson, L. and Ascher, L.M. (eds.). Anxiety and Stress Disorders: Cognitive-Behavioral Assessment and Treatment. New York: The Guilford Press. 191-212.

McKhann, G.M. 2001. A Neurologist Looks Ahead to 2025. *Cerebrum*, 3(3), 83-104.

Miller, N.E. 1969. Learning of Visceral and Glandular Responses. *Science*, 163, 434-445.

National Aeronautics and Space Administration (NASA). 2002. BioMolecular Systems Research Program. NASA AstroBionics Program. Retrieved April 7, 2002 from http://astrobionics.arc.nasa.gov/prog_bsrp.html.

National Aeronautics and Space Administration (NASA) and the National Cancer Institute (NCI). 2002. Biomolecular Sensor Development: Overview. Retrieved May 3, 2002 from http://nasa-nci.arc.nasa.gov/overview_main.cfm.

National Cancer Institute (NCI) and the National Aeronautics and Space Administration (NASA). 2001. "Fundamental Technologies for Development of Biomolecular Sensors." NASA/NCI Broad Agency Announcement (BAA) (N01-CO-17016-32). Retrieved April 7, 2002 from http://rcb.nci.nih.gov/appl/rfp/17016/Table%20of%20Contents.htm.

National Science and Technology Council Committee on Technology, Subcommittee on Nanoscale Science, Engineering and Technology. 2000. National Nanotechnology Initiative: The Initiative and its Implementation Plan. Vision: Advanced Healthcare, Therapeutics and Diagnostics: a. Earlier Detection and Treatment of Disease: Sensors. Washington, D.C.

Norris, S.L., and Currieri, M. Performance Enhancement Training Through Neurofeedback. 1999. In Evans, J.R. and Abarbanel, A. (eds.) Introduction to Quantitative EEG and Neurofeedback. San Diego: Academic Press. 223-240.

Prinzel, L.J., Pope, A.T., and Freeman, F.G. 2002. Physiological Self-Regulation and Adaptive Automation. *The International Journal of Aviation Psychology*, 12(2), 181-198.

Roco, M.C. and Bainbridge, W.S. (eds.). 2001. Societal Implications of Nanoscience and Nanotechnology. Dordrecht, Netherlands: Kluwer Academic Publishers.

IMPROVING QUALITY OF LIFE OF DISABLED PEOPLE USING CONVERGING TECHNOLOGIES

G. Wolbring, U. Calgary, and R. Golledge, UCSB

It is understood that NBIC should be used in a way that diminishes the discrimination against disabled people, advances their acceptance and integration into society, and increases their quality of life.

The Vision

1. NBIC has the potential to give disabled people, and this includes many elderly, the ability to choose between different modes of information output, whether visual, audio, print, or others, as all these modes can be offered routinely at the same time. It has the potential to change computer interface architecture so that disabled people, including those who are blind, sight-impaired, dyslexic, arthritic, immobile, and deaf, can access the Internet and its webpages as transparently and quickly as able-bodied people by means of, for example, holographic outputs; force-feedback, vibrotactile, vastly improved natural speech interfaces; and realtime close captioning. Multimodal access to data and representations will provide a cognitively and perceptually richer form of interaction for all persons, regardless of impairment, handicap, or disability. It will allow for more flexibility in the mode of working (from home or a company building or elsewhere) and representation (in person or virtual). Meetings like this workshop could easily take place within a 3-D virtual reality once the modes of interaction are available in real time and adaptable to different needs (see e.g., http://www.digitalspace.com/avatars/). Even private conversations during breaks could be easily arranged in this virtual reality. This virtual reality would be an alternative to travel. Multimodal input and output interfaces will allow human-computer (HC) interaction when sight is not available (e.g., for blind or sight-impaired users), when sight is an inappropriate medium (e.g., accessing computer information when driving a vehicle at high speeds), or when features and objects are occluded or distant.

2. NBIC has the potential to increase the quality of life of disabled people by allowing for alternative modes of transportation. One technique that could potentially increase quality of life immensely would be mobile teleportation devices. Teleportation would be linked to global positioning devices (see http://www.research.ibm.com/quantuminfo/teleportation/) so that someone could just teleport themselves where they have to go.

3. NBIC will allow for improving assistive devices for disabled people. For example, wheelchairs, which so far haven't changed much in the last 20 years, could be improved in several ways: nanomaterials could make them cheaper, lighter, and more durable; nanotechnology could be used to improve batteries or develop alternative energy generating devices (such as small fuel cells); NBIC could increase wheelchair capabilities (such as stair climbing) and make them more intelligent. The resulting device would allow a person sitting in it to move in any direction, horizontal or vertical, without regard to obstacles such as stairs. It have no need to physically attach to a surface for movement (it could hover). It would allow for the exploration of rough terrain such as the outdoors. This kind of personal moving/flying device could of course be developed for all people. NBIC also might lead to functional artificial limbs, which might even be better than existing human limbs. The same is true for the development of artificial devices for hearing, vision, and cognitive abilities such as comprehension and memory.

4. NBIC will greatly improve the functionality and design of houses, allowing voice command, intelligent applications, etc., that enable disabled (and elderly) people to be more independent.

5. NBIC has the potential to change the public space to make it much more user friendly and inclusive. Means will include IT advances to enable wearable

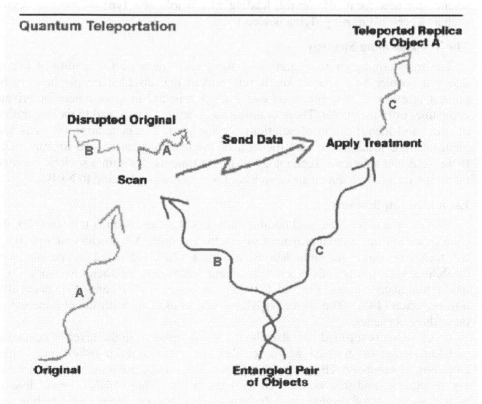

Figure C.18. On the quantum level this transport is achievable (Shahriar, Shapiro and Hemmer 2001). A mobile human teleportation device that can transport the person wherever the person wants to be would solve many accessibility and transportation problems.

computers for use in everyday living (e.g., finding when the next bus is due or where it is now); creation of smart environments (e.g., Remote Auditory Signage Systems [RASS] like talking signs, talking buses, etc., to facilitate wayfinding, business/object location identification, recognition of mass transit services, and intermodal transfer); use of IT and cognitive technology to develop voice-activated personal guidance systems using GPS and GIS; and multimodal interfaces to assist travel and environmental learning.

6. NBIC has the potential to improve communication on a global scale (e.g., universal translation devices), which would allow for a greater exchange of knowledge among people and a faster dissemination of advances in NBIC. The devices available today are not accurate and intelligent enough for use in day-to-day communication.

7. NBIC has the potential to help in the health management of disabled — and all — people.

The Role of Converging Technologies

The converging of technologies is needed if a systematic approach is to be undertaken to use technology for the benefit of disabled people. Often the same tool will have to rely on more than one technology to be workable (e.g., a wheelchair needs improved nanomaterials science for weight reduction and IT and cognoscience for new forms of control, leading to a whole new type of moving device such as a personal moving/flying device.)

The Transforming Strategy

The transforming strategy starts with the goal to increase the quality of life of disabled people. This goal makes it self-evident that disabled people have to be present at every brainstorming on every level, whether in government or private companies or in the public. These brainstorming activities will lead to the generation of ideas and identification of solutions for the goal. The generation of ideas and identifications leads to the identification of the technologies needed to implement these ideas and solutions. Technology is all the time used within a societal context; therefore, the societal dimension also has to be explored — leading to NBICs.

Estimated Implications

If the vision is fulfilled (and nothing indicates that the vision is not feasible), we should see a drop in unemployment of disabled people. A Canadian survey found the following three accommodations are most often identified by people with disabilities not in the labor force as being necessary for them to work: (1) modified/reduced hours (33%); (2) job redesign (27%); and (3) accessible transportation (14%). The above NBICs vision should help with the elimination of these three obstacles.

If the vision is fulfilled, we also should see an increase in the level of education and knowledge of disabled people (which in itself should translate into higher employment numbers). Higher levels of knowledge and employment would lead to higher income, and that would lead to better health. Thus, NBICs would lead to better integration of disabled people into society, making them more mobile and increasing their self-esteem. The disabled, including many elderly people, will feel less isolated and will participate more in society, which will lead to many other effects, including increased well-being.

Reference

Lloyd, S., M.S. Shahriar, J.H. Shapiro, P.R. Hemmer. 2001. Phys Rev Lett. Oct 15;87(16):167903 Long Distance, Unconditional Teleportation of Atomic States via Complete Bell State Measurements

Unison In 1998: A Canadian Approach to Disability Issues A Vision Paper Federal/Provincial/Territorial Ministers Responsible for Social Services.

D. ENHANCING GROUP AND SOCIETAL OUTCOMES

THEME D SUMMARY

Panel: J.S. Albus, W.S. Bainbridge, J. Banfield, M. Dastoor, C.A. Murray, K. Carley, M. Hirshbein, T. Masciangioli, T. Miller, R. Norwood, R. Price, P. Rubin, J. Sargent, G. Strong, W.A. Wallace

The third multidisciplinary theme is concerned with NBIC innovations whose benefits would chiefly be beyond the individual level, for groups, the economy, culture, or society as a whole. It naturally builds on the human cognition and physical capabilities themes and provides a background for the national security and scientific unification panels. In particular, it is focused on a nexus issue that relates logically to most technological applications discussed in this report and that connects all four NBIC scientific and technological realms — that is, how to enhance group human productivity, communication, and cooperation.

The starting point for enhancing group and societal outcomes was the workshop *Societal Implications of Nanoscience and Nanotechnology*, convened by the National Science Foundation September 28-29, 2000. Members of the 2001 workshop were all given copies of the earlier workshop report (Roco and Bainbridge 2001), and they considered how to build on the earlier nanotechnology foundation to develop a broader vision giving equal weight to biotechnology, information technology, and cognitive science, with a focus on enhancing human performance.

The report of the 2000 workshop stressed that the study of the societal implications of nanotechnology must be an integral part of the National Nanotechnology Initiative, and the same is true for future NBIC efforts. The term *societal implications* refers not merely to the impact of technology on society, but also to the myriad ways in which social groups, networks, markets, and institutions may shape development of the technology. Also, as the report recognized, "...sober, technically competent research on the interactions between nanotechnology and society will help mute speculative hype and dispel some of the unfounded fears that sometimes accompany dramatic advances in scientific understanding" (Roco and Bainbridge 2001, v). Similarly, involvement of the social and behavioral sciences in the convergence of NBIC disciplines will help maximize the gains that can be achieved in human performance.

Participants first considered a wide range of likely group and societal benefits of NBIC convergence, then developed the specific vision that they judge has the greatest potential and requires the most concentrated scientific effort to achieve.

There are many potential society-wide benefits of NBIC. Working together, the NBIC sciences and technologies can increase American productivity and lead to revolutionary new technologies, products and services. NBIC can significantly help us proactively deal with the environment, create new energy sources that will reduce our reliance on foreign oil, and ensure the sustainability of our economy. Multidisciplinary research could develop a secure national integrated data system for health data that relies on nano-bio interfaces to obtain, update, and monitor personal data. Combined with new treatments and preventive measures based on NBIC convergence, such a system will extend life and improve its quality. NBIC industries of the future will employ distributed manufacturing, remote design, and production management for individualized products; cognitive control through simulated human intelligence; and a host of other techniques that will promote progress. In addition, converging technologies promise advances in simultaneous group interaction by using cognitive engineering and other new strategies.

In the vast array of very significant potential benefits of NBIC, one stands out that would catalyze all the others and that would require a special, focused effort to achieve success in the 10-20 year time frame. The panel strongly asserted that work should begin now to create *The Communicator*, a mobile system designed to enhance group communication and overcome barriers that currently prevent people from cooperating effectively. A concentrated effort involving nanotechnology, biotechnology, information technology, and cognitive science could develop in one or two decades a mature system to revolutionize people's capability to work together regardless of location or context.

The Communicator: Enhancing Group Communication, Efficiency, and Creativity

The Communicator is envisioned as a multifaceted system relying on the development of convergent technologies to enhance group communication in a wide variety of situations, including formal business or government meetings, informal social interaction, on the battlefield, and in the classroom. This system will rely on expected advances in nanotechnology fabrication and emerging information technologies, tightly coupled with knowledge obtained from the biological and cognitive domains. The convergence of these technologies will enhance individual attributes and remove barriers to group communication such as incompatible communication technologies, users' physical disabilities, language differences, geographic distance, and disparity in knowledge possessed by group members.

At the heart of The Communicator will be nano/info technologies that let individuals carry with them information about themselves and their work that can be easily shared in group situations. Thus, each individual participant will have the option to add information to the common pool of knowledge, across all domains of human experience — from practical facts about a joint task, to personal feelings about the issues faced by the group, to the goals that motivate the individual's participation.

The Communicator will also be a facilitator for group communication, an educator or trainer, and/or a translator, with the ability to tailor its personal appearance, presentation style, and activities to group and individual needs. It will be able to operate in a variety of modes, including instructor-to-group and peer-to-peer interaction, with adaptive avatars that are able to change their affective behavior to fit not only individuals and groups, but also varying situations. It will operate in multiple modalities, such as sight and sound, statistics and text, real and virtual circumstances, which can be selected and combined as needed in different ways by different participants. Improving group interactions via brain-to-brain and brain-machine-brain interfaces will also be explored.

In total, a Communicator system with these attributes will contribute to help overcome inequality between people, isolation of the individual from the environment, injustice and deprivation, personal and cultural biases, misunderstanding, and unnecessary conflict. In the broadest sense, it will be a powerful enhancer of communication and creativity, potentially of great economic and social benefit.

Statements and Visions

The collective vision, called The Communicator here, draws together numerous applications and sciences. In particular, it connects cognitive science and the individual-centered behavioral sciences to the broad range of group-centered social sciences. In addition, this chapter includes contributions on engineering the science of cognition, engineering of the mind, and the NBIC role in the environment and in fundamentally new manufacturing processes and products. An example of such products is the future transport aircraft. A fresh approach to culture is advanced based on biological metaphors and information science methodologies. The statements and visions contributed by members of this working group naturally form a bridge back to the Roco and Bainbridge 2001 report on the societal implications of nanotechnology.

Reference

Roco, M.C. and W.S. Bainbridge, eds. 2001. *Societal Implications of Nanoscience and Nanotechnology.* Dordrecht, Netherlands: Kluwer.

STATEMENTS

COGNITION, SOCIAL INTERACTION, COMMUNICATION, AND CONVERGENT TECHNOLOGIES

Philip Rubin, National Science Foundation[1]

I am impressed with how my teenaged daughter and her friends marshal current technology for group communication. Most of their use of this technology, including AOL "Instant Messaging," email, cellphones, and transportation, is for social interaction.

The technological world 20 years from now will be a very different one. Prognostication is not my specialty and seems like a dangerous enterprise; however, I can talk about some things that we can do to help shape our future. Some of these are merely extensions of current technology and our current abilities, but the critical considerations I want to mention are well beyond our current capabilities. The unifying vision for these comments is the merging of cognition and communication.

Imagine a future without cellphones, laptops, PDAs, and other cumbersome devices. Going beyond the existing smart environments described by Reg Golledge

---

[1] The views expressed in this essay do not necessarily represent the views of the National Science Foundation.

and his colleagues (see Golledge essay in Chapter B and Loomis essay in Chapter C), we will soon be moving through a world in which we are continuously broadcasting, receiving, storing, synthesizing, and manipulating information. We will be embedded in dynamic, continually changing communicative clouds of data signals that communicate information about place, location, language, identity, persona, meaning, and intent. How will social and personal interaction be restructured in this new world? How can we use cognition to help us fly through these clouds effectively? I will leave the first question to experts like Sherry Turkle (see essay in Chapter B), who have thought long and hard about them, and will, instead, briefly mention where we need to go in the area of cognition.

The approaches that we will use for social and group communication in 20 years will rely on a variety of cognitive considerations. Here is a partial listing.

- **Intent.** *Neuro-nano* technology, such as *neural interfaces*, will enable us to provide the direct guidance of choice and selection of behaviors based on cognitive intent. This will allow for binary and graded choice directly under cognitive control.

- **Adaptation.** Communication and knowledge systems will learn and adapt, based upon an understanding of human behavior. Fundamental to this is a serious consideration of the adaptive landscapes that characterize this new communicative, social world and how they mesh with our cognitive capabilities.

- **Perception, analysis, and action.** Embedded and distributed systems and sensors will be enhanced by our fundamental understanding of human perceptual and analytic behavior and skills, including the following: auditory and visual scene analysis (Biederman 1995; Bregman 1994); the visual control of action (Loomis and Beall 1998; Turvey and Remez 1979; and Warren 1988); multimodality, including vision, audition, gesture, and haptic sensing and manipulation (Cassell et al. 2000; and Turvey 1996); spatial cognition (Golledge 1999); linguistic analysis, including statistically-based natural language processing and analysis (Biber, Conrad, and Reppen 1998; and Manning and Schutze 1999); and language use (Clark 1996).

- **Selection.** Cognitive selection, prioritization, and organization of information are essential if the information/communication clouds of the future are not to overwhelm us. Critical abilities to filter, organize, restrict, or enhance information will rely on cognitive selection, personal preference, and automatic adaptation that will evolve based on previous behavior, patterns, choices, and preferences.

- **Semantics.** Meaning will guide the performance of the systems of the future; it will be grounded by a variety of factors, including ties to the real world and its structure and requirements, biases, and personal and social needs. Semantically based systems will make communication more flexible, effective, and natural.

- **Self-organization and complexity.** Increasingly, approaches to understanding human cognition, perception, and behavior will rely on more sophisticated analytic, statistical, and conceptual tools. Examples include nonlinear dynamical systems; self-organization, complexity and emergent behavior; complex adaptive systems; agent-based modeling; naturalistic Bayesian-networks that include subjectively-based categorization and representation; and the like (Holland 1995; Kauffman 1995, 2000; Kelso 1997; Varela et al. 1991; and Waldrop 1992. See also essay by J. Pollack in Chapter B).

What is needed to make these changes happen? First, they rely on the presumed convergence of nano-, bio-, info-, and cognitive technologies. Obviously, some of these changes are already on the way, particularly in the realm of nanotechnology, information technology, communication systems, and engineering. Progress has been significantly slower on the cognitive end, for a variety of reasons. The problems to be tackled in areas such as cognition and perception are often broad and very complex. These difficulties have been compounded by the need for noninvasive approaches for probing and exploring the human cognitive system. Mind and behavior have usually been explored from the outside. In essence, the cognitive system has been treated as a "black box" that can be probed in a variety of ways. Often such approaches have been conducted independently of the constraints imposed both by human physiology and by the environment. Other techniques that are more invasive, such as lesion studies, work with a system that is not in its normal functioning state. The difficulties in probing this system hamper our understanding of it.

Recent technological advances have raised the possibility of obtaining additional data about neural functioning during normal cognitive activities that can help to inform and constrain our theorizing. New advances in functional neuroimaging, including fMRI, PET, and MEG, coupled with the detailed study of neural circuitry and the theoretical advances in a number of areas, hold great promise (Gazzaniga et al. 1998; Lyon and Rumsey 1996; Marantz et al. 2000; and Posner and Raichle 1997). Functional imaging has the potential to be the telescope that lets us observe the universe of the mind. The goal is not to localize behavior but to have a tool that can potentially aid in understanding a massively complex system and in exploring brain behavior. However, these techniques will not be adequate on their own. They must be used in the context of a basic understanding of human cognition, perception, learning, development, and so forth.

Unfortunately, the fundamental understanding of how cognition works in areas such as spatial and cognition perception (auditory, haptic, and visual) has been massively underestimated. These are complex problems that will require significant basic research. For example, we need to understand our interaction with the world before we can fully understand the role the brain plays in helping us navigate this world. Before we can fully understand the role of the brain in vision, we must have a better depiction of what is available in the world for us to see. Before we fully understand the role of the brain in language, we need a clear theoretical understanding of what language is, how it is structured and organized at a variety of levels. Considerable progress that has been made in such areas points to the promise of theory-based research coupled with emerging technologies for visualization and simulation.

The "intelligent" systems of the future that will be fundamental to group and social communication will be far removed from the expert systems and the ungrounded formal systems of the artificial intelligence (AI) of past years. Instead, they will rely on the gains made in the fundamental understanding of the psychology, biology, and neuroscience of human behavior and performance, including cognition, perception, action, emotion, motivation, multimodality, spatial and social cognition, adaptation, linguistic analysis, and semantics. These gains will be enhanced by consideration of human behavior as a complex adaptive biological system tightly coupled to its physical and social environment.

It remains to be seen whether the national support that is necessary to make substantial progress in these areas of cognition that hold such promise is forthcoming. However, if we hope to see truly convergent technologies leading to smart devices and the enhancement of human behavior, communication, and quality of life, we must tackle the difficult problems related to cognition on the large scale more commonly seen in areas such as computer science and engineering. Now is the time to seriously begin this effort.

References

Biber, D., S. Conrad, and R. Reppen. 1998. *Corpus linguistics: Investigating language structure and use*. Cambridge: Cambridge University Press.

Biederman, I. 1995. Visual object recognition. Chapter 4 in *An invitation to cognitive science, 2nd ed., Vol. 2, Visual cognition,* S.M. Kosslyn and D.N. Osherson, eds. Cambridge, MA: MIT Press.

Bregman, A.S. 1994. *Auditory scene analysis*. Cambridge, MA: MIT Press.

Cassell, J., J. Sullivan, S. Prevost, and E. Churchill. 2000. *Embodied conversational agents*. Cambridge, MA: MIT Press.

Clark, H.H. 1996. *Using language*. Cambridge: Cambridge University Press.

Gazzaniga, M.S., R.B. Ivry, and G.R. Mangun. 1998. *Cognitive neuroscience: The biology of the mind*. New York: W.W. Norton and Company.

Golledge, R.G., ed. 1999. *Wayfinding behavior: Cognitive mapping and other spatial processes*. Baltimore, MD: John Hopkins University Press.

X 1995. *Hidden order: How adaptation builds complexity*. New York: Addison-Wesley.

Kauffman, S. 1995. *At home in the universe: The search for the laws of self-organization and complexity*. Oxford: Oxford University Press.

_____. 2000. *Investigations*. Oxford: Oxford University Press.

Kelso, J., and A. Scott. 1997. *Dynamic patterns: The self-organization of brain and behavior*. Cambridge, MA: MIT Press.

Loomis, J.M. and A. Beall. 1998. Visually controlled locomotion: Its dependence on optic flow, three-dimensional space perception, and cognition. *Ecological Psychology* 10:271-285.

Lyon, G.R. and J.M. Rumsey. 1996. *Neuroimaging: A window to the neurological foundations of learning and behavior in children*. Baltimore, MD: Paul H. Brookes Publishing Co.

Manning, C.D., and H. Schutze. 1999. *Foundations of statistical natural language processing*. Cambridge, MA: MIT Press.

Marantz, A., Y. Miyashita, and W. O'Neil, eds. 2000. *Image, language, brain*. Cambridge, MA: MIT Press.

Posner, M.I. and M.E. Raichle. 1997. *Images of mind*. New York: W.H. Freeman and Co.

Turvey, M.T. 1996. Dynamic touch. *American Psychologist* 51:1134-1152.

Turvey, M.T. and R.E. Remez. 1970. Visual control of locomotion in animals: An overview. In *Interrelations of the communicative senses,* L. Harmon, Ed. Washington, D.C.: National Science Foundation.

Varela, F.J., E. Thompson, and E. Rosch. 1991. *The embodied mind: Cognitive science and human experience*. Cambridge, MA: MIT Press.

Waldrop, M.M. 1992. *Complexity: The emerging science at the edge of order and chaos.* New York: Simon and Schuster.

Warren, W.H. 1988. Action modes and laws of control for the visual guidance of action. In *Complex movement behaviour: The motor-action controversy,* O.G. Meijer and K. Roth, eds. Amsterdam: North-Holland .

ENGINEERING THE SCIENCE OF COGNITION TO ENHANCE HUMAN PERFORMANCE

William A. Wallace, Rensselaer Polytechnic Institute

The purpose of this paper is to provide a rationale for a new program whose purpose would be to integrate the science of cognition with technology to improve the performance of humans. We consider *cognition* to be "thinking" by individuals and, through consideration of emergent properties, "thinking" by groups, organizations, and societies. *Technology* is all the means employed by a social group to support its activities, in our case, to improve human performance. *Engineering* is the creation of artifacts such as technologies. Therefore, research concerned with engineering the science of cognition to improve human performance means research on the planning, design, construction, and implementation of technologies.

The purpose of such research should be to enhance performance, i.e., goal-directed behavior in a task environment, across all four levels of cognition: individual, group, organization, and society. In order to do so, we must consider the effective *integration* of cognition and technology as follows:

- integration of technology into the human central nervous system

- integration of important features of human cognition into machines

- integration of technologies (cognitive prosthetics) into the task environment to enhance human performance.

We see a synergistic combination of convergent technologies as starting with cognitive science (including cognitive neuroscience) since we need to understand the how, why, where, and when of thinking at all four levels in order to plan and design technology. Then we can employ nanoscience and nanotechnology to build the technology and biotechnology and biomedicine to implement it. Finally, we can employ information technology to monitor and control the technology, making it work.

ENGINEERING OF MIND TO ENHANCE HUMAN PRODUCTIVITY

James S. Albus, National Institute of Standards and Technology

We have only just entered an era in history in which technology is making it possible to seriously address scientific questions regarding the nature of mind. Prior to about 125 years ago, inquiry into the nature of mind was confined to the realm of philosophy. During the first half of the 20th century, the study of mind expanded to include neuroanatomy, behavioral psychology, and psychoanalysis. The last 50 years have witnessed an explosion of knowledge in neuroscience and computational theory. The 1990s, in particular, produced an enormous expansion of understanding of the molecular and cellular processes that enable computation in the neural substrate, and more is being learned, at a faster rate, than almost anyone can comprehend:

- Research on mental disease and drug therapy has led to a wealth of knowledge about the role of various chemical transmitters in the mechanisms of neurotransmission.

- Single-cell recordings of neural responses to different kinds of stimuli have shown much about how sensory information is processed and muscles are controlled.

- The technology of brain imaging is now making it possible to visually observe where and when specific computational functions are performed in the brain.

- Researchers can literally see patterns of neural activity that reveal how computational modules work together during the complex phenomena of sensory processing, world modeling, value judgment, and behavior generation.

- It has become possible to visualize what neuronal modules in the brain are active when people are thinking about specific things and to observe abnormalities that can be directly related to clinical symptoms (Carter 1998).

The Brain and Artificial Intelligence

In parallel developments, research in artificial intelligence and robotics has produced significant results in planning, problem-solving, rule-based reasoning, image analysis, and speech understanding. All of the fields below are active, and there exists an enormous and rapidly growing literature in each of the following areas:

- Research in learning automata, neural nets, fuzzy systems, and brain modeling is providing insights into adaptation and learning and knowledge of the similarities and differences between neuronal and electronic computing processes.

- Game theory and operations research have developed methods for decision-making in the face of uncertainty.

- Genetic algorithms and evolutionary programming have developed methods for getting computers to generate successful behavior without being explicitly programmed to do so.

- Autonomous vehicle research has produced advances in realtime sensory processing, world modeling, navigation, path planning, and obstacle avoidance.

- Intelligent vehicles and weapons systems are beginning to perform complex military tasks with precision and reliability.

- Research in industrial automation and process control has produced hierarchical control systems, distributed databases, and models for representing processes and products.

- Computer-integrated manufacturing research has achieved major advances in the representation of knowledge about object geometry, process planning, network communications, and intelligent control for a wide variety of manufacturing operations.

- Modern control theory has developed precise understanding of stability, adaptability, and controllability under various conditions of uncertainty and noise.

- Research in sonar, radar, and optical signal processing has developed methods for fusing sensory input from multiple sources and for assessing the believability of noisy data.

In the field of *software engineering*, progress is also rapid, after many years of disappointing results. Much has been learned about how to write code for software agents and build complex systems that process signals, understand images, model the world, reason and plan, and control complex behavior. Despite many false starts and overly optimistic predictions, artificial intelligence, intelligent control, intelligent manufacturing systems, and smart weapons systems have begun to deliver solid accomplishments:

- We are learning how to build systems that learn from experience, as well as from teachers and programmers.

- We understand how to use computers to measure attributes of objects and events in space and time.

- We know how to extract information, recognize patterns, detect events, represent knowledge, and classify and evaluate objects, events, and situations.

- We know how to build internal representations of objects, events, and situations and how to produce computer-generated maps, images, movies, and virtual reality environments.

- We have algorithms that can evaluate cost and benefit, make plans, and control machines.

- We have engineering methods for extracting signals from noise.

- We have solid mathematical procedures for making decisions amid uncertainty.

- We are developing new manufacturing techniques to make sensors tiny, reliable, and cheap.

- Special-purpose integrated circuits can now be designed to implement neural networks or perform parallel operations such as are required for low-level image processing.

- We know how to build human-machine interfaces that enable close coupling between humans and machines.

- We are developing vehicles that can drive without human operators on roads and off.

- We are discovering how to build controllers that generate autonomous tactical behaviors under battlefield conditions.

As the fields of brain research and intelligent systems engineering converge, the probability grows that we may be able to construct what Edelman (1999) calls a "conscious artifact." Such a development would provide answers to many long-standing scientific questions regarding the relationship between the mind and the body. At the very least, building artificial models of the mind would provide new insights into mental illness, depression, pain, and the physical bases of perception, cognition, and behavior. It would open up new lines of research into questions that hitherto have not been amenable to scientific investigation:

- We may be able to understand and describe intentions, beliefs, desires, feelings, and motives in terms of computational processes with the same degree of precision that we now can apply to the exchange of energy and mass in radioactive decay or to the sequencing of amino acid pairs in DNA.

- We may discover whether humans are unique among the animals in their ability to have feelings and start to answer the questions,

  - To what extent do humans alone have the ability to experience pain, pleasure, love, hate, jealousy, pride, and greed?

  - Is it possible for artificial minds to appreciate beauty and harmony or comprehend abstract concepts such as truth, justice, meaning, and fairness?

  - Can silicon-based intelligence exhibit kindness or show empathy?

  - Can machines pay attention, be surprised, or have a sense of humor?

  - Can machines feel reverence, worship God, be agnostic?

Engineering Intelligent Systems

The book *Engineering of Mind: An Introduction to the Science of Intelligent Systems* (Albus and Meystel 2001) outlines the main streams of research that we believe will eventually converge in a scientific theory that can support and bring about the engineering of mind. We believe that our research approach can enable the design of intelligent systems that pursue goals, imagine the future, make plans, and react to what they see, feel, hear, smell, and taste. We argue that highly intelligent behavior can be achieved by decomposing goals and plans through many hierarchical levels, with knowledge represented in a world model at the appropriate range and resolution at each level. We describe how a high degree of intelligence can be achieved using a rich dynamic world model that includes both *a priori* knowledge and information provided by sensors and a sensory processing system. We suggest how intelligent decision-making can be facilitated by a value judgment system that evaluates what is good and bad, important and trivial, and one that estimates cost, benefit, and risk of potential future actions. This will enable the development of systems that behave as if they are sentient, knowing, caring, creative individuals motivated by hope, fear, pain, pleasure, love, hate, curiosity, and a sense of priorities.

We believe that this line of research on highly intelligent systems will yield important insights into elements of mind such as attention, gestalt grouping, filtering, classification, imagination, thinking, communication, intention, motivation, and subjective experience. As the systems we build grow increasingly intelligent, we will begin to see the outlines of what can only be called mind. We hypothesize that mind is a phenomenon that will emerge when intelligent systems achieve a certain level of sophistication in sensing, perception, cognition, reasoning, planning, and control of behavior.

There are good reasons to believe that the computing power to achieve human levels of intelligence will be achieved within a few decades. Since computers were invented about a half-century ago, the rate of progress in computer technology has been astounding. Since the early 1950s, computing power has doubled about every three years. This is a compound growth rate of a factor of ten per decade, a factor of 100 every two decades. This growth rate shows no sign of slowing and, in fact, is accelerating: during the 1990s, computing power doubled every 18 months — a factor of ten every five years. Today, a typical personal computer costing less than $1000 has more computing power than a top-of-the-line supercomputer of only two decades ago. One giga-op (one billion operations per second) single-board computers are now on the market. There appears to be no theoretical limit that will slow the rate of growth in computing power for at least the next few decades. This means that within 10 years, a relatively inexpensive network of 10 single-board computers could have computational power approaching one tera-ops (one trillion, or 10^{12} operations per second). Within 20 years, 10 single-board computers will be capable of 10^{14} operations per second. This is equivalent to the estimated computational power of the human brain (Moravec 1999). Thus, it seems quite likely that within two decades, the computing power will exist to build machines that are functionally equivalent to the human brain.

Of course, more than raw computing power is necessary to build machines that achieve human levels of performance. But the knowledge of how to utilize this computing power to generate highly intelligent behavior is developing faster than most people appreciate. Progress is rapid in many different fields. Recent results from a number of disciplines have established the foundations for a theoretical framework that might best be called a "computational theory of mind." In our book, Meystel and I have organized these results into a reference model architecture that we believe can be used to organize massive amounts of computational power into intelligent systems with human-level capabilities. This reference model architecture consists of a hierarchy of massively parallel computational modules and data structures interconnected by information pathways that enable analysis of the past, estimation of the present, and prediction of the future.

This architecture specifies a rich dynamic internal model of the world that can represent entities, events, relationships, images, and maps in support of higher levels of intelligent behavior. This model enables goals, motives, and priorities to be decomposed into behavioral trajectories that achieve or maintain goal states. Our reference architecture accommodates concepts from artificial intelligence, control theory, image understanding, signal processing, and decision theory. We demonstrate how algorithms, procedures, and data embedded within this architecture can enable the analysis of situations, the formulation of plans, the choice of behaviors, and the computation of current and expected rewards, punishments, costs, benefits, risks, priorities, and motives.

Our reference model architecture suggests an engineering methodology for the design and construction of intelligent machine systems. This architecture consists of layers of interconnected computational nodes, each containing elements of sensory processing, world modeling, value judgment, and behavior generation. At lower levels, these elements generate goal-seeking reactive behavior; at higher levels, they enable perception, cognition, reasoning, imagination, and planning. Within each level, the product of range and resolution in time and space is limited: at low levels, range is short and resolution is high, whereas at high levels, range is long and resolution is low. This enables high precision and quick response to be achieved at low levels over short intervals of time and space, while long-range plans and abstract concepts can be formulated at high levels over broad regions of time and space.

Our reference model architecture is expressed in terms of the Realtime Control System (RCS) that has been developed at the National Institute of Standards and Technology and elsewhere over the last 25 years. RCS provides a design methodology, software development tools, and a library of software that is free and available via the Internet. Application experience with RCS provides examples of how this reference model can be applied to problems of practical importance. As a result of this experience, we believe that the engineering of mind is a feasible scientific goal that could be achieved within the next quarter century.

Implications for the Future

Clearly, the ability to build highly intelligent machine systems will have profound implications — in four important areas in particular: science, economic prosperity, military power, and human well-being, as detailed below.

Science

All of science revolves around three fundamental questions:

1. What is the nature of matter and energy?
2. What is the nature of life?
3. What is the nature of mind?

Over the past 300 years, research in the physical sciences has produced a wealth of knowledge about the nature of matter and energy, both on our own planet and in the distant galaxies. We have developed mathematical models that enable us to understand at a very deep level what matter is, what holds it together, and what gives it its properties. Our models of physics and chemistry can predict with incredible precision how matter and energy will interact under an enormous range of conditions. We have a deep understanding of what makes the physical universe behave as it does. Our knowledge includes precise mathematical models that stretch over time and space from the scale of quarks to the scale of galaxies.

Over the past half-century, the biological sciences have produced a revolution in knowledge about the nature of life. We have developed a wonderfully powerful model of the molecular mechanisms of life. The first draft of the human genome has been published. We may soon understand how to cure cancer and prevent AIDS. We are witnessing an explosion in the development of new drugs and new sources of food. Within the next century, biological sciences may eliminate hunger, eradicate most diseases, and discover how to slow or even reverse the aging process.

Yet, of the three fundamental questions of science, the most profound may be, "What is mind?" Certainly this is the question that is most relevant to understanding the fundamental nature of human beings. We share most of our body chemistry with all living mammals. Our DNA differs from that of chimpanzees by only a tiny percentage of the words in the genetic code. Even the human brain is similar in many respects to the brains of apes. Who we are, what makes us unique, and what distinguishes us from the rest of creation lies not in our physical elements, or even in our biological makeup, but in our minds.

It is only the mind that sharply distinguishes the human race from all the other species. It is the mind that enables humans to understand and use language, to manufacture and use tools, to tell stories, to compute with numbers, and to reason with rules of logic. It is the mind that enables us to compose music and poetry, to worship, to develop technology, and to organize political and religious institutions. It is the mind that enabled humans to discover how to make fire, to build a wheel, to navigate a ship, to smelt copper, refine steel, split the atom, and travel to the moon.

The mind is a process that emerges from neuronal activity within the brain. The human brain is arguably the most complex structure in the known universe. Compared to the brain, the atom is an uncomplicated bundle of mass and energy that is easily studied and well understood. Compared to the brain, the genetic code embedded in the double helix of DNA is relatively straightforward. Compared to the brain, the molecular mechanisms that replicate and retrieve information stored in the genes are quite primitive. One of the greatest mysteries in science is how the computational mechanisms in the brain generate and coordinate the images, feelings, memories, urges, desires, conceits, loves, hatreds, beliefs, pleasures, disappointment, and pain that make up human experience. The really great scientific question is "What causes us to think, imagine, hope, fear, dream, and act like we do?" Understanding the nature of mind may be the most interesting and challenging problem in all of science.

Economic Prosperity

Intelligent machines can and do create wealth. And as they become more intelligent, they will create more wealth. Intelligent machines will have a profound impact on the production of goods and services. Until the invention of the computer, economic wealth (i.e., goods and services) could not be generated without a significant amount of human labor (Mankiw 1992). This places a fundamental limit on average per capita income. Average income cannot exceed average worker productive output. However, the introduction of the computer into the production process is enabling the creation of wealth with little or no human labor. This removes the limit to average per capita income. It will almost certainly produce a new industrial revolution (Toffler 1980).

The first industrial revolution was triggered by the invention of the steam engine and the discovery of electricity. It was based on the substitution of mechanical energy for muscle power in the production of goods and services. The first industrial revolution produced an explosion in the ability to produce material wealth. This led to the emergence of new economic and political institutions. A prosperous middle class based on industrial production and commerce replaced aristocracies based on slavery. In all the thousands of centuries prior to the first industrial revolution, the vast majority of humans existed near the threshold of survival, and every major civilization was based on slavery or serfdom. Yet, less than 300 years after the beginning of the first industrial revolution, slavery has almost disappeared, and a large percentage of the world's population lives in a manner that far surpasses the wildest utopian fantasies of former generations.

There is good reason to believe that the next industrial revolution will change human history at least as profoundly as the first. The application of computers to the control of industrial processes is bringing into being a new generation of machines that can create wealth largely or completely unassisted by human beings. The next industrial revolution, sometimes referred to as the robot revolution, has been triggered by the invention of the computer. It is based on the substitution of electronic computation for the human brain in the control of machines and industrial processes. As intelligent machine systems become more and more skilled and numerous in the production process, productivity will rise and the cost of labor, capital, and material will spiral downward. This will have a profound impact on the structure of civilization. It will undoubtedly give rise to new social class structures and new political and economic institutions (Albus 1976).

The Role of Productivity

The fundamental importance of productivity on economic prosperity can be seen from the following equation:

Output = Productivity x Input

where

Input = labor + capital + raw materials

and

Productivity = the efficiency by which the input of labor, capital,
and raw material is transformed into output product

Productivity is a function of knowledge and skill, i.e., technology. Growth in productivity depends on improved technology. The rapid growth in computer technology has produced an unexpectedly rapid increase in productivity that has confounded predictions of slow economic growth made by establishment economists only a decade ago (Symposia 1988; Bluestone and Harrison 2000). In the future, the introduction of truly intelligent machines could cause productivity to grow even faster. Given only conservative estimates of growth in computer power, unprecedented rates of productivity growth could become the norm as intelligent machines become pervasive in the productive process.

Intelligent systems have the potential to produce significant productivity improvements in many sectors of the economy, both in the short term and in the long term. Already, computer-controlled machines routinely perform economically valuable tasks in manufacturing, construction, transportation, business, communications, entertainment, education, waste management, hospital and nursing support, physical security, agriculture and food processing, mining and drilling, and undersea and planetary exploration.

As intelligent systems become widespread and inexpensive, productivity will grow and the rate of wealth production will increase. Intelligent machines in manufacturing and construction will increase the stock of wealth and reduce the cost of material goods and services. Intelligent systems in health care will improve services and reduce costs for the sick and elderly. Intelligent systems could make quality education available to all. Intelligent systems will make it possible to clean up and recycle waste, reduce pollution, and create environmentally friendly methods of production and consumption.

The potential impact of intelligent machines is magnified by that fact that technology has reached the point where intelligent machines have begun to exhibit a capacity for self-reproduction. John von Neumann (1966) was among the first to recognize that machines can possess the ability to reproduce. Using mathematics of finite state machines and Turing machines, von Neumann developed a theoretical proof that machines can reproduce. Over the past two decades, the theoretical possibility of machine reproduction has been empirically demonstrated (at least in part) in the practical world of manufacturing:

- Computers are routinely involved in the processes of manufacturing computers
- Computers are indispensable to the process of designing, testing, manufacturing, programming, and servicing computers
- On a more global scale, intelligent factories build components for intelligent factories

At a high level of abstraction, many of the fundamental processes of biological and machine reproduction are similar. Some might object to a comparison between biological and machine reproduction on the grounds that the processes of manufacturing and engineering are fundamentally different from the processes of biological reproduction and evolution. Certainly there are many essential differences between biological and machine reproduction. But the comparison is not entirely far-fetched. And the results can be quite similar. Both biological and machine reproduction can produce populations that grow exponentially. In fact, machine reproduction can be much faster than biological. Intelligent machines can flow from a production line at a rate of many per hour.

Perhaps more important, machines can evolve from one generation to the next much faster and more efficiently than biological organisms. Biological organisms evolve by a Darwinian process, through random mutation and natural selection. Intelligent machines evolve by a Lamarckian process, through conscious design improvements under selective pressures of the marketplace. In the machine evolutionary process, one generation of computers often is used to design and manufacture the next generation of more powerful and less costly computers. Significant improvements can occur in a very short time between one generation of machines and the next. As a result, intelligent machines are evolving extremely quickly relative to biological species. Improved models of computer systems appear every few months to vie with each other in the marketplace. Those that survive and are profitable are improved and enhanced. Those that are economic failures are abandoned. Entire species of computers evolve and are superceded within a single decade. In other words, machine reproduction, like biological reproduction, is subject to evolutionary pressures that tend to reward success and punish failure.

The ability of intelligent systems to reproduce and evolve will have a profound effect on the capacity for wealth production. As intelligent machines reproduce, their numbers will multiply, leading to an exponential increase in the intelligent machine population. Since intelligent machines can increase productivity and produce wealth, this implies that with each new generation of machine, goods, and services will become dramatically less expensive and more plentiful, while per capita wealth will increase exponentially.

The Prospects for Technology Growth

It is sometimes argued that technology, and therefore productivity, cannot grow forever because of the law of diminishing returns. It is argued that there must be a limit to everything, and therefore, productivity cannot grow indefinitely. Whether this is true in an abstract sense is an interesting philosophical question. Whether it is true in any practical sense is clear: it is not. From the beginning of human civilization until now, it remains a fact that the more that is known, the easier it is to discover new knowledge. And there is nothing to suggest that knowledge will be subject to the law of diminishing returns in the foreseeable future. Most of the scientists who have ever lived are alive and working today. Scientists and engineers today are better educated and have better tools with which to work than ever before. In the neurological and cognitive sciences, the pace of discovery is astonishing. The same is true in computer science, electronics, manufacturing, and many other fields. Today, there is an explosion of new knowledge in almost every field of science and technology.

There is certainly no evidence that we are nearing a unique point in history where progress will be limited by an upper bound on what there is to know. There is no reason to believe that such a limit even exists, much less that we are approaching it. On the contrary, there is good evidence that the advent of intelligent machines has placed us on the cusp of a growth curve where productivity can grow exponentially for many decades, if not indefinitely. Productivity growth is directly related to growth in knowledge. Growth in knowledge is dependent on the amount and effectiveness of investment in research, development, and education. This suggests that, given adequate investment in technology, productivity growth could return to 2.5 percent per year, which is the average for the 20th century. With higher rates of investment, productivity growth could conceivably rise to 4 percent, which is the average for the 1960-68 time frame. Conceivably, with sufficient investment, productivity growth could exceed 10 percent, which occurred during the period between 1939 and 1945 (Samuelson and Nordhaus 1989).

If such productivity growth were to occur, society could afford to improve education, clean up the environment, and adopt less wasteful forms of production and consumption. Many social problems that result from slow economic growth, such as poverty, disease, and pollution, would virtually disappear. At the same time, taxes could be reduced, Social Security benefits increased, and healthcare and a minimum income could be provided for all. The productive capacity of intelligent machines could generate sufficient per capita wealth to support an aging population without raising payroll taxes on a shrinking human labor force. Over the next three decades, intelligent machines might provide the ultimate solution to the Social Security and Medicare crisis. Benefits and services for an aging population could be continuously expanded, even in countries with stable or declining populations.

Military Power

Intelligent systems technologies have the potential to revolutionize the art of war. The eventual impact on military science may be as great as the invention of gunpowder, the airplane, or nuclear weapons. Intelligent weapons systems are already beginning to emerge. Cruise missiles, smart bombs, and unmanned reconnaissance aircraft have been deployed and used in combat with positive effect. Unmanned ground vehicles and computer-augmented command and control systems are currently being developed and will soon be deployed. Unmanned undersea vehicles are patrolling the oceans collecting data and gathering intelligence. These are but the vanguard of a whole new generation of military systems that will become possible as soon as intelligent systems engineering becomes a mature discipline (Gourley 2000).

In future wars, unmanned air vehicles, ground vehicles, ships, and undersea vehicles will be able to outperform manned systems. Many military systems are limited in performance because of the inability of the human body to tolerate high levels of temperature, acceleration, vibration, or pressure, or because humans need to consume air, water, and food. A great deal of the weight and power of current military vehicles is spent on armor and life support systems that would be unnecessary if there were no human operators on board. A great deal of military tactics and strategy is based on the need to minimize casualties and rescue people from danger. This would become unnecessary if warriors could remain out of harm's way.

Intelligent military systems will significantly reduce the cost of training and readiness. Compared to humans, unmanned vehicles and weapons systems will require little training or maintenance to maintain readiness. Unmanned systems can be stored in forward bases or at sea for long periods of time at low cost. They can be mobilized quickly in an emergency, and they will operate without fear under fire, the first time and every time.

Intelligent systems also enable fast and effective gathering, processing, and displaying of battlefield information. They can enable human commanders to be quicker and more thorough in planning operations and in replanning as unexpected events occur during the course of battle. In short, intelligent systems promise to multiply the capabilities of the armed forces, while reducing casualties and hostages and lowering the cost of training and readiness (Maggart and Markunas 2000).

Human Well-being

It seems clear that intelligent systems technology will have a profound impact on economic growth. In the long run, the development of intelligent machines could lead to a golden age of prosperity, not only in the industrialized nations, but throughout the world. Despite the explosion of material wealth produced by the first industrial revolution, poverty persists and remains a major problem throughout the world today. Poverty causes hunger and disease. It breeds ignorance, alienation, crime, and pollution. Poverty brings misery, pain, and suffering. It leads to substance abuse. Particularly in the third world, poverty may be the biggest single problem that exists, because it causes so many other problems. And yet there is a well-known cure for poverty. It is wealth.

Wealth is difficult to generate. Producing wealth requires labor, capital, and raw materials — multiplied by productivity. The amount of wealth that can be produced for a given amount of labor, capital, and raw materials depends on productivity. The level of productivity that exists today is determined by the current level of knowledge embedded in workers' skills, management techniques, tools, equipment, and software used in the manufacturing process. In the future, the level of productivity will depend more and more on the level of knowledge embedded in intelligent machines. As the cost of computing power drops and the skills of intelligent machines grow, the capability for wealth production will grow exponentially. The central question then becomes, how will this wealth be distributed?

In the future, new economic theories based on abundance may emerge to replace current theories based on scarcity. New economic institutions and policies may arise to exploit the wealth-producing potential of large numbers of intelligent machines. As more wealth is produced without direct human labor, the distribution of income may shift from wages and salaries to dividends, interest, and rent. As more is invested in ownership of the means of production, more people may derive a substantial income from ownership of capital stock. Eventually, some form of people's capitalism may replace the current amalgam of capitalism and socialism that is prevalent in the industrialized world today (Albus 1976; Kelso and Hetter 1967).

Summary and Conclusions

We are at a point in history where science has good answers to questions such as, "What is the universe made of?" and "What are the fundamental mechanisms of life?" There exists a wealth of knowledge about how our bodies work. There are solid theories for how life began and how species evolved. However, we are just beginning to acquire a deep understanding of how the brain works and what the mind is.

We know a great deal about how the brain is wired and how neurons compute various functions. We have a good basic understanding of mathematics and computational theory. We understand how to build sensors, process sensory information, extract information from images, and detect entities and events. We understand the basic principles of attention, clustering, classification, and statistical analysis. We understand how to make decisions in the face of uncertainty. We know how to use knowledge about the world to predict the future, to reason, imagine, and plan actions to achieve goals. We have algorithms that can decide what is desirable and plan how to get it. We have procedures to estimate costs, risks, and benefits of potential actions. We can write computer programs to deal with uncertainty and compensate for unexpected events. We can build machines that can parse sentences and extract meaning from messages, at least within the constrained universe of formal languages.

As computing power increases and knowledge grows of how the brain converts computational power into intelligent behavior, the ability of machines to produce greater wealth (i.e., goods and services that people want and need) will enable many possible futures that could never before have been contemplated. Even under very conservative assumptions, the possibilities that can be generated from simple extrapolations of current trends are very exciting. We are at a point in history where some of the deepest mysteries are being revealed. We are discovering how the brain processes information, how it represents knowledge, how it makes decisions and controls actions. We are beginning to understand what the mind is. We will soon have at our disposal the computational power to emulate many of the functional operations in the brain that give rise to the phenomena of intelligence and consciousness. We are learning how to organize what we know into an architecture and methodology for designing and building truly intelligent machines. And we are developing the capacity to experimentally test our theories. As a result, we are at the dawning of an age where the engineering of mind is feasible.

In our book (Albus and Meystel 2001), we have suggested one approach to the engineering of mind that we believe is promising, containing the following elements:

- *a perception system* that can fuse *a priori* knowledge with current experience and can understand what is happening, both in the outside world and inside the system itself

- *a world-modeling system* that can compute what to expect and predict what is likely to result from contemplated actions

- *a behavior-generating system* that can choose what it intends to do from a wide variety of options and can focus available resources on achieving its goals

- *a value judgment system* that can distinguish good from bad and decide what is desirable

We have outlined a reference model architecture for organizing the above functions into a truly intelligent system, hypothesizing that in the near future it will become possible to engineer intelligent machines with intentions and motives that use reason and logic to devise plans to accomplish their objectives.

Engineering of mind is an enterprise that will prove at least as technically challenging as the Apollo program or the Human Genome project. And we are convinced that the potential benefits for humankind will be at least as great, perhaps much greater. Understanding the mind and brain will bring major scientific advances in psychology, neuroscience, and education. A computational theory of mind may enable us to develop new tools to cure or control the effects of mental illness. It will certainly provide us with a much deeper appreciation of who we are and what our place is in the universe.

Understanding the mind and brain will enable the creation of a new species of intelligent machine systems that can generate economic wealth on a scale hitherto unimaginable. Within a half-century, intelligent machines might create the wealth needed to provide food, clothing, shelter, education, medical care, a clean environment, and physical and financial security for the entire world population. Intelligent machines may eventually generate the production capacity to support universal prosperity and financial security for all human beings. Thus, the engineering of mind is much more than the pursuit of scientific curiosity. It is more even than a monumental technological challenge. It is an opportunity to eradicate poverty and usher in a golden age for all humankind.

References

Albus, J.S. and A.M. Meystel. 2001. *Engineering of mind: An introduction to the science of intelligent systems.* New York: John Wiley and Sons.

Albus, J.S. 1976. *Peoples' capitalism: The economics of the robot revolution.* Kensington, MD: New World Books. See also Peoples' Capitalism web page at http://www.peoplescapitalism.org.

Bluestone, B., and B. Harrison. 2000. *Growing prosperity: The battle for growth with equity in the twenty-first century.* New York: Houghton Mifflin Co.

Carter, R. 1998. *Mapping the mind.* University of California Press.

Edelman, G. 1999. Proceedings of International Conference on Frontiers of the Mind in the 21st Century, Library of Congress, Washington D.C., June 15

Gourley, S.R. 2000. Future combat systems: A revolutionary approach to combat victory. *Army* 50(7):23-26 (July).

Kelso, L., and P. Hetter. 1967. *Two factor theory: The economics of reality.* New York: Random House.

Maggart, L.E., and R.J. Markunas. 2000. Battlefield dominance through smart technology. *Army* 50(7).

Mankiw, G.N. 1992. *Macroeconomics.* New York: Worth Publishers.

Moravec, H. 1999. *Robot: Mere machine to transcendent mind.* Oxford: Oxford University Press.

Samuelson, P., and W. Nordhaus. 1989. *Economics*, 13th ed. New York: McGraw-Hill.

Symposia. 1988. The slowdown in productivity growth. *Journal of Economic Perspectives* 2 (Fall).

Toffler, A. 1980. *The third wave.* New York: William Morrow and Co.

von Neumann, J. 1966. Theory of self-reproducing automata (edited and completed by A. Burks). Urbana: University of Illinois Press.

MAKING SENSE OF THE WORLD: CONVERGENT TECHNOLOGIES FOR ENVIRONMENTAL SCIENCE

Jill Banfield, University of California, Berkeley

Through the combination of geoscience, biology, and nano- and information technologies, we can develop a fundamental understanding of the factors that define and regulate Earth's environments from the molecular to global scale. It is essential that we capture the complex, interconnected nature of the processes that maintain the habitability of the planet in order to appropriately utilize Earth's resources and predict, monitor, and manage global change. This goal requires long-term investments in nanogeoscience, nanotechnology, and biogeochemical systems modeling.

Introduction

Looking to the future, what are the greatest challenges our society (and the world) faces? Ensuring an adequate food supply, clean air, and clean water are problems intimately linked to the environment. Given the rate of accumulation of environmental damage, it seems appropriate to ask, can science and technology solve the problems associated with pollution and global change before it is too late? Where should we invest our scientific and technological efforts, and what might these investments yield?

One of the mysteries concerning environmental processes is the role of extremely small particles that, to date, have defied detection and/or characterization. We now realize that materials with dimensions on the nanometer scale (intermediate between clusters and macroscopic crystals) are abundant and persistent in natural systems. Nanoparticles are products of, and substrates for, nucleation and growth in clouds. They are also the initial solids formed in water, soils, and sediments. They are generated in chemical weathering and biologically mediated redox reactions, during combustion of fuel, and in manufacturing. For example, nanoparticles are by-products of microbial energy generation reactions that utilize inorganic ions (e.g., Mn, Fe, S, U) as electron donors or acceptors. They are highly reactive due to their large surface areas, novel surface structures, and size-dependent ion adsorption characteristics and electronic structures (including redox potentials). It is likely that they exert a disproportionately large, but as yet incompletely defined, influence on environmental geochemistry because they provide a means for transport of insoluble ions and present abundant novel, reactive surfaces upon which reactions, including catalytic reactions, occur.

It is widely accepted that the most rapid growth in knowledge in recent years has occurred in the field of biology. In the environmental context, the biology of single-celled organisms represents a critically important focus, for several reasons. First, microbes are extraordinarily abundant. They underpin many of the biogeochemical cycles in the environment and thus directly impact the bioavailability of contaminants and nutrients in ecosystems. They are responsible for the formation of reactive mineral particles and contribute to mineral dissolution. With analysis of these connections comes the ability to use microbes to solve environmental problems. Second, microorganisms are relatively simple, hence detailed analysis of how they work represents a tractable problem. Third, microbes have invented ways to carry out chemical transformations via enzymatic pathways at low temperatures. These pathways have enormous industrial potential because they provide energetically inexpensive routes to extract, concentrate, and assemble materials needed by society. Identification of the relevant microbial enzymatic or biosynthetic pathways requires analysis of the full diversity of microbial life, with emphasis on organisms in extreme natural geologic settings where metabolisms are tested at their limits.

Where does our understanding of microbes and nanoparticles in the environment stand today? Despite the fact that microbes dominate every habitable environment on Earth, we know relatively little about how most microbial cells function. Similarly, we have only just begun to connect the novel properties and reactivity of nanoparticles documented in the laboratory to phenomena in the environment. Although our understanding of these topics is in its infancy, science is changing quickly. The center of this revolution is the combination of molecular biology, nanoscience, and geoscience.

The Role of Converging Technologies

In order to comprehensively understand how environmental systems operate at all scales, convergence of biological, technological, and geoscientific approaches is essential. Three important tasks are described below.

Identification and Analysis of Reactive Components in Complex Natural Systems

Nanoparticles and microorganisms are among the most abundant, most reactive components in natural systems. Natural nanoparticles (often < 5 nm in diameter) exhibit the same novel size-dependent properties that make their synthetic equivalents technologically useful. The functions of some microbial cell components (e.g., cell membranes, ribosomes) probably also depend on size-related reactivity. A challenge for the immediate future is determination of the origin, diversity, and roles of nanoparticles in the environment. Similarly, it is critical that we move from detecting the full diversity of microorganisms in most natural systems to understanding their ranges of metabolic capabilities and the ways in which they shape their environments. These tasks require integrated characterization studies that provide molecular-level (inorganic and biochemical) resolution.

Massive numbers of genetic measurements are needed in order to identify and determine the activity of thousands of organisms in air, water, soils, and sediments. Enormous numbers of chemical measurements are also required in order to characterize the physical environment and to evaluate how biological and geochemical processes are interconnected. This task demands laboratory and field data that is spatially resolved at the submicron-scale at which heterogeneities are important, especially in interfacial regions where reactions are fastest. The use of robots in oceanographic monitoring studies is now standard, but this is only the beginning. Microscopic devices are needed to make *in situ*, fine-scale measurements of all parameters and to conduct *in situ* experiments (e.g., to assay microbial population makeup in algal blooms in the ocean or to determine which specific organism is responsible for biodegradation of an organic pollutant in a contaminated aquifer). These devices are also required for instrumentation of field sites to permit monitoring over hundreds of meters to kilometer-scale distances. Development of appropriate microsensors for these applications is essential.

Environmental science stands to benefit greatly from nanotechnology, especially if new sensors are developed with environmental monitoring needs in mind. In the most optimistic extreme, the sensors may be sufficiently small to penetrate the deep subsurface via submicron-scale pores and be able to relay their findings to data collection sites. It is likely that these extremely small, durable devices also will be useful for extraterrestrial exploration (e.g., Mars exploration).

Monitoring Processes in the Deep Subsurface

Many of the inorganic and organic contaminants and nutrients of interest in the environment may be sequestered at considerable depths in aquifers or geological repositories. Methods are needed to image the structure of the subsurface to locate and identify these compounds, determine the nature of their surroundings, and monitor changes occurring during natural or enhanced *in situ* remediation. Examples of problems for study include detection of nanoparticulate metal sulfide or uranium oxide minerals produced by biological reduction, possibly via geophysical methods; analysis of the role of transport of nanoparticulate contaminants away from underground nuclear waste repositories; and monitoring of the detailed pathways for groundwater flow and colloid transport.

Development of Models to Assist in Analysis of Complex, Interdependent Phenomena

After we have identified and determined the distributions of the reactive inorganic and organic nanoscale materials in natural systems, it is essential that we understand how interactions between these components shape the environment. For example, we anticipate development and validation of comprehensive new models that integrate predictions of particle-particle organic aggregation and crystal growth with models that describe how aggregates are transported through porous materials in the subsurface. These developments are essential for prediction of the transport and fate of contaminants during and after environmental remediation.

Environmental processes operate across very large scales on continents and in the oceans. Thus, remote collection of high-resolution data sets (e.g., by satellite-based remote sensing) can also be anticipated. The large quantities of data from direct and indirect monitoring programs will benefit from new methodologies for information management. Mathematical models are essential to guide cognition and to communicate the principles that emerge from the analyses. An example of an ecosystem model is shown in Figure D.1. Input from the cognitive sciences will be invaluable to guide development of supermodels of complex processes.

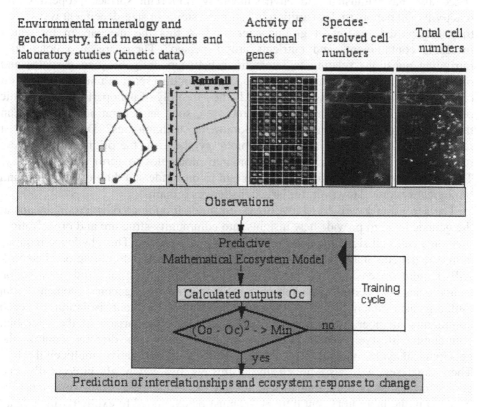

Figure D.1. Example of an ecosystem model that incorporates information about the physical and chemical environment with information about population size and structure and gene expression to analyze community interactions and predict response of the system to perturbations.

The Transforming Strategy

The first task toward an integrated understanding of the Earth's ecosystems is to identify and study the most important components. Focus on microorganisms is warranted based on their sheer abundance and metabolic versatility. The first two disciplinary partners, molecular biology and nanoscience, have already taken center stage with the integration of molecular biology and genome-enabled technologies (e.g., whole genome expression microarrays). In the next few years, these tools will allow us to decipher the full diversity of ways in which individual organisms grow, develop, reproduce, and evolve. These breakthroughs are critical to medicine, agriculture, and biologically assisted manufacturing and waste management.

Inorganic components also play key roles in natural systems. As noted above, exceedingly small particles intermediate in size between molecular clusters and macroscopic materials (nanoparticles) are abundant components of natural environments. Study of nanoparticle formation, properties, and stability is at the intersection of nanoscience, biology, chemistry, and geoscience. The unique characteristics of materials structured on the nanoscale have long been appreciated in the fields of materials science and engineering. It is now essential that we determine whether the nanoparticles in soils, sediments, water, the atmosphere, and space also have unusual and environmentally important surface properties and reactivity. Do nanoparticles partition in natural systems in size-dependent ways? Are they transported readily in groundwater, and is this the mechanism by which insoluble contaminants and nutrients are dispersed? There are also potentially intriguing questions relating to interactions between inorganic nanoparticles and organic molecules. For example, do nanoparticles in dust react in unusual ways with organic molecules (perhaps in sunlight)? Is the assembly of nanoparticles by organic polymers central to biomineralization processes, such as generation of bone? Can these interactions be harnessed for biomimetic technologies? Did reactions at nanoparticle surfaces play a role in prebiotic synthesis or the origin of life? Were nanoparticles themselves captured by organic molecules to form early enzymes? The answers to these questions are important to our understanding of inorganic and biological systems. However, far larger challenges remain.

The second task will be to investigate entire communities of microorganisms at the genetic level to provide new insights into community structure and organization, including cell-cell signaling and the partitioning of function. This challenge requires complete genetic analysis of all community members without cultivation. This task will require extension of current biological, computational, and information technologies to permit simultaneous reconstruction of genome content from multiorganism assemblages at the level of strains without isolation of each community member. Resulting data will also allow comparison of the microbial community lifestyle — characterized by the ability to directly control the geochemical cycles of virtually every element — to its alternative, multicellular life. These analyses will also unveil the pathways by which all biologically and geochemically important transformations are accomplished. This work must be initiated in the laboratory, but ultimately, must be expanded to explicitly include all environmental parameters and stimuli. Consequently, the task of understanding organisms in their environments stands before us as the third and longest-term task.

An additional component, geoscience, must be included in order to meet the challenge of molecularly resolved ecology. Environmental applications have lagged behind investigations of organisms in the laboratory because natural systems are extremely complicated. Critical environmental data include time-resolved measurements of the structure and organization of natural systems, organism population statistics, measurements of the levels of expression of all genes within communities of interacting species, and quantification of how these expression patterns are controlled by and control geochemical processes. This approach, which must ultimately include macroorganisms, will be essential for medical and agricultural, as well as environmental, reasons.

Education and Outreach

Analysis of complex systems, through integration of nanotechnology, nanoscience, geoscience, biology, ecology, and mathematics, will place special demands on the educational system. It will require training of a new generation of researchers with special experimental, communication, and quantitative reasoning skills. Because the task of ecosystem analysis is too large to be tackled in an individual project, it may be necessary to reconsider the structure of graduate student training programs. It is possible that traditional, carefully delineated, individual Ph.D. projects will be replaced by carefully integrated, collaborative Ph.D. research efforts that include individuals at all career levels. Changes such as this will have the added advantage of generating scientists that are able to work together to solve large, complicated problems.

The integration of science and technology to develop understanding of the environment should extend to all educational levels. For example, an effective use of nanotechnology may be to monitor processes in the vicinity of K-12 classrooms (e.g., bird migrations, air quality, pesticide degradation in soil) and to compare these data to those collected elsewhere. This may improve the public's appreciation of the Earth's environments as complex biogeochemical systems that change in definable and predictable ways as the result of human activities.

Conclusions

Molecularly resolved analyses of environmental systems will allow us to determine how increasingly complex systems, from the level of cells and microbial communities up to entire ecosystems at the planetary scale, respond to environmental perturbations. With this knowledge in hand, we can move toward rigorous determination of environmental state and prediction of ecosystem change.

High-resolution molecular- and nanometer-scale information from both inorganic and biological components of natural systems will dramatically enhance our ability to utilize microbial processes (such as light-harvesting molecules for solar cells or mineral-solubilizing enzymes for materials processing) for technological purposes. This may be of great importance if we are to reduce our dependence on energetically expensive manufacturing and existing energy resources. For example, bioleaching is an alternative to smelting, bioextraction is an alternative to electrochemistry, biosynthesis of polymers is an alternative to petroleum processing, biomineralization is an alternative to machine-based manufacturing. Ultimately, nano-bio-geo integration will allow us to tease apart the complex interdependencies between organisms and their surroundings so that we may ultimately gain sufficient

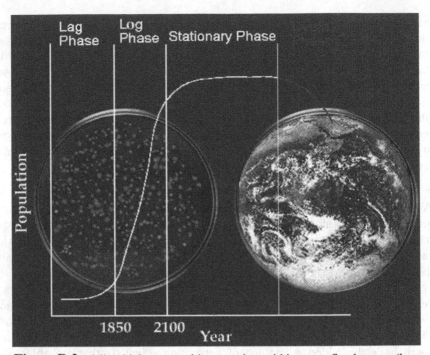

Figure D.2. Microbial communities growing within a confined space (here shown in a petri dish, left) have a cautionary tale to tell: overuse and/or unbalanced use of resources leads to build up of toxins, shortage of food, overpopulation, and death.

understanding of environmental systems to avoid the fate of microorganisms grown in a petri dish (Figure D.2).

FUNDAMENTALLY NEW MANUFACTURING PROCESSES AND PRODUCTS

M.C. Roco, National Science Foundation, Chair, National Science and Technology Council's Subcommittee on Nanoscale Science, Engineering, and Technology (NSET)[2]

Integration of NBIC tools is expected to lead to entirely new categories of materials, devices, and systems for use in manufacturing, construction, transportation, medicine, emerging technologies and scientific research. Fundamental research will be at the confluence of physics, chemistry, biology, mathematics and engineering. Nanotechnology, biotechnology and information technology will play an essential role in their research, design and production. For example, knowledge about the molecular level processes describing the growth and metabolism of living cells may be applied, through analogy, to development of man-made materials. New products may include pharmaceutical genomics, electronic devices with three-

---

[2] The views in this essay do not necessarily represent the views of the National Science Foundation or NSET.

dimensional architectures, large databases, software for realistic mutiphenomena/multiscale simulation of systems from basic principles; and quantitative studies in social sciences. Cognitive sciences will provide better ways to design and use the new manufacturing processes, products and services.

Several paradigm changes are expected in each of the four NBIC domains:

- Nanotechnology is expected to move from its current focus on scientific discovery towards technological innovation, leading to systematic manufacturing methods for mass production

- Biotechnology will move towards molecular medicine and pharmaceutical genomics, and biosystems will be integrated on an increased rate in advanced materials and systems

- In information technology, the quest for smallness and speed will be enhanced by the focus for new architectures, three-dimensional design, functionality and integration with application driven developments in areas such as biosystems and knowledge based technologies;

- Cognitive sciences will focus on explanation of the mind and human behavior from understanding of physico-chemical-biological processes at the nanoscale, and an increased attention will be given to human-technology co-evolution.

Broader range of research and development issues will need to be addressed. For example, one should consider the following challenges in nanoscale engineering:

- Three dimensional material/device/system spatial/temporal architectures
- Directed assembling/patterning/tem plating for heterogeneous nanosystems
- Hybrid and bio nanosystems for medicine and manufacturing
- Energy conversion and transfer
- Mutiphenomena, multiprocessors, multiscale design, including large scale atomistic modeling and simulation
- Integration of nanoscale into larger scales: creation and use of intermediary standard components
- Thermal and chemical stability
- Operational and environmental safety
- Reliability and reproductibity at the nanoscale

Fundamentally new technologies and products will develop. Examples are pharmaceutical genomics, neuromorphic technology, regenerative medicine, biochips with complex functions, molecular systems with multiscale architectures, realistic simulation of phenomena and processes from the nanoscale, and new flight vehicles.

A new infrastructure based on four NBIC research and development platforms will be necessary, to be available anywhere, in short time, to any industry and all those interested.

Adopting new indices such as a modified GNP to include changes in the human condition, impact on the environment, preparation of the infrastructure (including education) and other societal implications will be necessary in order to measure and better evaluate the outcomes. New holistic criteria, such as reducing the entropy of a system (i.e., less energy dissipation per computation and transmission of information; less material, energy, water, pollution in nanotechnology; and less change/degradation in biotechnology), may be considered.

Already NSF, NASA, EPA, DOD, DOE have several seed projects in developing R&D strategy based on unifying science and technology integration, creating infrastructure for research at the confluence of two or more NBIC domains, developing neuromorphic engineering, improving human performance, advancing changes in education including "learning how to learn", and preparing for societal implications of converging technologies.

VISIONARY PROJECTS

THE COMMUNICATOR: ENHANCEMENT OF GROUP COMMUNICATION, EFFICIENCY, AND CREATIVITY

Philip Rubin, Murray Hirschbein, Tina Masciangioli, Tom Miller, Cherry Murray, R.L. Norwood, and John Sargent

As envisioned, *The Communicator* will be a "smart," multifaceted, technical support system that relies on the development of convergent technologies to help enhance human group communication in a wide variety of situations, including meetings (both formal and informal), social exchanges, workplace collaborations, real-world corporate or battle training situations, and educational settings. This system will rely on expected advances in nanotechnology, fabrication, and a number of emerging information technologies, both software and hardware. In this system, these technologies will be tightly coupled with knowledge obtained from the biological and cognitive domains. The convergence of these technologies will serve to enhance existing attributes of individuals and remove barriers to group communication. This system will consist of a set of expanding implementations of these convergent technologies, growing more complex as the individual technologies mature over time. Some of these implementations are described below.

The initial goal of The Communicator is simple: to remove the kinds of barriers that are presently common at meetings where participants rely for communication on similar but slightly varying technologies. For example, it is standard for meeting participants to use software such as *PowerPoint* to present their ideas, but they often encounter technical difficulties moving between computers and computer platforms different from those on which they created their presentations. The transfer of information between systems during meetings is often hampered by varying media, connector differences, and incompatible data standards. At its simplest level, The Communicator would serve as an equalizer for communication in such situations, detecting the technological requirements of each participant and automatically resolving any differences in the presentation systems. The transfer and presentation of information would then become transparent.

Moving beyond this initial implementation, The Communicator would serve to remove more significant communication barriers, such as those related to physical disabilities or language differences. For example, the system, once apprised of a group member's hearing impairment, could tailor a presentation to that participant's needs by captioning the spoken or other auditory information. Similarly, it could produce auditory transcriptions of information presented visually in a group situation for any visually impaired member of the group. It could also provide simultaneous translation of meeting proceedings into a number of languages.

At the heart of The Communicator system are nano/info technologies that will allow individuals to carry with them electronically stored information about themselves that they can easily broadcast as needed in group situations. Such information might include details about preferences, interests, and background. Early implementations of this approach are do-able now.

An even more interesting and advanced implementation would consist of detection and broadcast of the physiological and affective states of group participants with the purpose of providing resources to individuals and tailoring interactivity in order to allow the group to more easily achieve its goals. Detection of participants' physiological and affective states would be determined by monitoring biological information (such as galvanic skin response and heart rate) and cognitive factors via pattern recognition (such as face recognition to detect facial emotion, and voice pitch analysis to detect stress levels). Based on determinations of the needs and physical and cognitive states of participants, The Communicator could tailor the information it supplies to each individual, providing unique resources and improving productivity. Participants would have the ability to define or restrict the kinds of information about themselves that they would be willing to share with other members of the group.

As an example of this implementation, in an international conference or tribunal, each participant could select simultaneous translation of the discourse. Through PDA-like devices or biopatches, the system could measure the empathy levels or stress levels of all negotiators. A personal avatar would serve as a "coach" for each individual, recalling past statements, retrieving personal histories, and functioning as a research assistant to prepare material for use in arguments and deliberations. The system would facilitate the building of consensus by identifying areas of nominal disagreement and searching for common values and ideas.

Beyond facilitation of group communication, The Communicator could also serve as an educator or trainer, able to tailor its presentation and to operate in a variety of modes, including peer-to-peer interaction and instructor/facilitator interaction with a group. The Communicator would function as an adaptive avatar, able to change its personal appearance, persona, and affective behavior to fit not only individuals or groups but also varying situations.

Design Components

Several important distinct components of The Communicator, introduced below, should be carefully designed to work together seamlessly.

The Individual Information Component

One key element of The Communicator system will be a model or record of each individual, including how each individual interacts with the environment and how s/he prefers to do so. This would include information like the language spoken, the preferred sensory channel, and limitations on input and output. This system should also include characteristics of the individual's cognitive capabilities: learning speed, preferences for learning modalities, areas of expertise, leisure activities, history of important social events, and other attributes that are relevant to a given task or situation. Ways that this element could be applied include the following:

- Using bioauthentication, the system could identify each individual in a group, including specific kinds of information about each individual. This could shorten the initial socialization process in a group setting.

- Users would be able to specify that they receive input translated into specific languages, including captioning or signing if needed.

- The system could determine what stress levels, information density, and learning rates work best for the individuals and the group as a whole.

- The system could provide support so that the individual learns in whatever modality works best for the topic at hand: auditory, haptic, text, images, virtual reality, and any specialized modality with which the user is comfortable. This would include such applications as using sounds to guide an individual through unknown territory if the person's vision and other senses are already monopolized by other inputs.

The Avatar Component

Another key element of The Communicator system will be avatars that can take on human appearance and behavior in a 3-D environment. They should be human-sized with full human fidelity, especially with respect to facial characteristics and emotion. The avatars should be able to assume any human form that is desired or most suitable (in terms of race, gender, and age, for example). The avatars' persona, mode of communication, and language should be able to be modified over time as the system learns the best method of communication or training for each individual.

The avatars should be life-like, so people will respond to them as though they are real. Avatars should be "in-a-box" and able to be placed and projected wherever needed, whether on a screen, as a hologram in the middle of a room, or through virtual reality viewers.

Possible applications include the following:

- Avatars could represent the human participants in a group to each other.

- They could also represent autonomous computerized agents that perform particular functions of the information and communication system.

- Avatars could be sent into dangerous situations, for example, to negotiate with a criminal holding a hostage.

- They could function as a resident nurse to the sick or as a companion to the elderly.

- An individual could perceive what his or her personal avatar encounters, e.g., "feeling" the presence of a biohazard or radiation in a dangerous environment while remaining immune to its harm.

- A training avatar (or a human tutor) could teach a person new skills by sharing the experience, for example, via a haptic suit, that could train a person in the physical movements required for dance, athletics, weaponry, or a refined manual skill such as surgery.

The Environmental Interface Component

A third key element of The Communicator system will be its interfaces with the surrounding "environmental network," creating the opportunity for enhanced, personalized communications and education. Characteristics of how humans interact with information and technology can be viewed as constraints, or they can be viewed as strengths that convergent technology can play to. For example, if an individual is good at detecting anomalies or patterns in data, the technology would enhance this capability. Perhaps the technology would provide a "rheostat" of sorts to increase or decrease the contrast in data differences. This interface is a two-way street. The environment knows who is present, and each user receives appropriate information in the preferred form.

- The transforming strategy would apply known neural assessment techniques along with standard educational objectives, progressing to full cogno-assisted individualized learning in a group setting or collaborative learning.

- The system would be useful for teleconferencing, since participants need not be in the same location.

- The system should make it possible to adjust the social structure of communications, from whole-group mode in which all parties receive all messages, to more structured communication networks in which subgroups and individuals play specialized roles.

Key design considerations of The Communicator include the following:

- Very high-speed communications are needed, whether cable or wireless.

- The human-computer interface should be a wearable system offering augmented reality in office, schoolroom, factory, or field situations.

Educational Applications

All communication involves learning, but an *Educator* version of The Communicator could be created that would enhance many kinds of education. The convergence of NBIC technologies can radically transform the teaching and learning process and maximize the sensory and cognitive abilities of students. Some examples of applications include assistance to the learning disabled, optimally timed and individually presented learning experiences, and learning in a collaboratively orchestrated environment.

Several strategies could be employed to implement the Educator vision, geared for either individuals or groups, the classroom or the field. In the K-12 educational experience, a personal avatar or "coach" could govern hands-on experiments in accomplishing such goals as learning reading, science, math, or foreign languages. It would "teach" students as a human teacher does but would optimize itself to the needs of the student. It would be patient, friendly, stern, or take on any appropriate behavior. It might be most suitable for younger students but could also be a mentor for adults. If needed, it could be a "copilot." In a work environment, it could not only teach prepared lessons but also monitor performance and instruct on how to improve it.

The system could merge the following technologies:

- biotechnology to assess the physiological and psychological state of the learner, sense moods and states of mind
- cognitive science and technology to present responsive and individualized presentations of material to the student through different modalities
- expert information technology to accumulate and supply educational information

Military training could employ The Educator to teach decision-making under stress in a battlefield game in which the battlefield is virtual and the soldier is the general. As the war game is played, the avatar could be the general's assistant, read out after the battle. In another scenario, the virtual battlefield could be in a real field where soldier-participants wear wireless PDA helmets. The system could also be used as a "decision-making under stress" teaching tool for corporate executives.

Educator avatars could assume a wide variety of images (male, female, young, old) and be capable of speaking in all languages (oral and otherwise); identifying individual learning styles and then adapting curricula to individual needs and using access to biological data to determine which methods are most effective for the assimilation and retention of knowledge. This could effectively improve education and training in all arenas from preschool through graduate school and across the corporate and military environments. It would equalize educational opportunities for all, enable learners to move through material at their own pace, and ensure that knowledge of learning styles would be retained and carried forward from year to year as children move from teacher to teacher or adults move from job to job.

Social Equalization

The adaptive capabilities of The Communicator would have the potential in group interactions to minimize the biases that arise from a variety of factors such as physical size and posture, gender, race, language, culture, educational background, voice tone and volume, and physical ability or disability. The result would be to maximize both individual and group performance. Examples include enhancing the performance of a poor learner, an athlete, or a soldier, and improving communication, collaboration, and productivity among people with a multitude of differences. Thus, the system would be not only a Communicator and Educator, but an *Equalizer* as well, enhancing human awareness, removing disabilities, and empowering all members of society.

On a more fundamental level, such a smart device could have a tremendous impact on the most disadvantaged people around the world, those who lack clean drinking water, adequate food supplies, and so on. Despite the lack of physical infrastructure like telephone cables, wireless Communicator technology could offer them the world of information in a form they can immediately use. Such knowledge will improve their agricultural production, health, nutrition, and economic status. No longer isolated from the global economic and cultural system, they will become full and valued participants.

Convergence

The Communicator system will incorporate each of the four NBIC technologies:

- *Nanotechnology* will be required to produce high-speed computational capabilities, wearable components that consume little energy, and pervasive sensors.

- *Biotechnology* will be fundamental to the interfaces, to monitoring the physical status of participants, and to the general design of human-friendly technologies.

- *Information technology* will be responsible for data management and transmission, translation across modalities and languages, and development of avatars and intelligent agents.

- *Cognitive science* will provide the understanding of effective learning styles, methods for elimination of biases, and the directions in which to search for common values and ideas that will be the foundation of a new form of social cooperation.

Some elements of The Communicator can be created today, but the full system will require aggressive research across all four of the convergent NBIC fields. Implementation of the entire vision will require an effort spanning one or two decades, but the payoff will be nothing less than increased prosperity, creativity, and social harmony.

ENHANCED KNOWLEDGE-BASED HUMAN ORGANIZATION AND SOCIAL CHANGE

Kathleen M. Carley, Carnegie Mellon University

Changes that bring together nanotechnology, information science, biology, and cognition have the potential to revolutionize the way we work and organize society. A large number of outcomes are possible. At the same time, existing social forms, legislation, and culture will limit and direct the potential outcomes. In a very real sense, technologies and societies, tools and cultures, capabilities and legislation will co-evolve. Without attempting to predict the future, a series of possible outcomes, issues, and research challenges are discussed. Particular emphasis is placed on issues of security and potentially radical change within groups, organizations, and society.

Data and Privacy

In the area of bioterrorism, a key issue is early detection or "biosurveillance." Early detection requires smart sensors at the biological level in the air, water, and ground and on humans. Early detection requires integrating this data with geographic, demographic, and social information. Even were the sensors to exist, there would still be a problem: Under current legislation and privacy laws, the data cannot be integrated and made readily accessible to practitioners and researchers. To develop and test data mining tools, knowledge management tools, and what-if policy simulators, access is needed to a wide range of data in real time; but, providing access to such data enables the users of these tools to "know" details of individual behavior.

In the area of organizations, a key issue is team design and redesign (Samuelson 2000). Team design and redesign requires accurate data of who knows what, can work with whom, and is currently doing what. Doing such a skill audit, network analysis, and task audit is a daunting task. Maintaining the information is even more daunting. Individuals are loathe to provide the information for fear of losing their basis of power or anonymity, or for fear of reprisal. However, much of the information is implicit in the locations that people occupy, their stress levels, webpages, curricula vitae, public conversations, and so on.

In the cases of both acquiring and maintaining individual data, all of the following can be used to enable better outcomes: nano-bio-sensors that are embedded in the body and that report on individual health, stress level, and location; intelligent surfaces that track who is present while reshaping themselves to meet the needs of and enhance the comfort of the users; auto-sensors that create a memory of what is said when people cough or sneeze; air and water sensors that sense contaminants; data-mining tools that locate information, simulation tools that estimate the change in social outcomes; information assurance tools and secure distributed databases. Indeed, such tools are critical to the collection, analysis, protection, and use of information to enhance group performance. The relatively easy problems here will be those that are dominated by technology, e.g., distributed database tools, data integration procedures, information assurance technology, and smart sensors. Those problems dealing with the need to change cultures, legislation, and ways of working will be more difficult. Privacy laws, for example, could mitigate the effectiveness of these tools or even determine whether they are ever developed. There are many critical privacy issues, many of which are well identified in the NRC report, *The Digital Dilemma* (http://www.nap.edu/catalog/9601.html). Views of knowledge as power will limit and impede data collection. Having such data will revolutionize healthcare, human resources, career services, intelligence services, and law enforcement. Having such data will enable "big-brotherism."

Were we able to overcome these two mitigating factors, then a key issue will become, "What will the bases for power be when knowledge is no longer a controlled commodity?" Since many organizations are coordinated and managed through the coordination and management of information, as knowledge is no longer controlled, new organizational forms should emerge. For example, a possible result might be the development of monolith corporations with cells of individuals who can do tasks, and as those tasks move from corporation to corporation, the cells would move as well. In this case, benefits, pay scales, etc., would be set outside the bounds of a traditional corporation. In this case, individual loyalty would be to the area of expertise, the profession, and not the company. Corporations would become clearinghouses linking agents to problems as new clients come with new problems.

Ubiquitous Computing and Knowledge Access

As computers are embedded in all devices, from pens to microwaves to walls, the spaces around us will become intelligent (Nixon, Lacey, and Dobson 1999; Thomas and Gellersen 2000). Intelligent spaces are generally characterized by the potential for ubiquitous access to information, people, and artificial agents, and the provision of information among potentially unbounded networks of agents (Kurzweil 1988). The general claim is that ubiquitous computing will enable everyone to have access to all information all the time. In such an environment, it is assumed that inequities will decrease. This is unlikely. While ubiquitous computing will enable more people to access more information more of the time, there will still be, short of major reforms, people with little to no access to computing. There will be excess information available, information making it difficult to discern true from false information. There will be barriers in access to information based on legislation, learning, and organizational boundaries. While information will diffuse faster, the likelihood of consensus being reached and being accurate given the information will depend on a variety of other factors such as group size, the complexity of the task and associated knowledge, initial distribution of information in the group, and so on. As a result, things may move faster, but not necessarily better.

Initial simulation results suggest that even when there are advanced IT capabilities, there will still be pockets of ignorance, certain classes of individuals will have privileged access to information and the benefits and power that derive from that, groups will need to share less information to be as or more effective, databases may decrease shared knowledge and guarantee information loss, and smaller groups will be able to perform as well or better than larger groups (Alstyne, M. v., and Brynjolfsson, E. 1996; Carley 1999). To address issues such as these, researchers are beginning to use multiagent network models. These models draw on research on social and organizational networks (Nohira and Eccles 1992), advances in network methodology (Wasserman and Faust 1994), and complex system models such as multiagent systems (Lomi and Larsen 2001). In these models, the agents are constrained and enabled by their position in the social, organizational, and knowledge networks. These networks influence who interacts with whom. As the agents interact, they learn, which in turn changes with whom they interact. The underlying networks are thus dynamic. The results suggest that organizations of the future might be flatter, with individuals coming and going from teams based on skills, that is, what they know, and not whom they know. As a result, social life will become more divorced from organizational life. Initial simulation results suggest that if information moves fast enough, decisions will become based not as much on information as on the beliefs of others; this should be particularly true of strategic decisions.

Socially Intelligent Technology

Major improvements in the ability of artificial agents to deal with humans and to emulate humans will require those artifacts to be socially intelligent. Socially intelligent agents could serve as intelligent tutors, nannies, personal shoppers, etc. Sets of socially intelligent agents could be used to emulate human groups/organizations to determine the relative efficacy, feasibility, and impact of new technologies, legislation, change in policies, or organizational strategy. At issue are questions of how social these agents need to be and what is the basis for social intelligence. It is relatively easy to create artificial agents that are more capable than a human for a specific well-understood task. It is relatively easy to create artificial agents that can, in a limited domain, act like humans. But these factors do not make the agents generally socially intelligent. One of the research challenges will be for computer scientists and social scientists to work together to develop artificial social agents. Such agents should be social at both the cognitive and precognitive (bio) level. Current approaches here are software- limited. They are also potentially limited by data; nanotechnology, which will enable higher levels of storage and processing, will also be necessary. That is, creating large numbers of cognitively and socially realistic agents is technically unfeasible using a single current machine. Yet, such agents need to exist on a single machine if we are to use such tools to help individuals manage change.

A key component of social intelligence is the ability to operate in a multiagent environment (Epstein and Axtell 1997; Weiss 1999). However, not all multiagent systems are composed of socially intelligent agents. For a machine to be socially intelligent, it needs to be able to have a "mental" model of others, a rich and detailed knowledge of realtime interaction, goals, history, and culture (Carley and Newell 1994). Socially intelligent agents need transactive memory, i.e., knowledge of who knows whom (the social network), who knows what (the knowledge network), and who is doing what (the assignment network). Of course this memory need not be accurate. For agents, part of the "socialness" also comes from being limited cognitively. That is, omniscient agents have no need to be social, whereas, as agents become limited — boundedly rational, emotional, and with a specific cognitive architecture — they become more social.

One of the key challenges in designing machines that could have such capabilities is determining whether such machines are more or less effective if they make errors like humans do. What aspects of the constraints on human cognition, such as the way humans respond to interrupts, the impact of emotions on performance, and so on, are critical to acquiring and acting on social knowledge? While we often see constraints on human cognition as limitations, it may be that social intelligence itself derives from these limitations and that such social intelligence has coordinative and knowledge benefits that transcend the limitations. In this case, apparent limits in individuals could actually lead to a group being more effective than it would be if it were composed of more perfect individual agents (Carley and Newell 1994).

A second key challenge is rapid development. Computational architectures are needed that support the rapid development of societies of socially intelligent agents. Current multiagent platforms are not sufficient, as they often assume large numbers of cognitively simple agents operating in a physical grid space as opposed to complex intelligent, adaptive, learning agents with vast quantities of social knowledge operating in social networks, organizations, and social space. Moreover, such platforms need to be extended to enable the co-evolution of social intelligence at the individual, group, and organizational level at differing rates and to account for standard human processes such as birth, death, turnover, and migration.

A third challenge is integrating such systems, possibly in real time, with the vast quantities of data available for validating and calibrating these models. For example, how can cities of socially intelligent agents be created that are demographically accurate, given census data?

Socially Engineered Intelligent Computer Anti-Viruses and DDOS Defenses

Computer viruses have caused significant financial losses to organizations (CSI 2000). Even though most organizations have installed anti-virus software in their computers, a majority of them still experience infections (ICSA 2000). Most anti-virus software can not detect a new virus unless it is patched with a new virus definition file. New virus countermeasures have to be disseminated once a new virus is discovered. Studies of viruses demonstrate that the network topology and the site of the initial infection are critical in determining the impact of the virus (Kephart 1994; Wang 2000; Pastor-Satorras 2001). What is needed is a new approach to this problem. Such an approach may be made possible through the use of socially intelligent autonomous agents.

The Web and the router backbone can be thought of as an ecological system. In this system, viruses prey on the unsuspecting, and distributed denial of service attacks (DDOS) spread through the networks "eating" or "maiming" their prey. Viruses are, in a sense, a form of artificial life (Spafford 1994). One approach to these attacks is to propagate another "species" that can in turn attack these attackers or determine where to place defenses. Consider a computer anti-virus. Computer anti-viruses should spread fixes and safety nets, be able to "eat" the bad viruses and restore the machines and data to various computers without, necessarily, the user's knowledge. Such anti-viruses would be more effective if they were intelligent and able to adapt as the viruses they were combating adapted. Such anti-viruses would be still more effective if they were socially intelligent and used knowledge about how people and organizations use computers and who talks to whom in order to assess which sites to infiltrate when. We can think of such anti-viruses as autonomous agents that are benign in intent and socially intelligent.

Social Engineering

Combined nano-, bio-, info-, and cogno-technologies make it possible to collect, maintain, and analyze larger quantities of data. This will make it possible to socially engineer teams and groups to meet the demands of new tasks, missions, etc. The issue is not that we will be able to pick the right combination of people to do a task; rather, it is that we will be able to pick the right combination of humans, webbots, robots, and other intelligent agents, the right coordination scheme and authority scheme, the right task assignment, and so on, to do the task while meeting particular goals such as communication silence or helping personnel stay active and engaged. Social engineering is, of course, broader than just teams and organizations. One can imagine these new technologies enabling better online dating services, 24/7 town halls, and digital classrooms tailored to each student's educational and social developmental level.

The new combined technologies are making possible new environments such as smart planes, "living" space stations, and so on. How will work, education, and play be organized in these new environments? The organizational forms of today are not adequate. Computational organization theory has shown that how groups are organized to achieve high performance depends on the tasks, the resources, the IT, and the types of agents. You simply do not coordinate a group of humans in a board room in the same way that you would coordinate a group of humans and robots in a living space station or a group of humans who can have embedded devices to enhance their memory or vision.

Conclusion

These areas are not the only areas of promise made possible by combining nano-, bio-, info-, and cogno-technologies. To make these and other areas of promise turn into areas of advancement, more interdisciplinary research and training is needed. In particular, for the areas listed here, joint training is needed in computer science, organizational science, and social networks.

References

Alstyne, M.v., and E. Brynjolfsson. 1996. Wider access and narrower focus: Could the Internet Balkanize science? *Science* 274(5292):1479-1480.

Carley, K.M. forthcoming, Smart agents and organizations of the future. In *The handbook of new media*, ed. L. Lievrouw and S. Livingstone.

_____. forthcoming. Computational organization science: A new frontier. In Proceedings, Arthur M. Sackler Colloquium Series on Adaptive Agents, Intelligence and Emergent Human Organization: Capturing Complexity through Agent-Based Modeling, October 4-6, 2001; Irvine, CA: National Academy of Sciences Press.

_____. forthcoming, Intra-Organizational Computation and Complexity. In *Companion to Organizations,* ed. J.A.C. Baum. Blackwell Publishers.

Carley, K.M., and V. Hill. 2001. Structural change and learning within organizations. In *Dynamics of organizations: Computational modeling and organizational theories,* ed. A. Lomi and E.R. Larsen. MIT Press/AAAI Press/Live Oak.

Carley, K.M. 1999. Organizational change and the digital economy: A computational organization science perspective. In *Understanding the Digital Economy: Data, Tools, Research,* ed. E. Brynjolfsson, and B. Kahin. Cambridge, MA: MIT Press.

Carley, K.M., and A. Newell. 1994. The nature of the social agent. *J. of Mathematical Sociology* 19(4): 221-262.

CSI. 2000. CSI/FBI computer crime and security survey. *Computer Security Issues and Trends.*

Epstein, J., and R. Axtell. 1997. *Growing artificial societies.* Boston, MA: MIT Press.

ICSA. 2000. ICSA Labs 6th Annual Computer Virus Prevalence Survey 2000. ICSA.net.

Kephart, J.O. 1994. How topology affects population dynamics. In *Artificial life III,* ed. C.G. Langton. Reading, MA: Addison-Wesley.

Kurzweil, R. 1988. *The age of intelligent machines.* Cambridge, MA: MIT Press.

Lomi, A., and E.R. Larsen, eds. 2001. *Dynamics of organizations: Computational modeling and organizational theories.* MIT Press/AAAI Press/Live Oak.

Nixon, P., G. Lacey, and S. Dobson, eds. 1999. Managing interactions in smart environments. In Proceedings, 1st International Workshop on Managing Interactions in Smart Environments (MANSE '99), Dublin, Ireland, December 1999.

Nohira, N. and R. Eccles, eds. 1992. *Organizations and networks: Theory and practice.* Cambridge, MA: Harvard Business School Press.

Pastor-Satorras, R., and A. Vespignani. 2001. Epidemic dynamics and endemic states in complex networks. Barcelona, Spain: Universitat Politecnica de Catalunya.

Samuelson, D. 2000. Designing organizations. *OR/MS Today.* December: 1-4. See also http://www.lionhrtpub.com/orms/orms-12-00/samuelson.html.

Spafford, E.H. 1994. Computer viruses as artificial life. *Journal of Artificial Life.*

Thomas, P. and H.-W. Gellersen, eds. 2000. Proceedings of the International Symposium on Handheld and Ubiquitous Computing: Second International Symposium, HUC 2000, Bristol, UK, September 25-27, 2000 .

Wang, C., J.C. Knight, and M.C. Elder. 2000. On computer viral infection and the effect of immunization. In Proceedings, IEEE 16th Annual Computer Security Applications Conference.

Wasserman, S. and K. Faust. 1994   *Social Network Analysis.* New York: Cambridge University.

Weiss, G., ed. 1999. *Distributed artificial intelligence.* Cambridge, MA: MIT Press.

A VISION FOR THE AIRCRAFT OF THE 21ST CENTURY

S. Venneri, M. Hirschbein, M. Dastoor, National Aeronautics and Space Administration

The airplane will soon be 100 years old. Over that period of time, it has evolved from the cloth and wood biplanes of the 1920s to the first all-metal single-wing aircraft of the 1930s, to the 100-passenger commercial transports of the 1950s, to the modern jet aircraft capable of reaching any point in the world in a single day. Nevertheless, the design of the modern airplane really has not changed much in the last 50 years. The grandfather of the Boeing 777 was the Boeing B-47 bomber designed in the late 1940s. It had a sleek, tubular aluminum fuselage, multiple engines slung under swept wings, a vertical tail, and horizontal stabilizers. Today, the fuselage is lighter and stronger, the wings more aerodynamic, and the engines much more efficient, but the design is a recognizable descendent of the earlier bomber.

The aircraft of the 21st century may look fundamentally different (Figure D.3).

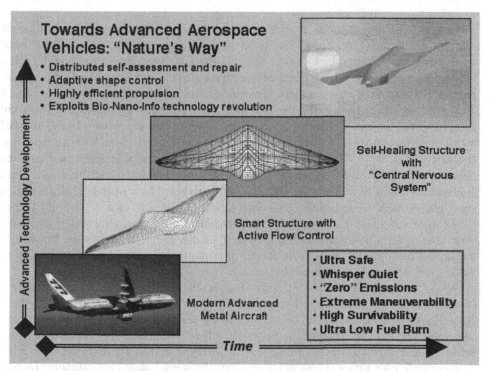

Figure D.3. Towards advanced aerospace vehicles: "Nature's Way."

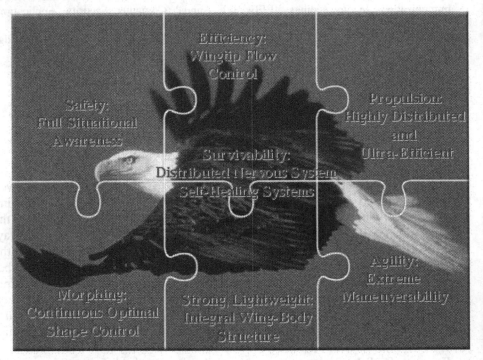

Figure D.4. Inspiration for the next generation of aircraft.

NASA is beginning to look to birds as an inspiration for the next generation of aircraft — not as a "blueprint," but as a biomimetic mode (Figure D.4). Birds have evolved over the ages to be totally at home in the air. Consider our national bird, the eagle. The eagle has fully integrated aerodynamic and propulsion systems. It can morph and rotate its wings in three dimensions and has the ability to control the air flow over its wings by moving the feathers on its wingtips. Its wings and body are integrated for exceptional strength and light weight. And the wings, body, and tail work in perfect harmony to control aerodynamic lift and thrust and balance it against the force of gravity. The eagle can instantly adapt to variable loads and can see forward and downward without parallax. It has learned to anticipate the sudden drag force on its claws as it skims the water to grab a fish and how to stall its flight at just the right moment to delicately settle into a nest on the side of a cliff. The eagle is made from self-sensing and self-healing materials. Its skin, muscle, and organs have a nervous system that detects fatigue, injury, or damage, and signals the brain. The eagle will instantly adapt to avoid further trauma, and tissues immediately begin to self-repair. The eagle is designed to survive.

NASA is pursuing technology today that is intended to lead toward just such a biomimetically inspired aircraft (Figure D.5). Advanced materials will make them lighter and more efficient to build. Advanced engines will make them fast and efficient. The airframe, engine, and cockpit will be "smarter." For decades, aircraft builders have worked to build wings that are stronger and stiffer. However, the wing that is needed for take-off and landing is not the wing needed for cruising. During

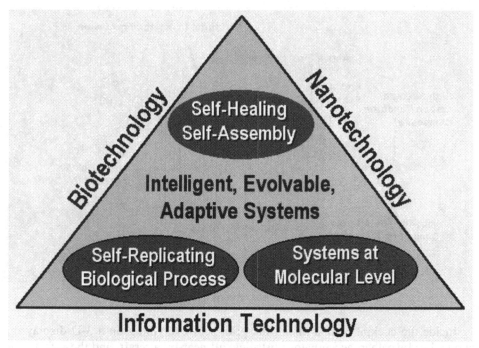

Figure D.5. Revolutionary technology vision as applied to future aircraft.

take-off and landing, the wing needs to be highly curved from leading edge to ` trailing edge to produce enough lift at low speed. But this also produces a lot of drag. Once airborne, the wing needs to be flat for minimal drag during cruise. To change the wing shape, NASA has employed leading-edge slats — an articulated "nose" that runs along the length of the wing — and multipiece flaps that can drop the trailing edge of the wing by 60 degrees. All of this requires gears, motors, and hydraulic pumps.

Imagine a bird-like wing of the future. It is not built from multiple, mechanically connected parts. It is made from new smart materials that have imbedded sensors and actuators — like nerves and sinew. The sensors measure the pressure over the entire surface of the wing and signal the actuators how to respond. But even the sensors are smart. Tiny computing elements detect how the aircraft responds to sensor signals. They eventually learn how to change the shape of the wing for optimal flying conditions. They also detect when there is damage to a wing and relay the extent and location to the pilot. And, like an injured bird, the wing adjusts its response to avoid further damage. This will not only be a very efficient and maneuverable airplane, but a very safe one.

Like the wings, the engines of this plane have integral health-management systems. Temperatures, pressures, and vibrations are all continuously monitored and analyzed. Unique performance characteristics are automatically developed for each engine, which then continually operates as efficiently as possible, and very safely. Long before a part fails, damage is detected and protective maintenance scheduled.

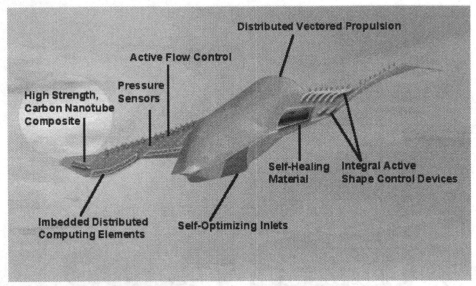

Figure D.6. NASA's dream of a future flight vehicle.

Inside the cockpit compartment, the pilot sees everything on a 3-D display that shows local weather, accentuates obstacles, all near-by aircraft, and the safest flight path. The on-board clear air turbulence sensor uses lasers to detect unsteady air well ahead of the aircraft to assure a smooth ride. When approaching a major airport, the lingering vortices that were shed from the wingtips of larger aircraft and that can upset a smaller one, can be easily avoided. This is a long-term vision, but emerging technology can make it real.

A key to achieving this vision is a fusion of nanoscale technology with biology and information technology (Figure D.6). An example is intelligent multifunctional material systems consisting of a number of layers, each used for a different purpose. The outer layer would be selected to be tough and durable to withstand the harsh space environment, with an embedded network of sensors, electrical carriers, and actuators to measure temperature, pressure, and radiation and to trigger a response whenever needed. The network would be intelligent. It would automatically reconfigure itself to bypass damaged components and compensate for any loss of capability. The next layer could be an electrostrictive or piezoelectric membrane that works like muscle tissue with a network of nerves to stimulate the appropriate strands and provide power to them. The base layer might be made of biomolecular material that senses penetrations and tears and flows into any gaps. It would trigger a reaction in the damaged layers and initiate a self-healing process.

Carbon nanotube-based materials are an example of one emerging technology with the potential to help make this a reality. They are about a hundred times stronger than steel but one-sixth the weight of steel. They can have thermal conductivities seven times higher than the thermal conductivity of copper with 10,000 times greater electrical conductivity. Carbon nanotube materials may also have piezoelectrical properties suitable for very high-force activators. Preliminary NASA studies indicate that the dry weight of a large commercial transport could be reduced by about half compared to the best composite materials available today. The application of high-temperature nanoscale materials to aircraft engines may be equally dramatic. Through successful application of these advanced lightweight materials in combination with intelligent flow control and active cooling, thrust-to-weight ratio increases of up to 50 percent and fuel savings of 25 percent may be possible for conventional engines. Even greater improvement can be achieved by developing vehicle designs that fully exploit these materials. This could enable vehicles to smoothly change their aerodynamic shape without hinges or joints. Wings and fuselages could optimize their shape for their specific flight conditions (take-off, cruise, landing, transonic, and high-altitude).

In the long-term, the ability to create materials and structures that are biologically inspired provides a unique opportunity to produce new classes of self-assembling material systems without the need to machine or process materials. Some unique characteristics anticipated from biomimetics are hierarchical organization, adaptability, self healing/self-repair, and durability. In the very long term, comparable advances in electrical energy storage and generation technology, such as fuel cells, could completely change the manner in which we propel aircraft. Future aircraft might be powered entirely electrically. In one concept, thrust may be produced by a fan driven by highly efficient, compact electric motors powered by advanced hydrogen-oxygen fuel cells. However, several significant technological issues must still be resolved in order to use hydrogen as a fuel, such as efficient generation and storage of hydrogen fuel and an adequate infrastructure necessary for delivering the fuel to vehicles (Figure D.7).

None of this is expected to happen quickly. Over the next decade we will likely see rapid development of advanced multifunctional, nanotechnology-based structural materials, such as carbon nanotube composites. Integrated health monitoring systems — for airframe and engine — may be developed, and deformable wings with imbedded actuators may also be developed. The cockpit will likely begin to become more of an extension of the pilot with greater use of senses other than sight

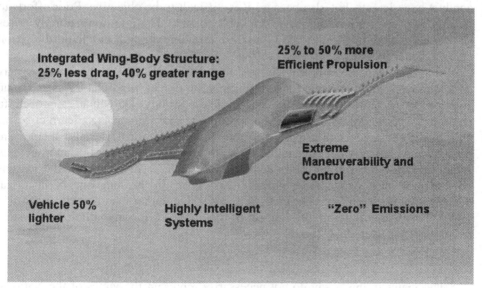

Figure D.7. Attributes of a future flight vehicle.

to provide "situational awareness" of the aircraft and its operating environment. In two to three decades, we may see the first "bio/nano/thinking/sensing" vehicles with significant use of nanotechnology-based materials, fully integrated exterior-interior flow control, and continuously deformable wings. By then, the aircraft may also have a distributed control/information system — like a nervous system — for health monitoring, some level of self-repair, and cockpits that create a full sensory, immersive environment for the pilot.

MEMETICS: A POTENTIAL NEW SCIENCE

Gary W. Strong and William Sims Bainbridge, National Science Foundation [3]

In the "information society" of the 21st century, the most valuable resource will not be iron or oil but culture. However, the sciences of human culture have lacked a formal paradigm and a rigorous methodology. A fresh approach to culture, based on biological metaphors and information science methodologies, could vastly enhance the human and economic value of our cultural heritage and provide cognitive science with a host of new research tools. The fundamental concept is the *meme*, analogous to the gene in biological genetics, an element of culture that can be the basis of cultural variation, selection, and evolution.

The meme has been characterized both as a concept that could revolutionize the social sciences as the discovery of DNA and the genetic code did for biology, and as a concept that cannot produce a general theory of social evolution because requirements for Darwinian evolution do not map into the social domain (Aunger 2000). There is a lot we do not understand about human behavior in groups, its relation to learning, cognition, or culture. There is no general theory that situates cognition or culture in an evolutionary framework, Darwinian or otherwise. It is also hard to conduct science in the social domain, not just because it is difficult to conduct experiments, but also because it is difficult to be objective. Prior efforts to "Darwinize" culture have a long and ignoble history. The question naturally arises as to what is new that might allow progress this time around, or should discretion take the better part of valor?

While any debate tends to sharpen the debate issues, in this case it may prematurely close off a search for a scientific definition of important terms and of an appropriate contextual theory. For example, a strictly Darwinian approach to cultural or social evolution may not be appropriate since humans can learn concepts and, in the same generation, pass them on to their offspring. Because memes are passed from one individual to another through learning, characteristics an individual acquires during life can be transmitted to descendents. This is one of the reasons why memes may evolve more rapidly than genes. In the language of historical debates in biology, culture appears to be Lamarckian, rather than Darwinian (Strong 1990). This would imply a different set of requirements for an evolutionary system that are not yet well understood.

---

[3] The views in this essay do not necessarily represent the views of the National Science Foundation.

As another example, we are only now discovering that many of the genes of an organism code for "chaperone" proteins that do not have "meaning" in a particular biological function, but, rather, play a role in molecular recycling and enabling the proteomic networks of molecules to interact in an orderly fashion (Kim et al. 1998). We do not yet understand how a balance is kept within a cell between the evolutionary need for variety and the need to preserve order in systems. Nevertheless, it is likely that in a fast-changing Lamarckian system, such processes become even more important. On the socio-cultural level, religious ideologies appear to have chaperone roles that may help keep individuals focused on important daily activities rather than getting caught up in unsolvable dilemmas and becoming unable to act. Even so, such ideologies cannot become so strict as to eliminate important variety from an evolutionary system. This tradeoff between order and disorder may operate like a regulator for social change (Rappaport 1988).

While there is no known Federal grants program focused on memetics, nor any apparent, organized research community, there are likely a number of existing and completed research projects that impact on the domain. These probably are found in a variety of disciplines and do not use a common vocabulary. For example, a few archaeologists apply evolutionary theory in their work (Tschauner 1994; Lyman and O'Brien 1998), and some cultural anthropologists explore the evolution of culture in a context that is both social and biological (Rindos 1985; Cashdan 2001; Henrich 2001). However, most archaeologists avoid theoretical explanations altogether, and cultural anthropology is currently dominated by a humanist rather than a scientific paradigm. So, even though starting a research program in this area would not have to begin from scratch, there would be much work to do. The biggest roadblock would be getting researchers from various disciplines to collaborate over a common set of interests.

At a first approximation, there are three different realms in which biological genetics is valuable to humanity. First, it contributes to the progress of medicine, because there is a genetic aspect to all illnesses, not only to those diseases that are commonly labeled "genetic" or "inherited." Second, it provides valuable tools for agriculture, most recently including powerful techniques of genetic engineering to design plants and animals that are hardier, more nutritious, and economically more profitable. Third, it answers many fundamental scientific questions about the nature and origins of biological diversity, thus contributing to human intellectual understanding of ourselves and the world we live in. Cultural memetics would have three similar realms of applications, as described below.

Cultural Pathology

Culture is not just art, music, language, clothing styles, and ethnic foods. Importantly, it also includes the fundamental values, norms, and beliefs that define a society's way of life. Thus, the classic problem of social science has been to understand how and why some people and groups deviate from the standards of society, sometimes even resorting to crime and terrorism. Recent attention on closed groups has once again raised the question, "Why do people believe weird things?" — to borrow from a recent book title (Schermer 2002). The problem of social order thus depends upon the dynamic interactions between cultures, subcultures, and countercultures.

For decades, various anthropologists have considered whether or not there is a cultural equivalent of the human genome underlying differences of belief and behavior across groups or whether cultural context differentially expresses elements from a common repertoire available to all humans. One way to approach the issue might be to study culture with methodologies similar to those of bioinformatics.

A key bioinformatics construct is the genomic code, the cultural equivalent of which has been widely discussed under the concept of "meme" (Dawkins 1976). Cross-cultural signals are often undetected or misidentified, and cultural miscommunication is commonplace, leading one to suspect the existence of such codes and their differentiation among social groups. Levi-Strauss (1966) refers to cultural concepts, or artifacts, as "things to think with." Such shared concepts may, however, be more a form of externalized representation, or "cognitive Post-It Notes," with important information processing functionality for a social group.

The prevalence of fundamentalist cultural and religious movements, for example, suggests that there may be an equivalent of the "auto-immune" response at the cultural level. Religion appears to be what Talcott Parsons (1964) called an "evolutionary universal," essential to the functioning of societies and prominent in every long-lasting culture. Within the realm of religion, diversification also appears to be universal, and it may be vain to hope that all people can eventually come to share compatible religious beliefs (Stark and Bainbridge 1987). At the present time, it is crucial to understand that revitalization or nativistic movements appear to be universal in times of great social change (Wallace 1956). Such movements tend toward increased orthodoxy and the involvement of charismatic leaders. Anthropologists have studied such movements from the time of the "Ghost-Dance" cults of native North Americans at the end of the 19th century to the rise of militant groups in Islam today (La Barre 1972).

"World-views" may be self-regulating, in this respect, each dominant ideology naturally stimulating the evolution of counter-ideologies. Just when Western Civilization rejoiced that it had vanquished Nazism and Marxism, and the "end of history" was at hand, radical Islam emerged to challenge its fundamental values (El-Affendi 1999). Quite apart from the issue of terrorist attacks from radical fringes of Islam, the entire Muslim religious tradition may have an evolutionary advantage over western secularism, because it encourages a higher birth rate (Keyfitz 1986). An inescapable natural law may be at work here, comparable to that which regulates the constantly evolving relations between predators and prey in the biological realm, ensuring that there is always a rival culture, and complete victory is impossible (Maynard Smith 1982). However, deep scientific understanding of the memetic processes that generate radical opposition movements may help government policymakers combat them effectively. It may never be possible to eradicate them entirely, but with new scientific methods, we should be able to prevent them from driving our civilization to extinction.

A science of memetics, created through the convergence of many existing disciplines, would likely give a basis for understanding the relationship between social groups and globalization — a topic of enormous recent interest. Fundamentalist groups are no longer "fringe" as they practice tactics to deal with variety and change, and they have become a topic not only for cultural anthropologists but also for law enforcement and governments in general. Certain "ideas" may have the force of a social virus that spreads as quickly and can have as deleterious effects on a population as do biological viruses (Boyd and Richerson 1985; Dennett 1995; Sagan 1997). It is important to examine such theories and to consider whether or not people are naturally vulnerable to "hacking" in the concept domain, as their computer networks are vulnerable in cyberspace. At the same time, memetics can help us understand the forces that promote cooperation between people and sustain culturally healthy societies (Axelrod 1990).

Memetic Engineering

Since long before the dawn of history, human beings have influenced the evolution of plants and animals, by domesticating them, breeding them, and now by engineering their genetic structure directly (Diamond 1997). Over the same span of millennia, humans became progressively more sophisticated in the processes by which they generate and transmit new culture, leading to the advanced electronic media of today. However, while agriculture in recent centuries has employed genetic science and technology of advancing complexity to domesticate plants and animals, the culture-based industries have not yet made use of memetic science.

It is important to realize that the term *culture* is defined very broadly by anthropologists and other social scientists. It is not limited to high artistic culture (symphonies, oil paintings, and great poetry), popular culture (rock music, best-selling novels, and dress styles), or intellectual culture (academic philosophies, schools of scholarship, and scientific theories). It also includes the practices of skilled professions, from surgery to litigation, financial accounting to bridge building, dentistry to uranium mining, and from auto mechanics to rocket science. The habitual patterns of behavior in families, neighborhoods, corporations, and government agencies are also forms of culture. We can say that culture refers to any pattern of thought and behavior that is shared through learning, rather than being rooted in biological inheritance.

We take for granted the assumption that government agencies like the National Science Foundation, National Institutes of Health, Defense Advanced Research Projects Agency, and Department of Energy should conduct fundamental scientific research that will ultimately be of benefit to manufacturing and transportation industries and to the military. At the same time, debates range over how heavily government should be involved in supporting culture through agencies like National Endowment for the Arts or National Endowment for the Humanities. But here we are discussing something very different from grants to support the work of artists and humanists. Rather, we refer to fundamental scientific research on the dynamics of culture, that will be of benefit to culture-creating and communication industries, and to national security through relations with other countries and through an improved ability to deal successfully with a wide range of nongovernmental organizations and movements.

If manufacturing creates the hardware of modern economies, the culture industries create the software. Both are essential to prosperity, and in the modern world, both should be grounded in solid scientific knowledge. If we understood better how human beings actually innovate, whether in music or the engineering design of consumer products, we could help them do it better. If we had a better map of culture, analogous to the Linnean system that classifies biological organisms into species and genera, we could help people find the culture they want and we could locate "uninhabited" cultural territories that could profitably be colonized by growing industries. Many of the social problems faced by contemporary American society seem to have substantial cultural aspects, so the findings of scientific memetics would be extremely valuable for both the government agencies and private organizations that have to deal with them.

As the Human Genome Project drew to its conclusion, it became clear to everyone that "mapping the human genome" was only part of the work. Also necessary was studying the great genetic diversity that exists from person to person around the planet, and discovering the biochemical pathways through which each gene was expressed in the phenotypic characteristics of the individual. Comparable work will be required in cultural memetics. For any given cultural trait, there may exist a number of distinct alternatives, like alleles in biological genetics, the mutational forms of a gene. The characteristics of varied individuals are the complex result of different alleles interacting across numerous genes. Categorization of culture from a memetic perspective will identify these alleles, and memetic engineering could make extensive use of techniques for combining these cultural traits in new ways (Bainbridge 1985).

Understanding how memes are expressed in actual human behavior will require advances in cognitive science that will have spin-off benefits in education and the culture industries. For example, research on how language is encoded both memetically and cognitively will contribute to better language instruction in schools and more effective commercial and governmental translation across languages. As in any major scientific endeavor, there may be a large number of unexpected benefits, but the gains we can identify now already more than justify the development of memetic science on economic grounds alone.

A Science of Culture

Participants in the Convergent Technologies (NBIC) conference recommended a new scientific initiative, analogous to the Human Genome Project that charted the human genetic code, which they called the Human Cognome Project — unraveling the secrets of the human cognitive genome. Any attempt to solve the riddles of the human mind will have to be far more than an exercise in brain neurology; most importantly, it will have to attack the mysteries of the cultural genome.

One major benefit of a program in memetics would be to better understand culture as an evolutionary process in its own context, whether as a Darwinian, Lamarckian, or as yet unknown system (Boyd and Richerson 1985). The knowledge gained could create a framework for a scientific rebirth in social and cultural domains. While opinions vary, it would not be too harsh to suggest that several social sciences seem to have stalled, some of them achieving very little progress in recent decades. The same thing occasionally happens in physical sciences. For example, planetary astronomy had practically stalled in the two or three decades prior to the launch of the first interplanetary space probes. Similarly, cancer research has achieved progress only very slowly over the past century, but the Human Genome Project offers new hope of breakthroughs. Memetic science could provide just the intellectual boost and potent research methodology needed by such diverse fields as Anthropology, Political Science, and Sociology.

Development of new theories and methods will require cooperation between hundreds of scientists in perhaps a dozen fields, so here with our limited perspectives we can suggest only a few of the possibilities. Perhaps there are a number of common features of natural codes, including both cultural and biological codes:

- *The "independence" feature*: Natural code elements tend to have arbitrary meaning (C.S. Peirce's symbols, as opposed to icons or indices) facilitating abstraction and reuse.

- *The "combinatorial advantage" feature*: The number of potential representations is much larger in combinations of elements than in one-to-one element coding — perhaps because evolutionary selection favors representational richness available by combination sets.

- *The self-regulation of natural codes*: Dependency upon a code results in a constraint for new input to be interpreted in terms of the code; change is thereby limited to evolution of the code over time.

Work on applying language modeling to genomic sequences at Carnegie Mellon University has suggested that genomes differentiate species by having distributions that include rare occurrences and where such rare occurrences can often be species-unique. This work suggests that some species-unique sequences have an unusual generative power, such as those playing an important role in fold initiation of proteins. Perhaps cultural codes also contain some rare occurrences that serve to differentiate cultures and are heavily associative, or generative, within the culture.

The study of cultural codes, such as suggested here, has not proceeded as rapidly as other fields such as bioinformatics. Perhaps there are reasons of politics and objectivity that have lowered the expectation of resources available for doing such research. Cultural codes may be easier and more politically feasible to study in the short-run in culturally primitive groups or other large-brained species. Bottlenose dolphins, for example, participate in fluid, short-term social associations, and their vocal plasticity as well as their behavior appears to be related to their fission/fusion social structure (Reiss et al. 1997). Perhaps dolphins' fluid social groups provide external cognitive representations (perhaps via "mirror neurons") in a manner similar to the totems of primitive human cultural groups.

Several systematic research methodologies need to be developed. One breakthrough that seems within reach would be the memetic equivalent of the Linnean system for classifying species, genera, and other kinds of biological clades. In recent years, information science has developed a range of techniques, such as latent semantic analysis and semantic concept space technology (Harum et al. 1996). United with cognitive science, these methods should go a long way to identifying the structure of the cultural genome and the mechanisms by which it changes or sustains itself. Through the development of memetic science, we will want to look to genetics for inspiration and selectively import both theories and methods from biology when appropriate.

The scientific study of culture is both possible and pregnant with knowledge of human behavior. Thus, it deserves to be given more resources, especially in light of current events. These events include not only the terrorism of September 11, 2001, but also the dot-com crash and the failure of nations as diverse as Argentina,

Indonesia, and Japan to sustain their economic development. Memetic science could help us deal with challenges to American cultural supremacy, discover the products and services that will really make the information economy profitable, and identify the forms of social institutions most conducive to social and economic progress.

A Transforming Strategy

The most obvious barrier to the emergence of a successful science of memetics is the lack of a unified scientific community to create it. We suggest that three kinds of major projects would be needed to establish the nucleus for this vital new field:

1. *Professional conferences, scientific journals, and a formal organization devoted to memetics.* A scientific community needs communication. Because memetics spans biology, information science, cognitive science, and cultural studies, the people who will create it are strewn across many different disciplines that hold their annual meetings at different times in different cities. Thus, a series of workshops and conferences will be essential to bring these people together. Out of the conferences can emerge publications and other mechanisms of communication. An electronic communication network at the highest level of scientific quality needs to be established.

2. *Data infrastructure, in the form of multiuse, multiuser digital libraries incorporating systematic data about cultural variation, along with software tools for conducting scientific research on it.* Some work has already been accomplished of this kind, notably the decades-long efforts to index the findings of cultural anthropological studies of the peoples of the world, accessible through World Cultures Journal (http://eclectic.ss.uci.edu/~drwhite/ worldcul/world.htm), and cross-cultural questionnaire surveys such as The World Values Survey (http://wvs.isr.umich.edu/). However, existing data were not collected with memetic analysis in mind. They typically ignore most dimensions of modern cultures, and they lack information about the networks of communication between individuals and groups that are fundamental to memetic mutation and diffusion. Thus, entirely new kinds of cultural data infrastructure are needed, to provide the raw material for memetic science.

3. *Specific major research projects assembling multidisciplinary teams to study distinct cultural phenomena that are most likely to advance fundamental memetic science and to have substantial benefits for human beings.* Because culture is highly diverse, it is essential to support multiple projects in different domains. This strategy would connect data infrastructure projects with teams of scientists oriented toward answering specific but profound scientific questions. One recent suggestion that has merit on both scientific and practical grounds is to create an distributed digital library devoted to all aspects of Islamic culture, with special attention to understanding how it evolves and divides. Another worthwhile project would be to link existing linguistic data archives, for example represented by the Linguistic Data Consortium, then transform them into a laboratory for studying the constant process of change that goes on within and across languages. A very different project, with a wide range of intellectual and economic benefits, would be an institute to study the transformation of engineering and manufacturing by the

development of nanotechnology, gaining fundamental scientific understanding of the innovation process, to improve the methods by which new
technologies are developed.

References

Aunger, R., ed. 2000. *Darwinizing culture: The status of memetics as a science*. Oxford: Oxford
University Press.

Axelrod, R. 1990. *The evolution of cooperation*. New York: Penguin.

Bainbridge, W.S. 1985. Cultural genetics. In *Religious movements*, ed. R. Stark. New York:
Paragon.

Boyd, R. and P.J. Richerson. 1985. *Culture and the evolutionary process*. Chicago: University
of Chicago Press.

Dawkins, R. 1976. *The selfish gene*. Oxford: Oxford University Press.

Dennett, D.C. 1995. *Darwin's dangerous idea*. New York: Simon and Schuster.

Diamond, J. 1997. *Guns, germs, and steel: The fates of human societies*. New York: Norton.

El-Affendi, A. 1999. Islam and the future of dissent after the "end of history." *Futures*
31:191-204.

Harum, S.L., W.H. Mischo, and B.R. Schatz. 1996. Federating repositories of scientific
literature: An update on the Digital Library Initiative at the University of Illinois at
Urbana-Champaign. *D-Lib Magazine*, July/August, www.dlib.org.

Keyfitz, N. 1986. The family that does not reproduce itself. In *Below-replacement fertility in
industrial societies: Causes, consequences, policies*, ed. K. Davis, M.S. Bernstam, and R.
Campbell. (A supplement to *Population and Development Review*).

Kim, K.K., R. Kim, and S.-H. Kim. 1998. Crystal structure of a small heat-shock protein.
Nature 394:595-599.

Levi-Strauss, C. 1966. *The savage mind*. Chicago: University of Chicago Press.

Lyman, R.L., and M.J. O'Brien. 1998. The goals of evolutionary archaeology: History and
explanation. *Current Anthropology* 39:615-652.

Maynard Smith, J. 1982. *Evolution and the theory of games*. New York: Cambridge
University Press.

Reiss, D., B. McCowan, and L. Marino. 1997. Communicative and other cognitive
characteristics of bottlenose dolphins. *TICS* 140-145.

Strong, G.W. 1990. Neo-Lamarckism or the rediscovery of culture. *Behavioral and Brain
Sciences* 13: 92.

Parsons, T. 1964. Evolutionary universals in society. *American Sociological Review* 29: 339-
357.

Rappaport, R. 1988. *Ecology, meaning, and religion*. Richmond, California: North Atlantic
Books.

Sagan, C. 1997. *The demon-haunted world: Science as a candle in the dark*. New York:
Ballantine Books.

Schermer, M. 2002. *Why people believe weird things: Pseudoscience, superstition, and other
confusions of our time*. New York: H. Holt.

Stark, R. and W.S. Bainbridge. 1996. *A theory of religion*. New Brunswick, NJ: Rutgers
University Press.

Tschauner, H. 1994. Archaeological systematics and cultural evolution: Retrieving the
honour of culture history. *Man* 29:77-93.

Wallace, A.F.C. 1956. Revitalization movements. *American Anthropologist* 58:264-281.

E. National Security

Theme E Summary

Panel: R. Asher, D.M. Etter, T. Fainberg, M. Goldblatt, C. Lau, J. Murday,
W. Tolles, G. Yonas

The fourth NBIC theme examines the ways in which the United States and modern civilization can meet the intelligence and defense challenges of the new century. In a world where the very nature of warfare is changing rapidly, national defense requires innovative technology that (a) projects power so convincingly that threats to the United States are deterred, (b) eliminates or minimizes the danger to U.S. warfighters from foe or friendly fire, and (c) reduces training costs by more than an order-of-magnitude through augmented reality and virtual reality teaching aids.

Investment in convergent nanotechnology, biotechnology, information technology, and cognitive science is expected to result in innovative technologies that revolutionize many domains of conflict and peacekeeping. We are entering an era of network-centric combat and information warfare. Increasingly, combat vehicles will be uninhabited, and robots or other automated systems will take on some of the most hazardous missions. Effective training will make extensive use of augmented or virtual reality. Nanotechnology will offer reliable means for detecting and protecting against chemical and biological agents. Convergence of many technologies will enhance the performance of human warfighters and defenders, in part through monitoring health and instituting prophylaxis and through magnifying the mental and physical capabilities of personnel.

The Defense Science and Technology Strategy (Department of Defense 2000) seeks to ensure that the warfighters today and tomorrow have superior and affordable technology to support their missions and to give them revolutionary war-winning capabilities. There is special focus on information assurance with emphasis on security; battlespace awareness with emphasis on sensor webs, miniaturized platforms, netted information, and cognitive readiness; force protection with emphasis on chemical/biological defense; and support for the warfighter.

In the recent past, new technologies have dramatically enhanced American ability to both prepare for and execute military actions. By implementing advances in information technologies, sensors, and simulation, we have strengthened our ability to plan and conduct military operations, quickly design and produce military systems, and train our forces in more realistic settings. These technologies are central to greater battlefield awareness, enabling our forces to acquire large amounts of information, analyze it quickly, and communicate it to multiple users simultaneously for coordinated and precise action. As former Defense Secretary William J. Perry has noted, these are the technological breakthroughs that are "changing the face of war and how we prepare for war."

Numerous special programs, reports, and presentations address these goals. The Department of Defense has designated nanoscience as a strategic research area in order to accelerate the expected benefits (Murday 1999). Various conferences and studies have been devoted to assessing nanotechnology status and needs for defense (Department of Defense 2000; National Research Council 2002). Attention has also been paid to anticipating more global societal consequences of those efforts in support of national security (Roco and Bainbridge 2001).

National Security Goals for NBIC

This conference panel identified seven goals for NBIC augmentation of national security, all of which require the close integration of several of the nanotechnology, biotechnology, information technology, and cognition fields of endeavor. The seven goals, listed below, are sufficiently diverse that there is no common strategy beyond the need for interdisciplinary integration. The net result of accomplishing the stated goals would reduce the likelihood of war by providing an overwhelming U.S. technological advantage, would significantly reduce the cost of training military manpower, and would significantly reduce the number of lives lost during conflict.

1. **Data linkage, threat anticipation, and readiness.** Miniaturized, affordable sensor suites will provide information from previously inaccessible areas; high-speed processing will convert the data into information; and wide-bandwidth communication pipelines with digital security will distribute information rather than data to all who need it.

2. **Uninhabited combat vehicles.** Automation technology (including miniaturization of sensing, augmented computation and memory, and augmented software capability) will enable us to replace pilots, either fully autonomously or with pilot-in-the-loop, in many dangerous warfighting missions. The uninhabited air vehicle will have an artificial brain that can emulate a skillful fighter pilot in the performance of its missions. Tasks such as take-off, navigation, situation awareness, target identification, and safe return landing will be done autonomously, with the possible exception of circumstances requiring strategic or firing decisions. Without the human g-force constraint and the weight of human physical support equipment (oxygen, ejection system, armor, etc.), the planes will be more maneuverable. Tanks, submarines, and other combat vehicles will experience similar benefits.

3. **Warfighter education and training.** A partnership between nanotechnology and information technology holds the promise for relatively inexpensive, high-performance teaching aids. One can envision a virtual-reality teaching environment that is tailored to the individual's learning modes, utilizes contexts stimulating to that individual, and reduces any embarrassment over mistakes. The information exchange with the computer can be fully interactive, involving speech, vision, and motion. Nanodevices will be essential to store the variety of necessary information and to process that information in the millisecond time frames necessary for realtime interaction.

4. **Chemical/biological/radiological/explosive (CBRE) detection and protection.** Microfabricated sensor suites will provide ample, affordable, error-free forewarning of chemical, biological, radiological, or explosive threat. For those who must work in a contaminated environment, individual protection (masks and clothing) will induce heat stresses no greater than conventional uniforms while providing full protection. Decontamination and

neutralization procedures will be effective against agents, yet will be relatively benign to people and the environment. Monitors will provide information on warfighter physiological status and initiate any necessary prophylaxis.

5. **Warfighter systems.** The warfighter is subjected to periods of intense stress where life or death decisions must be made with incomplete information available, where the physiology of fatigue and pain cloud reason, and where supplemental technology must compete with the 120 pounds of equipment weight s/he must carry. NBIC technologies can address all of these aspects of warfighting. Nanotechnology holds the promise to provide much greater information, connectivity, and risk reduction to the warfighter. The continued miniaturization of electronic devices will provide 100 times more memory with less bulk and weight (a terabit of information in a cm^2). Processing speeds will increase to terahertz rates. Displays will be flexible and paper-thin, if not replaced by direct write of information on the retina. High-bandwidth communication will be netted. Prolific unattended sensors and uninhabited, automated surveillance vehicles under personal warfighter control will be providing high data streams on local situations. Weapons will automatically track targets and select precise firing times for greater accuracy. The marriage of semiconductors and biology will provide physiological monitors for alertness, chemical or biological agent threats, and casualty assessment. The small size of the nanodevices will limit the volume, weight, and power burdens.

6. **Non-drug treatments for enhancement of human performance.** Without the use of drugs, the union of nanotechnology and biotechnology may be able to modify human biochemistry to compensate for sleep deprivation and diminished alertness, to enhance physical and psychological performance, and to enhance survivability rates from physical injury.

7. **Applications of brain-machine interface.** The convergence of all four NBIC fields will give warfighters the ability to control complex entities by sending control actions prior to thoughts (cognition) being fully formed. The intent is to take brain signals (nanotechnology for augmented sensitivity and nonintrusive signal detection) and use them in a control strategy (information technology), and then impart back into the brain the sensation of feedback signals (biotechnology).

Statements and Visions

Defense applications are intended for the highly competitive environments of deterrence, intelligence gathering, and lethal combat, so it is essential to be technologically as far ahead of potential opponents as possible. The United States and its closest allies represent only a small fraction of the world population, and in the asymmetrical conflicts of the early 21st century, even a small number of dedicated enemies can cause tremendous damage. Thus, the overview statements and future visions written by participants in the national security working group address very high-priority areas where the United States and its allies can achieve and maintain great superiority. The statements and visions cover areas from enhancing soldier performance (M. Goldblatt) and combat readiness (D.M. Etter) to future roles of NBIC for fighting terrorism (J. Murday, T. Fainberg, C. Lau) and equipping soldiers (R. Asher, J. Murday, T. Fainberg, C. Lau).

References

Department of Defense. 2000. *Defense science and technology strategy 2000*. Washington D.C.: Department of Defense.

Murday, J.S. 1999. Science and technology of nanostructures in the Department of Defense. *Journal of Nanoparticle Research* 1:501-505.

National Research Council. 2002. *Implications of emerging micro and nano technologies*. Washington D.C.: National Academies Press.

Roco, M.C., and W.S. Bainbridge, eds. 2001. *Societal implications of nanoscience and nanotechnology*. Dordrecht, Netherlands: Kluwer Academic Publishers.

STATEMENTS

COGNITIVE READINESS: AN IMPORTANT RESEARCH FOCUS FOR NATIONAL SECURITY

Delores M. Etter, United States Naval Academy

Cognitive readiness is a critical research area for the Department of Defense. Soldiers must not only be ready physically for the myriad of roles that they have in the world today, but they must also be ready cognitively. This cognitive readiness extends from handling stress and sleep deprivation, through training "anytime, anyplace," through additional information provided by augmented reality, and through realtime physical monitoring during operations. This range of cognitive readiness requires a serious investment in research covering a wide range of areas. This paper will present some of the focus of existing research and some of the paths for future research in this area as it applies to national security.

Critical Focus Areas for DOD S&T

Approximately three years ago the senior directors in the Office of the Deputy Under Secretary of Defense for Science and Technology selected five areas as especially critical in DOD's research program. These five research areas are the following: chemical and biological defense, hardened and deeply buried targets, information assurance, smart sensor web, and cognitive readiness. Today, these five areas seem to be obvious priorities, but three years ago that was not the case. These areas had existing research programs that were supported by the military service research programs and the defense agencies. The identification of these five areas by the Office of the Secretary of Defense gave them a corporate priority. Additional funds were provided to start new programs, to coordinate existing programs, and to support workshops to bring together new players who worked in various aspects of the areas.

The Department's focus on chemical and biological defense has been a clear priority for DOD over the last few years. The need for this research results from proliferation of inexpensive weapons of both chemical and biological agents. DOD's research has four key areas of priority: detection of the agents, protection from the agents, decontamination of equipment and people after exposure, and an understanding of the dispersion of the agents from a modeling and simulation perspective.

Concern over hardened and deeply buried targets comes from the fact that underground facilities are often used to conceal missiles and weapons of mass destruction. DOD's research program includes priorities in overhead imagery to attempt to locate the targets, sensor research to determine what activities are being carried out underground, delivery systems to neutralize facilities if necessary, and computational modeling activities to understand the structures and activities within them.

Cyberterrorism is a real part of today's world. Attacks come from hackers, from terrorists, and from insiders. Dealing with information warfare is critical to assure that our information is protected and is not compromised. Research in information assurance involves designs of new firewalls, malicious code detectors, encryption techniques, and correlation technologies.

Smart sensor web is a concept that provides complete situation awareness to the individual soldier in the field. It is based on integrating information from areas such as realtime imagery, micro weather information, and moving targets. The research includes physical model understanding, dynamic data bases, microsensors, wireless communications, and the next-generation Internet.

Cognitive readiness addresses human optimization. The challenges to the human include sustained operations, environmental ambiguity, and information overload. Research programs address topics such as physiological monitoring, embedded training, learner-centric instruction, and augmented reality. Figure E.1 shows the wide range of areas covered by cognitive readiness.

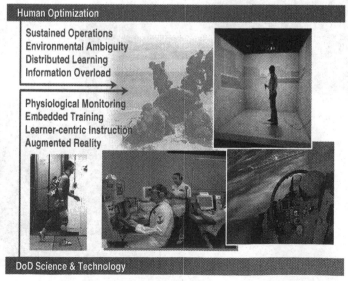

Figure E.1. Cognitive readiness research.

Cognitive Readiness Framework

The DOD has a multidisciplinary focus on the human dimension of joint warfighting capabilities. This cross-Service framework ensures that research addresses the following requirements:

- warfighters are mentally prepared for accomplishing their missions
- warfighters are performing at their optimum
- tools and techniques for preparing warfighters are the most effective and affordable
- tools and techniques that warfighters use are the most effective and affordable

The changing military environment compels a focus on cognitive readiness. Issues that affect this aspect of military readiness come from many directions. Soldiers have many different threats and changing missions that extend from peacekeeping to warfighting. Budget reduction brings personnel drawdowns in the military, and that brings demographic changes. In addition, military systems are becoming more complex, and soldiers need to handle new technologies. Figure E.2 illustrates the range of these interactions that soldiers must handle.

Four domains from science and technology research have been defined for cognitive readiness:

- *Sociology and personnel.* This domain deals with family, group, and culturally defined issues; selection and classification; and leadership.
- *Health and welfare.* This domain includes mental acuity, fatigue, physiological readiness, quality of life, and morale.

Figure E.2. Changing military environment.

- *Human systems integration*. This domain covers human-centered design, decision aids, and dynamic function allocation.

- *Education and training*. This domain includes using new technologies for teaching/learning and to develop specific tasks, skills, and/or procedures.

The following three examples demonstrate the wide range of research necessary to support cognitive readiness. *Augmented reality* involves bringing the information world to the soldier in real time. *Biomedical monitoring* combines sensors for measuring the physical readiness of soldiers to realtime monitoring to judge performance capability. *Survival technologies* present different areas of research to protect soldiers physically so that they are mentally and physically ready to perform their missions.

Example 1: Augmented Reality

Consider an urban environment. Soldiers need to know immediate answers to questions such as the following:

- How do I get to this building?

- What building is in front of me?

- Where is the main electric circuit in this building?

- What is the safest route to this building?

- Are there hidden tunnels under the streets?

- Street signs are missing – where am I?

- Have sniper locations been identified?

The area of augmented reality is an area in which technology is used to augment, or add, information for the soldier. For example, augmented reality could amplify natural vision by projecting information on a soldier's visor, or perhaps projecting it directly on the soldier's retina. This additional information added to the natural view could identify warnings for sniper locations and mines. Hidden infrastructure and utilities such as subways, service tunnels, and floor plans could be displayed. Virtual information such as simulated forces could be displayed to provide new training simulations. Figure E.3 gives an example of the type of information that would be very helpful if it were shown over an image to augment the information available to a soldier.

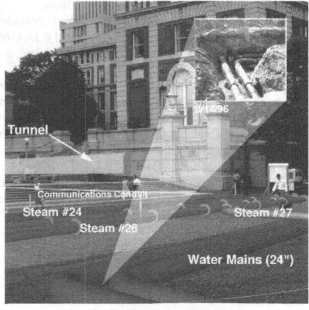

Figure E.3. Augmented reality.

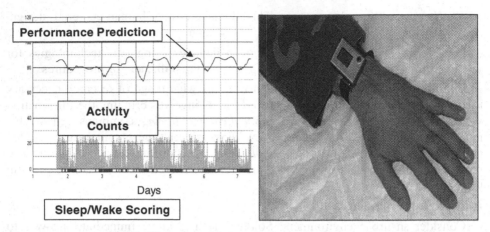

Figure E.4. Sustaining performance: managing sleep.

Example 2: Biomedical Status

Biomedical status monitoring is the medical equivalent of the Global Positioning System (GPS). It uses sensors for vital signs, electrolytes, stress hormones, neurotransmitter levels, and physical activity. In essence, it locates the soldier in physiological space as GPS does in geographic space.

The biomedical status monitoring program is integrated into several DOD programs, including Land Warrior, Warrior's Medic, and Warfighter Status Monitor. These programs allow dynamic operational planning with biomedical input that supports pacing of operations at sustainable tempo. It also allows commanders to anticipate and prevent casualties due to heat stress, dehydration, performance failures from sleep deprivation, and combat stress casualties. Not only can casualties be detected, but initial treatment can be guided.

Figure E.4 gives an example of a wrist monitor that predicts performance by monitoring sleep. Sleep is determined by the lack of motion of the wrist monitor. The graph in the figure predicts performance based on the amount of rest that the soldier has had.

Sensors can also help prevent casualties by monitoring soldiers in MOPP gear – the equipment worn to work in hazardous environments. The sensors can include core temperature (collected from a sensor that is swallowed by the soldier), skin temperature, heart rate, and activity rate. The combination of these sensors can be used to determine when a soldier needs to take a break in order to prevent possible injury or death.

Figure E.5 illustrates the hypothetical use of these biomedical status monitoring devices when they are combined with wireless communication systems. Individual soldier status can be monitored not only by soldiers working side by side, but also by central units that can be mobile or transmitted to satellite systems. Future sensors may also be embedded bionic chips.

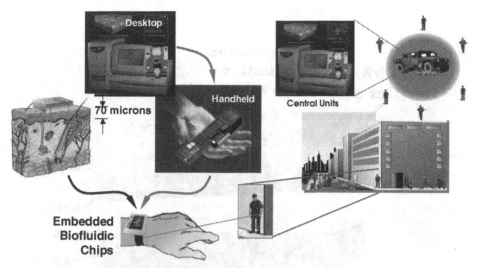

Figure E.5. Wrist-mounted remote biological assay.

Example 3: Survival Technologies

A number of new survival technologies are being developed to provide human protection in many different ways. Ballistics protection, shown in Figure E.6, is being studied using new high-performance fibers, composite materials, advanced ceramics, and metals. The analysis of new materials requires enhanced predictive modeling of the effects of ballistic weapons with these new materials. Another challenge is integrating the new materials into uniform systems.

Innovative research in chemical/biological protection for soldiers is investigating selectively permeable membranes that would provide an outer coating for uniforms. The coating would not allow aerosols or liquids to penetrate from outside the

Figure E.6. Ballistics protection.

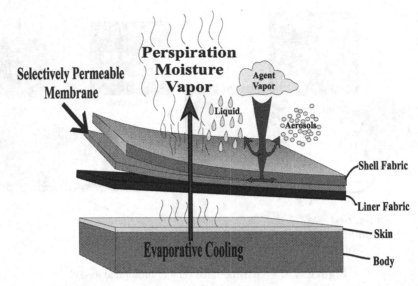

Figure E.7. Selectively permeable membranes for uniforms.

material. Additional research is being done in elastomeric protective materials and lightweight carbonless materials. Figure E.7 shows a diagram of some interactions between various layers of the material.

Directed-energy eye protection (protection from lasers) is a challenge because of the various frequencies of lasers. Some current systems are considering robust dielectric stacks on polycarbonate, enhanced-eye-centered holograms, operational dye technology, and nonlinear optical effects.

New materials are providing possibilities for multifunctional materials. Examples include aramid co-polymer chemistry and flame-retardant chemistry. Some of the possibilities for microencapsulation may provide phase-change materials — materials that change to match the environment of the soldier. This would provide a chameleon-like uniform.

Finally, systems integration will play an important part of combining many of the new capabilities such as microelectronics, improved lightweight sensors, and advanced materials. The work on high-resolution flat panel displays will provide wearable computer screens, and that will significantly reduce the weight of equipment that soldiers need to carry.

Conclusions

This article has briefly provided some of the reasons why cognitive readiness is such an important area to national security and identified some of the research that is being supported in this area. Successful research will require partnerships that bring together researchers from universities, government agencies, industry, and international coalitions. The benefits have far ranging possibilities that will address cognitive readiness not only of soldiers, but of general populations as well.

Acknowledgements

Significant contributions to this article were provided by Mr. Bart Kuhn from the Office of the Deputy Under Secretary of Defense for Science and Technology.

References

DOD. 1999 (Feb. 12). *Warrior protection systems.* Defense Science and Technology Seminar on Emerging Technologies, Sponsored by DOD, DUSD (S&T).
_____. 2000 (Oct. 13). Future warrior systems. Defense Science and Technology Seminar on Emerging Technologies.
_____. 2001 (Feb.). Defense science and technology strategy and plans. Washington, D.C.: DOD, DUSD (S&T).

DARPA'S PROGRAMS IN ENHANCING HUMAN PERFORMANCE

Michael Goldblatt, Defense Advanced Research Projects Agency

The Defense Advanced Research Projects Agency (DARPA) was established in 1958 as the first U.S. response to the Soviet launching of Sputnik. Since that time, DARPA's mission has been to assure that the United States maintains a lead in applying state-of-the-art technology for military capabilities and to prevent technological surprise from her adversaries.

With the infusion of technology into the modern theater of war, the human has become the weakest link, both physiologically and cognitively. Recognizing this vulnerability, DARPA has recently begun to explore augmenting human performance to increase the lethality and effectiveness of the warfighter by providing for super physiological and cognitive capabilities.

Metabolic Engineering

The Metabolic Engineering Program seeks to develop the technological basis for controlling metabolic demands on cells, tissues, and organisms. The initial phase of the program is focusing on the successful stabilization and recovery of cells and tissues from stress states representative of military operational conditions, with specific focus on blood and blood products (Figure E.8).

When successful, the application of this technology to combat casualties will result in greater salvage

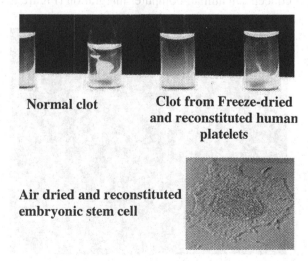

Normal clot **Clot from Freeze-dried and reconstituted human platelets**

Air dried and reconstituted embryonic stem cell

Figure E.8. Develop methods for controlled metabolism in cells, tissues, organs, and organisms needed by the U.S. military population.

of human life and limb from the battlefield, through the availability of cell-based therapy for hemorrhage, shock, and critical wounds. Additionally, stabilized cells and tissues will provide a stable substrate for prepositioning and large-scale manufacture of needed cellular and tissue products.

Exoskeletons for Human Performance Augmentation

The goal of the human performance augmentation effort is to increase the speed, strength, and endurance of soldiers in combat environments. The program will develop technologies, such as actively controlled exoskeletons, to enable soldiers to handle more firepower, wear more ballistic protection, and carry more ammunition and supplies, etc., in order to increase the lethality and survivability of ground forces in all combat environments (Figure E.9).

Two of the critical issues for exoskeletons are power for actuation and biomechanical control integration. The program is developing efficient, integrated power and actuation components to generate systems with duration that are operationally significant. Hence, researchers are exploring the use of chemical/hydrocarbon fuels (with very high energy density and specific energy) for energy conversion and mechanical actuation (as opposed to other energy storage media such as batteries or compressed air). An understanding of biomechanics, feedback, and control are also critical to building an integrated system that provides seamless compatibility with human kinetics, especially under battlefield stress.

Augmented Cognition

The DARPA Augmented Cognition program promises to develop technologies capable of extending the information management capacity of warfighters. This knowledge empowerment will be accomplished in part by exploiting the growth of computer and communication science and accelerating the production of novel concepts in human-computer integration (Figure E.10).

ISMS–Robot Supporting Human   Motion Capture System

Figure E.9. Incorporate and advance technologies to remove the burden of mass (120+ lbs.) and increase the soldier's strength, speed, endurance, and overall combat effectiveness.

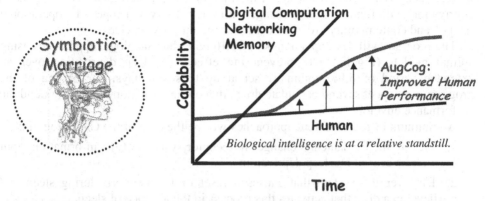

Figure E.10. Maintain a person's cognitive state at an optimal arousal level, then the person will have enhanced memory and the ability to perform optimally even under conditions of interruptions; this will improve and enhance the quality of military decisionmaking.

The mission of the Augmented Cognition program is develop and demonstrate quantifiable enhancements to human cognitive ability in diverse, stressful, operational environments. Specifically, this program will measure its success by its ability to enable a single individual to successfully accomplish the functions currently carried out by three or more individuals.

The program will explore the interaction of cognitive, perceptual, neurological, and digital domains to develop improved performance application concepts. Success will improve the way 21st century warriors interact with computer-based systems, advance systems design methodologies, and fundamentally reengineer military decisionmaking.

Continuous Assisted Performance (CAP)

The goal of this program is to discover new pharmacologic and training approaches that will lead to an extension in the individual warfighter's cognitive performance capability by at least 96 hours and potentially for more than 168 hours

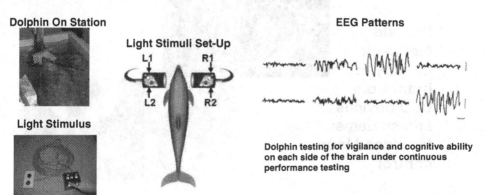

Figure E.11. Develop multifaceted approaches to prevent the degradation of cognitive performance caused by sleep deprivation in order to extend personnel "duty cycle."

without sleep. The capability to resist the mental and physiological effects of sleep deprivation will fundamentally change current military concepts of "operational tempo" and contemporary orders of battle for the military services.

The program will develop a number of different pharmacologic approaches using animal models (Fig. E.11) to prevent the effects of sleep deprivation over an extended period of time, nominally set at up to seven days. At the end of the program, we expect several candidate drugs that alone, or in combination, extend the performance envelope.

A minimum of four different approaches will be the core of the CAP program:

1. Prevent the fundamental changes in receptor systems of the information input circuits caused by sleep deprivation.

2. Discover the system that causes a reset of the network during sleep and develop a drug that activates this process in the absence of sleep.

3. Stimulate the normal neurogenesis process that is part of learning and memory, thereby increasing the reserve capacity of the memory circuits.

4. Determine if individuals resistant to sleep deprivation use a different strategy in solving problems and, if so, then develop a training approach that makes this possible for everyone.

Brain-Machine Interface

This program uses brain-machine interfaces to explore augmenting human

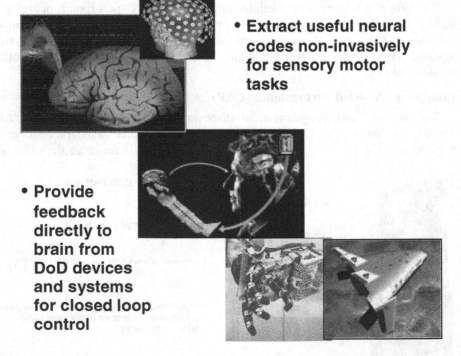

• **Extract useful neural codes non-invasively for sensory motor tasks**

• **Provide feedback directly to brain from DoD devices and systems for closed loop control**

Figure E.12. Augment human performance by harnessing brain activity to command, control, actuate, and communicate with the world directly through brain integration and control of peripheral devices and systems.

performance by extracting neural codes for integrating and controlling peripheral devices and systems. The program attacks the technological challenges across many disciplines and will require assembly of interdisciplinary teams to achieve the ambitious goal of having humans interact with and control machines directly from brain activity.

Three of the significant challenges that the program will explore are as follows:

1. fundamental extraction of patterns of neuronal code as they relate to motor activity and the proprioceptive feedback necessary for executing motor commands

2. non-invasive access to the necessary brain activity (access a 500-micron square area where temporal spike train outputs can be measured)

3. design and fabrication of new machines (elasticity, compliance, force dynamics) that could be optimally controlled by the brain.

NBIC FOR HOMELAND DEFENSE: CHEMICAL / BIOLOGICAL / RADIOLOGICAL / EXPLOSIVE (CBRE) DETECTION/PROTECTION

James Murday, Naval Research Laboratory

The coupling of nanoscale sensors for chemical/biological/radiological/explosive protection (CBRE) with improvements in information technology and physiology can critically impact national security programs by providing sensitive, selective, and inexpensive sensor systems that can be deployed for advance security to the following kinds of locations:

- transportation modes (security protection for air, bus, train/subway, etc.)

- military (for protection of facilities and equipment)

- federal buildings (government offices, U.S. embassies, all other federal buildings)

- U.S. Customs (for border crossings, international travel, etc.)

- civilian businesses (in large and small cities)

- the environment (public water supplies, waste treatment plants, natural resource areas, reservoirs, etc.)

- schools (to prevent weapons, explosives such as pipe bombs, etc.)

Improvements in detection systems, coupled with new approaches to protection, promise potential impact that is vast and critical.

Role of Converging Technologies

Converging NBIC technologies will integrate the biology, chemistry, electronics, engineering, materials, and physics research communities to establish the interdisciplinary nanoscience knowledge and expertise needed to exploit nanofabrication and nanostructures in the development of the following:

- miniaturized, intelligent sensor systems with revolutionary CBRE performance

- new high-surface-area, templated adsorbents for personnel/collective protection systems
- nanofibers for effective clothing with minimal heat loading
- catalytic materials effective against agent while relatively benign to humans and environment
- mechanisms to disrupt biological agent viability

Nanotechnology will provide innovative new hardware. Information technology will provide the effective transformation of new data into information. Biotechnology will provide new insights into human physiology and prophylaxes. Together, these three technologies can lead to effective new protection systems against the CBRE weapons of mass destruction.

Transforming Strategy to Reach Vision

Short-Term (1-5 Year) Transition Opportunities

To be successful in the 1-5 year timeframe, opportunities must have already demonstrated proof-of-principle and have existing commercial interest. Specific examples include those shown in Table E.1.

Table E.1. Examples of Commercialized Nanotechnologies

Investigator	Institute	Technology	Company
Mirkin	Northwestern	nanoAu biological sensing	Nanosphere, Inc.
Lieber	Harvard	nanotube sensors	Nanosys
Snow	NRL	nanoAu chemical sensing	MicroSensor Systems
Klabunde	Kansas State	nanocluster agent catalysis	Nanoscale Materials
Thundat	ORNL	cantilever bio/chem sensing	Protiveris
Smalley	Rice	CNT for adsorbents	CTI
Doshi		polymer nanofibers	eSpin

SBIR and STTR funding can accelerate the transformation of the existing science discovery into technology ready for commercial attention.

Mid-term (5-10 Year) Transition Opportunities

Those areas where an investment in nanoscience holds the promise for paradigm-breaking approaches to detection/protection/neutralization with commercial product transition in the 5-10 year timeframe include the following:

Sensing

- transduction/actuation mechanisms for greater sensitivity/selectivity
- biotic/abiotic interfaces to marry semiconductors with *in vivo* biology
- environmental energy sources to minimize battery requirements

Protection

- high-surface-area materials with templated structure for selective adsorption

-   controlled porosity for separation
-   nanofibers for clothing with improved adsorption/neutralization of agent
- neutralization/decontamination
    -   nanostructures to disrupt biological function
    -   catalytic nanostructures
- Therapeutics
    -   Encapsulated drugs for targeted release
    -   MEMS "capsules" for controlled drug release

Long-Term (10-20 Year) Transition Opportunities

Investment in the science base long-term is believed to be important for ultimate integration of many components into a complex system (e.g., sensor suites) and for providing sufficient insights into a complex system (e.g., cell and spore physiology) to enable innovative technologies. Examples include the following:

- Laboratory on a chip — incorporation of multiple separation and detection technologies at sub-micron scales on a single chip in order to obtain inexpensive, rapid detection technology with low false positive/negative events
- Cell-based sensing — development of sensing technology that responds to unknown new threats by measuring the response of living systems that can mimic human biochemistry
- Nanoelectromechanical systems (NEMS) — extension of the MEMS technologies another three orders smaller in order to incorporate significantly more capability

Estimated Implications

Since the United States presently can dominate any military confrontation, it is highly likely that the nation will continue to suffer from terrorist actions such as the World Trade Center and the subsequent anthrax distribution. The application of convergent technologies to national defense has the potential for revolutionary new capability to counter the threats.

References

U.S. National Science and Technology Council. 2002. Chemical, biological, radiological, explosive: Detection and protection. In *National Nanotechnology Initiative, The Initiative and its Implementation Plan. Detailed Technical Report Associated with the Supplemental Report to the President's FY 2003 Budget.* Chapter 10, New Grand Challenges in Fiscal Year 2003. White House: Washington, DC.

DOD. 2000 (March). *Chemical and biological defense program. Annual report to Congress.* Fort Belvoir, VA: Department of Defense, Defense Technical Information Center.

FUTURE ROLES FOR SCIENCE AND TECHNOLOGY IN COUNTERTERRORISM

Tony Fainberg, Defense Threat Reduction Agency, Department of Defense

The natural reaction among scientists, engineers, and technical experts following the atrocities of September 11 was the fervent wish to apply their knowledge, abilities, and creativity in order to contribute to the defeat of current and future terrorist threats to the United States and its international friends and allies.

Indeed, there is ample opportunity for directing technical advances to this end. However, it should be emphasized that much can be accomplished nearly independently of technical innovations. Security procedures need to be improved in many venues. The most talked-about area today is aviation security; for example, the need to know who has access to airplanes at airports is pressing. Background checks to this end are now being instituted and, although enabled by advances in computer technologies of various sorts, can already be accomplished, given bureaucratic acquiescence. But although technical applications can enable these checks, the main barriers to doing so in the past have been cost, inconvenience, and concerns about intrusion on privacy. Another example is in the area of explosives detection. Excellent equipment for detecting explosives in baggage had been developed and manufactured as long ago as 1994. Since 1997, this equipment has been deployed and further developed, but it could be deployed in such a way as to cover the whole civil aviation system rather than just 10 percent of it. Under the current, new imperatives, these and a number of other matters can and will be solved through national resolve rather than advanced technology. Especially for the near-term, there is much that can be done to reduce our vulnerabilities (indeed, much is being done), without developing a lot that is new in the way of science and technology.

But, although science and technology are not the only answers to the diverse and menacing terrorist threat, they are part of the answer and will increasingly become so in the future. New integrated systems and approaches will be necessary both to increase the robustness of our society against bioattacks and to face newer threats, which themselves may be developed through the use of science and technology.

I will try to lay out some thoughts about where we might conceivably look for new tools to deal with threats that have occurred or that we can easily imagine occurring. My emphasis is on technologies that could begin to produce useful results in the mid-term (say, 2-3 years to 10 years), particularly those areas that are within the scope of this workshop's focus on the convergence of nanoscience/ nanotechnology, biotechnology/biomedicine, information technologies, and cognitive science.

Aviation Security[1]

One main problem in the area of aviation security that might be addressed by some of the NBIC technologies would be trying to find out (a) who the people are who have access to aircraft and (b) what their intentions are.

A second problem lies in the timely detection of chemical or biological agents, particularly in airports, and in what to do about the alarms, false and real. Chemical detectors are fairly good right now, although like everything else, they can be improved, especially regarding false alarms. One does need to program them to look for the particular agents of interest. The issues then are cost, where to deploy, and how to deal with false alarms. I will touch more on biosensors in the following section.

Infotech is the technical key to determining who the people are who have access to aircraft, and it also offers the first clues to their intentions. The people with access are those who work at airports, including screeners, and the passengers and crew. One problem is to distill information from various databases, most domestic, some international, to ferret out those individuals who are known or suspected to be threats. There will be resistance to sharing from those possessing the information on highly sensitive databases. At the minimum, a means must be found for providing only the essential information to the parties controlling access to the aircraft.

Biometrics, including facial recognition technologies, can in principle provide an additional identification tool, beyond the usual name, a minimal amount of personal data, and, perhaps, a picture. However, none of this is any use unless one has the individual of concern in one's database already. In the case of the 19 hijackers, from publicly available information, only three would have triggered any sort of alert. These were due to overstaying visas or having had minor run-ins with the law.

For those with access to aircraft, a serious background check needs to access databases that go back longer than a few months or even years: I would assert that it is necessary to track someone's credentials for eight years or more to get a clear enough picture of their potential for criminal conduct. And one constantly needs to verify that those granted access are actually the ones who have been approved for access. We don't want access given to someone who steals an ID, for example. Here, too, infotech and biometrics can only help with part (a substantial part, true) of the job. Procedural security changes are required to protect the civil aviation system adequately from the "insider" threat.

Regarding those who actually board a flight, it would be nice to know whether they have malevolent intentions that pose a risk to others. This is where some technological futurism might possibly be of use. Remote detection of heart rate, adrenaline on the skin, and perhaps other chemicals connected with the "fight or flight" reaction, is imaginable, and some efforts have been proceeding in these areas for years. Voice stress analysis is another possibility, although to my knowledge, there are no highly convincing data that this would provide a reliable trigger for the purposes considered here. And, in the neurological/cognitive realm, on an even

---

[1] For comparison with current work, the research and development plans for aviation security within the Federal Aviation Administration may be downloaded from the site, http://www.faa.gov/asd/red98.htm.

more futuristic note, would there be clues one could obtain from a remote (at a meter or two) electroencephalogram that would be useful?

I am somewhat skeptical of all of these possibilities, but the problem is serious enough, in my view, to justify some work in these areas. At the least, one could easily imagine useful by-products for public health and neurological research. Experimental data are needed to learn how reliable (if at all) such indicators would be in a civil aviation context. The obvious issues of effectiveness, false positives, and false negatives will be determinant: a simple demonstration of some vague effect is insufficient.

One needs to bear in mind that the consequences for an individual of triggering the system may not necessary be immediate incarceration for life. A trigger may simply indicate the need to examine carefully just what the individual has brought onto the plane. One might also want to correlate alarms from different individuals on the same flight. False positives, while they need to be controlled, can be tolerated at a moderately low level (say, less than a percent).

Information technologies could obviously be applied to the issue of monitoring or controlling a hijacked plane automatically or from the ground, as has been discussed openly in the press. All this is feasible with current processing, communications, and information technologies and appears to need little in further new research. Whether this approach (especially controlling flight) is a good idea or not, is another question. Pilots tend to think it is not.

Biodefenses

Sensors[2] (Refs)

It would be useful if highly sensitive, specific, broad-spectrum sensors, capable of detecting biological or chemical agents before they could threaten human life, were placed in many environments: transportation nodes, vehicles, public buildings, even homes. They should also be rapid (seconds to a few minutes at the most) and have manageable false alarm rates. What is manageable in this case is rather less than what is manageable in controlling airplane boarding. A false alarm rate of one per year per detector might be barely manageable in some contexts, unless one has the ability to run quick follow-up tests for verification. Even considering only public buildings, probably the most likely civilian target category for attack, the problem is still extremely challenging.

Biotechnology and nanotechnology (or, at least, microtechnology) converge here. There have been efforts in this area for years. I refer particularly to the "lab-on-a-chip" concept, which is being developed and used by national laboratories and private companies. For the purpose of protecting against terrorism (and serious work is going on in this area), one may envision arrays of perhaps up to 1000 by 1000 sites on a small chip, each one populated by a DNA sample from a particular

---

[2] Descriptions of government research and development work in chemical and biological detectors may be found in U.S. Department of Energy, *Chemical and Biological National Security Program, FY00 Annual Report* (Washington, DC: U.S. Department of Energy 2000) and U.S. Department of Defense, *Nuclear /Biological/ Chemical (NBC) Defense, Annual Report to Congress,* (Washington, DC: U.S. Department of Defense 2000).

pathogen. If one can sample well enough and devise a PCR process to be fast enough, one might imagine that highly specific detection would be possible. The rub is the time required: current prototypes that do DNA analysis typically require on the order of an hour to process a sample and have a rather small number of pathogens to which they are sensitive. The hope is to reduce this time to minutes or less.

If major improvements in biosensors are, indeed, possible within a few years, the applications in the public health arena are easy to imagine. If a national medical surveillance network is assembled, as some researchers envision (notable among them, Alan Zellicoff of Sandia) and many advocate, the use of an even broader pathogen-detection chip (if cheap enough) could have enormous benefits, both for monitoring and for individual treatment. The spin-offs would more than justify the expense incurred in the main counterterrorist thrust. This is an area I consider extremely fertile for more research and development, perhaps more than any other in the counterterrorist field, and one that needs even more attention than it is currently receiving.

Decontamination

Sensors would have obvious uses for decontamination after an attack. But what about decontaminating the air in buildings? There are current technologies that could be useful, as a matter of course, in buildings with high levels of circulation. Ultraviolet radiation, electron discharges, and nuclear radiation all come to mind as possibilities. As retrofits to current buildings, the cost would generally be prohibitive except for high-value targets. But if reasonable cost options were feasible, new buildings could incorporate such measures. This is an engineering issue and one that I suggest is worthy of some study.

Vaccines and Therapeutics

Vaccines and therapeutics are areas that have, of course, been pursued for a long time: centuries, in fact. Nowadays, the terrorist threat gives new impetus to these pursuits. Especially regarding vaccines, the lack of a strong market has made the large drug companies uninterested in working very hard in this area, and I assert that there is therefore a major role for the government.

A major new field is antiviral drugs, which is highly relevant to terrorism, since many putative agents, from smallpox to the hemorrhagic fevers, are viruses. To an outsider, this looks like a burgeoning subject of study, one poised on the cusp of serious breakthroughs. Major efforts need to be placed here. In this field of bioresearch, as well as many others, the stimulation of work for counterterrorist or defense ends will have many spin-offs for public health that are perhaps more valuable than the original purpose of the work.

Another approach is to look for methods to counter the chemistry and mechanics of infections, to look for commonalities in the way that different agents wreak havoc on multicellular organisms, and to counter the pathogen attack in a generic way. The Department of Energy, DARPA, and, indeed the whole field of microbiology actively work these areas of research. To an outside observer, again, the approach seems intriguing and promising. What I would suggest here is coordination of such work that particularly applies to microorganisms of interest as agents of bioattacks.

A totally different field that has received some attention lately, but probably not enough, is the area of edible vaccines.[3] Synthetically coding for receptor sites on the protein coats of pathogens, and then inserting these DNA strings into a plant genome has produced interesting early results. Workers at the Boyce-Thompson Plant Research Institute at Cornell, in collaboration with researchers at Baylor University, have found immune response in human subjects generated by eating the potatoes that result from such genetic manipulation. Since we have experienced such difficulties in producing a vaccine in large quantity just for anthrax, a totally different path might be in order. Side effects should be minimized by this technique. One could even imagine, eventually, a cocktail, a V-8, of tomatoes, bananas, or some other food, bred to protect against a variety of pathogens. The doses could be easily distributed and delivered, and, in remote or poor areas, would need a minimum of refrigeration and require no needles. Possibly, none of this will work out: maybe the required doses of food will just be too great or will have to be re-administered too often to be practical. But, it seems to me that this is interesting enough to investigate with more vigor than is currently the case.

Other Areas

Time and space severely limit what can be described in an extremely short paper. I will just touch upon other areas that appear to me to be important in combating terrorism. All would involve nanotechnologies and information sciences, falling under the NBIC rubric, since they would probably require advances in computing power to be most effective.

One can try to apply information technology and social sciences in an effort to discern patterns of behaviors in nasty organizations. If one were to focus on correlating a large volume of diverse data that include the characteristics, motivations, and actions, could one achieve any predictive value? Predicting a specific event at a specific time is clearly unlikely, but perhaps a result could be generalized cues that would enable intelligence services to look more closely at a given time at a given group. DARPA is pursuing such avenues, as are, no doubt, other branches of the government.[4] I would not call this cognition per se, but this type of effort does try to encompass, in part through behavioral sciences, how certain types of people might think in specific situations.

Finally, I would like to point to the issue of integrating architectures, applied to many counterterrorist areas. This, too, involves cognition and information science and technology. As a simple example, the security at an airport would greatly benefit from some integration of all the security activities that go on there, including alarms, alarm resolution, personnel assignments, equipment status, and so on.

On a much more complex level, the response to a major terrorist act, involving weapons of mass destruction, would benefit enormously from a generalized C4ISR (command, communications, control, computers, information, surveillance, and

---

[3] http://www.sciam.com/2000/0900issue/0900langridge.html, also in *Scientific American*, Sept. 2000.

[4] http://schafercorp-ballston.com/wae/, accessed last on 27 December 2001, contains a description of a DARPA project entitled Wargaming the Asymmetric Environment.

reconnaissance) architecture. How does one put the response all together, among so many federal, state, and local agencies? How does urgent information get down to the street quickly and accurately? How is it shared rapidly among all those who urgently need to know? How does one communicate most effectively to inform the public and elicit the most productive public reaction to events? How can one best guide the decisions of high-level decisionmakers in responding effectively to the attack? How are their decisions most effectively implemented? True, we can always muddle through; we always have. But a comprehensive architecture for emergency response could make an enormous difference in how well the society will respond, and it could minimize casualties. And cognitive science and information technology together could greatly help in devising such architectures. Much talk and much work is proceeding in this area, especially in the past two months. My impression, however, is that thinking by newcomers to the counterterrorist field — who have the expertise in operations research, information technology, and cognitive sciences — would be highly productive.

NANOTECHNOLOGY AND THE DEPARTMENT OF DEFENSE

Clifford Lau, Office of the Deputy Under Secretary of Defense for Research

The Department of Defense (DOD) recognized the importance of nanotechnology well before the National Nanotechnology Initiative (NNI). DOD investment in nanoscience dated back to the early 1980s when the research sponsored by DOD began to approach the nanometer regime. Nanoscience and nanotechnology is one of six research areas identified by DOD as strategically important research areas. After careful evaluation and coordination with other federal agencies within the Interagency Working Group on Nanotechnology, the DOD investment was organized to focus on three nanotechnology areas of critical importance to DOD: Nanomaterials by Design, Nano-Electronics/Magnetics/Optoelectronics, and Nanobiodevices. DOD has traditionally provided leadership in nanotechnology research, particularly in the areas of nanoelectronics, chemistry, and materials. The research sponsored by DOD will provide the scientific foundation for developing the nanotechnology to enhance our warfighting capabilities.

DOD Impact

It is anticipated that nanotechnology would impact practically all arenas of warfighting in DOD, including command, control, communications, computers, intelligence, surveillance, and reconnaissance (C4ISR). In addition to providing much greater capability in computing power, sensors, and information processing, nanotechnology will also save more lives of our men and women in uniform by the development of lightweight protective armors for the soldiers. The value of nanotechnology to DOD includes, but is not limited to, the following:

a) *Chemical and biological warfare defense*. Nanotechnology will lead to the development of biochemical sensors to monitor the environment in the battlefield. Chemical and biological warfare agents must be detected at very low levels in real time. Nanotechnology will dramatically improve detection sensitivity and selectivity, even to the point of responding to a few molecules of the biochemical agent. Nanostructures are showing the potential for decontamination and neutralization as well.

b) *Protective armor for the warrior.* Nanotechnology will lead to the development of extremely strong and lightweight materials to be used as bullet-stopping armors.

c) *Reduction in weight of warfighting equipment.* Nanotechnology will reduce the volume and weight of the warfighting equipment a soldier/marine must carry in the battlefield by further miniaturization of the sensor/information systems. Development in nanoelectronics and portable power sources based on nanotechnology will provide much-needed capability in information dominance in sensing, communication, situational awareness, decision support, and targeting.

d) *High-performance platforms and weapons.* By providing small structures with special properties that can be embedded into larger structures, nanotechnology will lead to warfighting platforms of greater-stealth, higher-strength, and lighter-weight structural materials. In addition to higher performance, new materials manufactured by nanotechnology will provide higher reliability and lower life-cycle cost. One example, already in fleet test by the Navy, utilizes nanostructured coatings to dramatically reduce friction and wear. In another example, nanocomposites where clay nanoparticles are embedded in polymer matrices have been shown to have greater fire resistance and can be used onboard ships.

e) *High-performance information technology (IT).* Nanotechnology is expected to improve the performance of DOD IT systems by several orders of magnitude. Current electronics devices will reach a limit at 100 nm size in another five years. Continued advances in IT will require further advances in nanotechnology. Information dominance in network centric warfare and the digital battlefield is critical to DOD in winning the wars of the future.

f) *Energy and energetic materials.* The DOD has a unique requirement for energetic materials. Fast-release explosives and slow-release propellants must have high energy density while retaining stability. Nanoparticles and nano-energetic materials have shown greater power density than conventional explosives. Nanopowdered materials have also shown promise for improved efficiency in converting stored chemical energy into electricity for use in batteries and fuel cells.

g) *Uninhabited vehicles and miniature satellites.* Nanotechnology will lead to further miniaturization of the technology that goes into uninhabited vehicles and miniature satellites. The Uninhabited Air Vehicles (UAVs) will have greater range and endurance due to the lighter payload and smaller size. Uninhabited Combat Air Vehicles (UCAVs), will have greater aerial combat capabilities without the g-force limitations imposed on the pilot. Uninhabited Underwater Vehicles (UUVs) will be faster and more powerful due to miniaturization of the navigation and guidance electronics.

DOD Programs

Because of the large potential for payoffs in enhancing warfighting capabilities, nanotechnology continues to be one of the top priority research programs within the Department of Defense. In the Office of the Secretary of Defense, the DURINT (Defense University Research Initiative on Nanotechnology) will continue to be funded out of the University Research Initiative (URI) program. All three services and DARPA have substantial investments in nanotechnology 6.1 basic research. New 6.2 applied research programs are being planned to transition the research results to develop the nanotechnology for DOD.

ADVANCED MILITARY EDUCATION AND TRAINING

James Murday, Naval Research Laboratory

The U.S. military annually inducts 200 thousand new people, 8 percent of its person power. Further, the anticipated personnel attrition during warfare requires extensive cross-training. With public pressure to reduce casualties, there is increasing utilization of high technology by the military. Warfighters must be trained in its use, recognizing that the education level of the average warfighter is high school. These circumstances present the military with an education and training challenge that is exacerbated by the fact that personnel are frequently in remote locations — onboard ship or at overseas bases — remote from traditional education resources. The entire K-12 education in the United States has similar problems, so any program that successfully addresses military training needs will certainly provide tools to enhance K-12 education as well.

The convergence of nano-, bio-, info- and cognitive technologies will enable the development of a highly effective teacher's aide — an inexpensive (~$100) virtual learning center that customizes its learning modes (audio, visual, and tactile) to individuals and immerses them into a custom environment best suited for their rapid acquisition of knowledge.

Role of Converging Technologies

Nano. Nanotechnology holds the promise for relatively inexpensive, high-performance teaching aides. One can envision a virtual-reality teaching environment that is tailored to the individual's learning modes, utilizes contexts stimulating to that individual, and reduces any embarrassment over mistakes. The information exchange with the computer can be fully interactive — speech, vision, and motion. Nanodevices will be essential to store the variety of necessary information or imagery and to process that information in the millisecond timeframes necessary for realtime interaction.

Bio. Biotechnology will be important to provide feedback on the individual's state of acuity and retention.

Info. Information technology must develop the software to enable far more rapid information processing and display. Since military training must include teaming relationships, the software must ultimately accommodate interaction among multiple parties. Innovations are also needed to enable augmented-reality manuals whereby an individual might have realtime heads-up display of information that cues repair and maintenance actions.

Cogno. Effective learning must start with an understanding of the cognitive process. People have different learning modes — oral, visual, tactile. They respond to different motivators — individual versus group — and different contexts — sports for the male, social for the female, to use two stereotypes. Human memory and decision processes depend on biochemical processes; better understanding of those processes may lead to heightened states of acuity and retention.

Transforming Strategy to Reach the Vision

Under its Training and Doctrine Command (TRADOC, http://www-tradoc.army.mil/), the U.S. Army has a Training Directorate that endeavors to introduce more effective training and education methods. A collaborative program between the National Nanotechnology Initiative, the National Information Technology Initiative, NSF (science and math), the Department of Education (K-12 teaching), and TRADOC might lead to the most rapid progress toward this goal. The entertainment industry must also be included, since it has been a driver behind much of the recent technological progress

Estimated Implications

This opportunity has benefit for education and training of students at all age levels, not just the military. Further, all technology benefits from larger markets to lower the unit cost. A low-cost instruction aide as described above, especially in mathematics and science, could bypass the problem of preparing adequately knowledgeable K-12 teachers. Success at this project could revolutionize the nation's approach to education.

VISIONARY PROJECTS

HIGH-PERFORMANCE WARFIGHTER

James Murday, Naval Research Laboratory

If one were to look for situations where the confluence of nano, bio, info and cogno technologies would make a critical difference, the military warfighter would certainly be seriously considered a leading example. The warfighter is subjected to periods of intense stress where life or death decisions (cogno) must be made with incomplete information (info) available, where the physiology of fatigue and pain cloud reason (bio), and where supplemental technology (nano) must compete with the 120 pounds of equipment weight s/he must carry.

The confluence of the NBIC technologies will provide the future U.S. warfighter with the capability to dramatically out-fight any adversary, thereby imposing inhibitions to using warfare with the United States as a means to exert power and reducing the risk of U.S. casualties if war does occur.

Role of Converging Technologies

Nanotechnology holds the promise to provide much greater information, connectivity, and risk reduction to the warfighter. The continued miniaturization of electronic devices will provide 100 times more memory (a terabit of information in a cm^2). Processing speeds will increase to terahertz rates. Displays will be flexible and paper thin, if not replaced by direct write of information on the retina. High-bandwidth communication will be netted. Prolific unattended sensors and uninhabited, automated surveillance vehicles under personal control will provide high data streams on local situations. The marriage of semiconductors and biology will provide physiological monitors for alertness, chemical/biological agent threat, and casualty assessment. Nanofibers and nanoporous adsorbents will protect against CB threats while minimizing heat burdens and providing chameleon-like color adaptation for camouflage. The small size of the nanodevices will limit the volume, weight, and power burdens.

Presuming nanotechnology delivers the hardware, advances must be made to create information out of the manifold data streams. The soldier must stay alert to the environment, heads-up or retinal displays are essential, as well as the traditional flat, flexible (paper-like) media. Voice dialogue with the computer is essential to keep hands free for other functions. GPS-derived location, high-precision local maps (cm^2 voxels — potentially three-dimensional representations that include information about building structures, underground tunnels, and the like); language translators (for interrogation of the local citizenry); automated weapons that track target location and control the precise moment to fire: all of these capabilities will require new software.

Biotechnology promises considerable advances in monitoring and controlling the physiological condition of a warfighter. New innovations are likely to include sensitive, selective transduction of biological events into signals compatible with electronic devices; new approaches to the neutralization of biological and chemical agents without aggressively attacking other constituents in the local environment; and possible harnessing of body chemistry as a source of local power.

The nano-, info-, biotechnology items above are aids toward more effective learning and decision making. Rapid, effective cognition is critically dependent on body physiology, and on the manner information is organized and delivered (audio, visual, tactile) (Figure E.13).

Transforming Strategy to Reach the Vision

Nanoscale science, engineering, and technology will provide the understanding critical to rapid progress in the development of new, higher-performance, information technology nanodevices, of high performance materials, and of sensors/activators for biological systems. In a simplified, but useful, perspective, nanoscience will underpin the information technology and biotechnology components of a warfighter system program. The National Nanotechnology Initiative (NNI) will provide a broad-based program in nanoscience; it remains a challenge to couple that program most effectively with information technology and biotechnology.

Information Technology (ITI) is also a U.S. national initiative. The coordinating offices for both the NNI and ITI programs have been collocated in order to

Nano-Technology for the Future Warrior

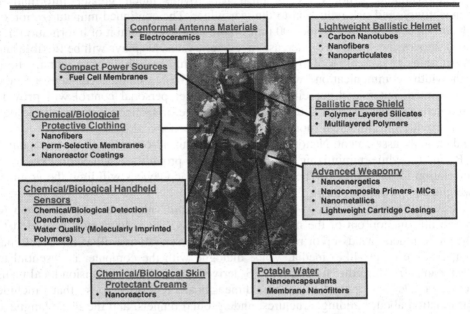

Figure E.13. Soldier system of the future (courtesy Dr. Andrzej W. Miziolek, U.S. Army Research Laboratory, AMSRL-WM-BD, Aberdeen Proving Ground, MD).

encourage close collaboration. The Information Technology Initiative identifies areas where advances in device capability would be most effective and works to advance modeling and simulation (high-performance computing) so that theoretical contributions to nanoscience will be an equal partner with the experimental efforts. The Nanotechnology Initiative must accelerate progress in those areas where new, cost-effective technology will lead to the most significant impact on information systems.

Biotechnology is effectively a third U.S. national imitative if one includes the NIH budget for health and medicine. A principal challenge here is acceleration of chemical, physical, materials, and engineering contributions to biotechnology. Biology must also better identify the biochemical basis for alertness, acuity, and memory retention.

The large investments already present in nano-, info- and biotechnology should be coordinated and coupled with efforts in cognition. DARPA, NASA, NIH, and NSF already have major programs that seek to integrate nano-, bio- and info- research. Within the DOD, the Army and Marines have the lead efforts in technologies to impact the individual warfighter. The Army is presently competing a University-Affiliated Research Center (UARC) on

Figure E.14. Wearable device for non-drug treatments.

the topic, "Institute for Soldier Nanotechnologies," that potentially can integrate the essential components of this opportunity.

Estimated Implications

Technology has led to dramatic improvements in fighting capability, but not for the individual soldier or marine. While air and sea power certainly have a major role in attacking any opponent, in any major conflict, soldiers and marines will be engaged in ground combat. Utilizing the convergent NBIC technologies, we have the opportunity to improve significantly the ability to control the local situation at minimal risk of personal casualty.

References

Nanotechnology for Soldier Systems Conference, Cambridge, Massachusetts, July 7-9, 1998.
Natick Soldier Center (NSC). 2002. Mission: maximize the individual warfighter's survivability, sustainability, mobility, combat effectiveness. Website: http://www.natick.army.mil/soldier/. Aberdeen, MD: U.S. Army Soldier and Biological Chemical Command (SBCCOM) and Nattick, MA: Nattick Labs.

NON-DRUG TREATMENTS FOR ENHANCEMENT OF HUMAN PERFORMANCE

Robert Asher, Sandia National Laboratories

Human performance enhancement may require modifications to the biochemical aspects of the human. Maintained alertness, enhanced physical and psychological performance, and enhanced survivability rates in serious operations all require modifications to the biochemical aspect of the human. DARPA is in the process of developing drugs to enhance performance when a person has been sleep-deprived. Drug companies spend an average of $800 million to develop new drugs that may have negative side effects. An alternative is to develop non-drug approaches to human performance enhancement. As an example, it is common medical practice to immerse a person in a hot bath preceding heart operations to build up stress proteins that will give greater survivability when s/he receives blood products.

Consider the use of externally applied, non-dangerous electromagnetic fields to increase the rate of production of body biochemicals that enhance human performance. DARPA has a proposal to increase the rate of stress protein production before a soldier goes into combat. The intent is to increase the survivability rate when the soldier is wounded and needs to receive blood products. Beyond that, one can envision increasing the rate of production of ATP, which will yield higher energy levels by natural means, will help ion pumping to aid in nerve recovery and contraction of muscles, and will speed recovery from combat stress. What other changes can be engineered by a specifically shaped electromagnetic pulse that might enhance human performance without pharmaceuticals? This investigation may spawn a new industry in which the human is enhanced by externally applied electromagnetic pulses so shaped as to enhance specific biochemical changes within the body without drugs or in combination with drugs, with fewer side effects. For instance, nanoparticles might be formulated to release drug dosages only when irradiated with electromagnetic pulses focused at certain sites, allowing treatments to specific areas without the whole body being affected by the drug therapy.

Role of Converging Technologies

All of the NBIC technologies have a role in the goals of non-drug enhancement of human performance:

Nano. Develop and understand the nano aspects of the use of electromagnetic field interactions with cellular structures. Develop and understand how treatments may be developed by nano particle interactions only at specific sites where the electromagnetic fields are focused. Investigate whether electromagnetics can be used as a power source to conduct mechanical actions at the sites.

Bio. Develop a detailed understanding of the effects of electromagnetics on cells and neuronal networks, including the full range of scales, from micro effects on proteins to macro effects on neuronal networks.

Info. Develop methods to shape optimal electromagnetic pulses to carry messages to the cells and neurons.

Cogno. Understand how electromagnetics can be used to enhance cognitive performance as well as physiological performance.

Transforming Strategy to Reach Vision

The strategies to achieve these goals are as follows:

- Develop a program that will explore the use of electromagnetics for enhancement of human performance. This program will be multidisciplinary in orientation, utilizing
  - electromagnetics as the actuation mechanism for the treatments
  - biotechnology in the understanding of cellular interaction with the electromagnetic fields
  - nanotechnology to help engineer solutions that may include specific site treatments released by a focused electromagnetic field
  - information technology in that the pulses need to be so shaped as to cause desired interconnected cell electromagnetic responses of cognition by external fields
- Fund work towards the goal of understanding in detail the effects of electromagnetics on cellular systems and on cognition.
- Consider cellular electrochemical and structural changes and actions imposed by electromagnetics.
- Fund work towards electromagnetic and biochemical dynamical modeling of cellular systems in order to both understand electromagnetic and biochemical aspects, as well as to optimize the shape of electromagnetic pulses to impose desired cell changes without inducing side effects.
- Fund experimental basic work in understanding the effects of electromagnetics on cells.

Estimated Implications

The impact on society of such a program can be great, as this might yield treatments to enhance human performance without the use of drugs and provide new exciting treatments for ailments that require site-specific treatments. A new industry can be born from this work. It may also lead to treatments that will enhance human cognition.

BRAIN-MACHINE INTERFACE

Robert Asher, Sandia National Laboratories

Increasingly, the human is being asked to take in multisensory inputs, to make near-instantaneous decisions on these inputs, and to apply control forces to multitask and control machines of various sorts. The multitasking, multisensor environment stresses the human, yet, more and more s/he being asked to operate in such an environment. As an example, the visionary project on uninhabited combat vehicles discusses an increased workload in piloting combat vehicles. DARPA has a brain-machine interface program about to start. This program has as its goal human ability to control complex entities by sending control actions without the delay for muscle activation. The major application for this program is control of aircraft. The intent is to take brain signals and use them in a control strategy and then to impart feedback signals back into the brain.

The DARPA program could be extended to include a broader range of potential impact by including the possibility of other applications: learning and training, automobile control, air traffic control, decision-making, remote sensing of stress, and entertainment. Learning and training might be implemented as information coded into brain signals and then input into the person. Air traffic control in increasingly busy skies can use such capability: the controller has multiple inputs from multiple aircraft. These can be input into his brain in a 3-D aspect and an alertness signal used to "wake him up" when his attention drifts beyond acceptable limits. Not only intellectual data might be passed from one person to another without speaking, but also emotional and volitional information. Decision-making may become more precise as emotional, fatigue, and other cognitive states can be appraised prior to making a critical decision.

The potential impact on automobile safety is great. The driver can have quicker control of his automobile (Figure E.15), allowing for safer driving while reducing the car-to-car spacing on congested highways. This would help alleviate highway congestion and the need for more highways. Furthermore, it would allow for safer driving as driver attention can be measured and the driver "alerted" or told in some manner to pay attention to his or her driving when attention wanders beyond safe margins. It can allow for detection of driver impairment so that the vehicle may be made either not to start or to call emergency.

Direct connection into the brain could yield a revolution in entertainment, as people may be "immersed," MATRIX-style, into the midst of a movie or educational show. Can you imagine the impact of being immersed in a fully 3-D audio-visual simulation of the battle of Gettysburg?

Figure E.15.   Hands-off control of an automobile through
a device for reading and implanting brain waves.

Role of Converging Technologies

Nano. The brain-machine interface effort will require nanotechnologies in order
to make the required experimental measurements and to implement the devices for
both receiving brain electromagnetic signals and transmitting signals back into the
brain.

Bio. This is a highly biological, neuroscience effort, which requires detailed
understanding and measurements of the brain's electromagnetic activity. It requires
a significant measurement protocol.

Cogno. This effort by its very nature will directly affect the cognitive aspects of
the individual by externally applied electromagnetic fields by implanting
information for the individual. Thus, this effort can lead to increased learning and
other cognitive results.

Transforming Strategy to Reach the Vision

To achieve these goals, enter a partnership with DARPA to fund additional
technologies and applications that would enhance the brain-machine interface effort.
Work should be focused on the goals of using the technologies for cognitional
aspects, understanding memory, and learning brain function to be able to design
devices to increase their capabilities.

Estimated Implications

This effort would yield a technological revolution, in applications from
computers to entertainment. It would give the United States a global competitive
advantage while yielding solutions to specific domestic problems such as air traffic
control and highway safety in increasingly crowded environments. It will
revolutionize education. This effort will yield devices that may be applied to a
number of activities and be sufficiently small as to be wearable in a car or at home.

NANO-BIO-INFO-COGNO AS ENABLING TECHNOLOGY FOR UNINHABITED COMBAT VEHICLES

Clifford Lau, Office of the Deputy Under Secretary of Defense for Research

It is envisioned that in 20-30 years, when the research and development are successfully completed, nano-bio-info-cogno (NBIC) technology will enable us to replace the fighter pilot, either autonomously or with the pilot-in-the-loop, in many dangerous warfighting missions. The uninhabited air vehicle will have an artificial "brain" that can emulate a skillful fighter pilot in the performance of its missions. Tasks such as take-off, navigation, situation awareness, target identification, and safe return landing will be done autonomously, with the possible exception of person-in-the-loop for strategic and firing decisions. Removing the pilot will result in a more combat-agile aircraft with less weight and no g-force constraints, as well as reduce the risk of pilot injury or death. The fighter airplane will likely derive the greatest operational advantages, but similar benefits will accrue to uninhabited tanks, submarines, and other military platforms.

Role of Converging Technologies

The convergent NBIC technologies, although at the early stage of basic research, are anticipated to have an impact on practically all arenas of warfighting and peacekeeping and thus are vitally important to national security. For instance, today's fighter airplanes are loaded with sensors, avionics, and weapon systems. The complexity of these systems and the information they provide place tremendous workload on the pilot. The pilot must fly the fighter airplane in hostile environment, watch the cockpit displays, be aware of the situation, process the sensor information, avoid anti-air missiles, identify and destroy the targets, and return safely. No wonder there is information overload on the pilot, in spite of the many decision aid systems. Furthermore, fighter pilots are highly valued and trained warriors, and the country cannot afford to lose them from anti-air fire. The need for autonomous or semi-autonomous air vehicles to accomplish surveillance and strike missions is clear (Figure E.16).

Nano. Nanotechnology will continue to the current trend in miniaturization of

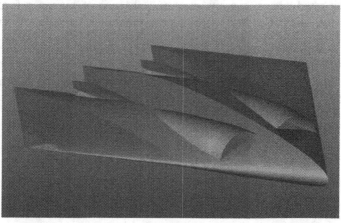

Figure E.16. Uninhabited combat air vehicle (UCAV).

sensors, electronics, information processors, and computers. Miniaturization will reduce the weight, size, and power of the on-board systems in the air vehicle and will increase information processing power.

Bio. Brain research will help us to understand how pilots process the massive amount of information coming from the sensors and intelligence. That understanding will allow us to design an artificial "brain" to process the information and to control the air vehicle autonomously.

Info. Research in information technology will enable us to design specialized systems that do not require writing millions of lines of code, such as the adaptive learning strategy used by the brain. Storage and retrieval of massive amounts of data and information fusion to allow the system to make decisions will also be an important aspect of this research.

Cogno. Understanding the principles behind cognition is extremely important in the design of an autonomous system with the capabilities of target recognition and situation awareness. For autonomous air vehicles, it is particularly important to recognize the intent of encounters with friendly or unfriendly aircraft in its vicinity.

Transforming Strategy to Reach Vision

The DOD presently has a number of projects working toward uninhabited combat aircraft. The challenges to meet this goal are considerable. An NBIC program centered at universities would provide both the scientific discovery and the trained students that will be necessary for those projects to succeed quickly. In order to achieve the vision stated above, it is necessary to plan a coordinated and long-term research program considering the above strategies on how to get there. It is important to integrate the current research efforts on nanotechnology with the other research areas to form a multidisciplinary research program. A university-based basic research program addressing the needed science must be interactive with the DOD programs addressing system design and manufacture.

Estimated Implications

Removal of the pilot from assault and fighter aircraft will reduce the risk of injury or death to highly trained warfighters. American public opinion makes this a clear priority. In addition, the lighter weight (no pilot, oxygen system, ejection system, man-rated armor, canopy, etc.) and absence of human g-force constraints will make the aircraft either more maneuverable or capable of more extended missions.

References

National Academy Press. 2000. Review of ONR's uninhabited combat air vehicles program. Washington, D.C.

Lazarski, A.J. (Lt. Col.). ND. Legal implications of the uninhabited combat aerial vehicle. http://www.aipower.maxwell.af.mil/airchronicles/cc/lazarski.html.

DATA LINKAGE AND THREAT ANTICIPATION TOOLS

Tony Fainberg, Defense Threat Reduction Agency

The United States will be subject to asymmetric military threats from lesser powers. On 11 September 2001, this observation moved from the theoretical to the real. To deal adequately with the future, the United States must develop an intelligence system to anticipate threats from adversary states or sub-state actors.

Role of Converging Technologies

The suggested approach is to use the power of *information technology* to assemble, filter, and analyze data about the adversary. First, it will be necessary to acquire a large volume of data regarding each potential enemy organization. Data linkage among many databases would be needed, including some from open source material and others from intelligence sources. The data would include the group's characteristics, its people, funds, and the movement of each, the motivations of the people, relevant current events, significant dates, and some way of encoding the cultural perspectives of the organization. In addition to information technology, the approach also requires *nanotechnology*, due to the large amount of data that need to be handled and analyzed. Further, some sociological analysis (for the group) and psychological profiling would be required, as well as country and culture experts. This requires broad social science input. Understanding how the adversary analyzes and makes decisions involves modeling his *cognition* processes. An automated translation capability would be helpful in the data mining, since frequently there may not be enough analysts familiar with the necessary languages to keep up with the data input.

Transforming Strategy to Reach Vision

DARPA's Information Technology Office is pursuing similar methodologies, as have, no doubt, other branches of the government. It is possible that increased computing power, better application of the social sciences, plus more sophisticated integration of the information and modern decision algorithms might produce significantly better predictive tools. The National Science Foundation is in an excellent position to sponsor research in this area, as well as to coordinate similar programs of other agencies through interagency workshops.

Estimated Implications

The resulting decision tool or decision aid would probably not be able to predict a specific event at a specific time; however, it could possibly function to cue intelligence services to look more closely at the adversary when it gives an alarm and might also be useful for cueing heightened security alerts.

F. UNIFYING SCIENCE AND EDUCATION

THEME F SUMMARY

Panel: D.L. Akins, Y. Bar-Yam, J.G. Batterson, A.H. Cohen, M.E. Gorman, M. Heller, J. Klein-Seetharaman, A.T. Pope, M.C. Roco, R. Reddy, W. Tolles, R.S. Williams, D. Zolandz

The fifth and final NBIC theme explores the transformations of science and scientific education that will enable and be enhanced by technological convergence. The panel especially focused on the ways that education can transform science and unifying science (based on the unity of nature and using cause-and-effect explanation) can transform education, for the vast improvement of both. As a number of reports from the National Research Council (NRC 1996-2000) and comparable organizations attest, the future of society depends on continued scientific progress, which in turn depends upon science education. Converging scientific principles and technologies will raise the importance of this issue to a higher level.

Four factors demand significant changes in the science education received by students at all levels:

1. Many poorly understood social factors work against science in the educational system, and ways must be found to counter these anti-science forces using new S&T trends (NSF 2000).

2. Rapid progress in cognitive, biological, information, and nanoscale sciences offers new insights about how people learn that can guide effective reforms in curriculum, evaluation, and organizational structuring.

3. New education techniques and tools will be made available by converging technologies, and we need to prepare to take advantage of them.

4. Few mid-career professional scientists have the practical opportunity to redirect their careers to any significant extent, so unification of the sciences must largely begin in school.

Currently, scientific and engineering education is highly fragmentary, each part constrained by the boundaries of one particular discipline. In the future, the knowledge taught will be based on unifying concepts offered by nano, bio, info, and cognitive sciences throughout the educational establishment. Natural, engineering, social, and humanity sciences will converge. The corresponding basic concepts of unifying science will be introduced at the beginning of the teaching process in K-12, undergraduate, and graduate education. New tools will be developed by convergent technologies to provide high-quality, anywhere-anytime educational opportunities. NBIC science and engineering education will be made available to the majority of students and as continuing education to all interested adults.

No single discipline can describe or support the converging technologies by itself. Different disciplines may play a leading role in different applications. Interfaces are beginning to develop among the four NBIC domains, linking them in pairs, trios, and as a full quartet, in parallel with in-depth development within each field. The optimal process will not develop naturally: a systematic program must be created to encourage it.

Within academia, significant challenges must be overcome. Many teachers lack sufficient depth in their knowledge of mathematics and science, and not enough of the best students are attracted to science and technology. Also, qualified personnel who do understand science and technology generally get better-paying jobs outside the field of teaching.

What Can NBIC Do for Education?

The unification of the sciences is gaining momentum and will provide a knowledge base for education. The concepts on fundamental building blocks of matter employed in nanoscience can be applied in different disciplines, thus providing a multidisciplinary opportunity to introduce breadth while advancing depth. This creates the opportunity for integration across learning — moving from reductionism to integration. It also introduces the challenge of creating a common language for talking about the big picture.

Technologies that arise from the NBIC convergence will provide new tools and modalities for teaching. Some of these will be sensory, including visual, auditory, and tactile. Others will take advantage of better understanding of how the brain works. Still others will be logistic and include delivery of teaching and educational resources any time and anywhere. For advanced levels of scientific training, this will create opportunities at new research frontiers.

Across all levels, there will be opportunities to involve groups of people who have tended previously to be excluded from high-quality science education. We have a responsibility to achieve substantial inclusion and outreach, especially across race and gender. The entire 21st century workforce will be involved in the convergent technologies revolution. NBIC-related applications will be an excellent way to promote systemic, problem-based learning from the earliest educational levels.

What Can Education Do for NBIC?

Universities epitomize the ideal of uniting the intellectual heritage of mankind, so they are a relatively hospitable environment for scientific and technological convergence. Other kinds of educational institutions can also play crucial roles in bringing the scientific and technical disciplines together. In the economy, certain markets become trading zones where a great diversity of products, services, and institutions converge. Scientific trading zones will have to be created, perhaps anchored in university-based research centers or in joint academic-industrial partnerships, that will allow students and scientists to develop the necessary communication skills for trading ideas across disciplines.

The educational system can provide a stimulus for drawing recruits into the NBIC community. Classrooms can become a proving ground for exploring new technologies designed to facilitate learning and communication. Similarly, the educational system can be a developmental laboratory for testing useful technological directions in NBIC.

Many new educational approaches will have to be tried in order to see which are most effective in achieving technological convergence. For example, universities may offer retraining for scientists who already have doctorates and may already have extensive experience in industry or research laboratories. Perhaps young scientists will engage in post-doctoral work in a second field. NBIC will benefit from changes in life-long learning at all levels, including in both white-collar and blue-collar occupations. NBIC concepts must be adopted early, in advance of technological developments that would require a qualified workforce.

NBIC is likely to be both creative and destructive at all levels of the scientific, economic, and social establishment, for example, creating new industries and companies, with the inescapable result that some older ones will decline or even become extinct. Thus, it will be important to educate society about the potential unintended consequences of technological innovation. Maximizing the societal benefits of a new technology is essential for it to enjoy full public support (Roco and Bainbridge 2001).

NBIC Education for the Twenty-First Century

To enhance human performance most successfully, science and engineering education will have to evolve and, in some respects, radically reinvent itself. The knowledge taught will be based on concepts offered by nano, bio, info, and cognitive sciences, and these concepts will be introduced at the beginning of the K-12 teaching process. High-quality science education will be made available to the majority of students.

Special efforts must be made to stimulate communication between disciplines and develop in scientists the communication skills for doing so, so that conversations between them can be made focused and productive. Achievement of good interdisciplinary communication will synergistically enhance the knowledge and progress of all disciplines. Since mathematical tools represent a common language among and between disciplines, mathematics should be taught in greater depth and be a common focus among most scientific disciplines. At the same time, mathematics textbooks must use problems from science and engineering as examples.

Concerted efforts must be supported to write cross-disciplinary educational materials, using a variety of media at the university level that help with the language problems across traditional fields. A positive, inclusive social environment must be promoted that encourages creative growth of converging technologies. Improved pedagogy and accessibility are fundamental ingredients for the realization of converging technologies, incorporating the cultural differences that exist between students and between different technical fields.

At the college and graduate school levels, we may need a new program for multidisciplinary fellowships that would make it possible for students to move among professors and disciplines related to NBIC. A fellowship might travel with a student from one department or school to another and temporarily into a research integration or industry unit. Students might be allowed to define their own cross-disciplinary proposals, then funding would be provided directly to them rather than to an institution or mentor.

Depth in graduate studies is necessary and should not be compromised. However, if specific disciplines deliberately associate themselves with neighboring disciplines that use similar tools and models, breadth and a holistic perspective will come more easily to all.

Creating new educational curricula and methodologies will require problem-driven, system-oriented research and development. Cognitive scientists can analyze learning styles using NBIC and provide appropriate assistance. Better education is needed for teachers, including sufficiently funded research experiences and credit for in-service experiences in industry and research laboratories.

NBIC concepts should be introduced as early as possible. For example, basic concepts and problems of nanoscience could be taught in elementary schools. NBIC terms and concepts could be placed into childhood educational reading materials starting from the earliest levels. Virtual reality environments and websites could offer many kinds of exciting instructional materials. Practical demonstration kits could facilitate interactive learning. Research scientists could frequently visit schools to offer demonstrations and serve as role models.

NBIC courses and modules can be integrated to some extent into existing curricula and school settings, but novel alternatives will also have to be explored. Every way of making science and technology more interesting for young people would be helpful, such as using games to teach math and logic. To achieve these goals, it will be essential for educators, including members of school boards, curriculum development committees, and designers of standardized tests, to identify and encourage champions in K-12 schools. National standards for educational achievement will be indispensable tools to address the most challenging and promising NBIC areas.

In 15 years, we anticipate that education will be based to a significant extent on unifying principles in science and technology that are easier to understand and more valuable for the learner. The new NBIC science content will have been introduced and be available in about 50 percent of the public schools. A variety of new pedagogical tools will be widely available, based on new learning methods, using learning-enhancing devices developed by neuroscience in cooperation with information technology. The process of learning at home or school, either individually or in groups, will be faster and better because of the new methods, tools, and processes.

Statements and Visions

As in the other working groups, participants in the Science and Education group prepared statements offering strategies for transforming the current situation with respect to scientific unification and visions of what could be accomplished in 10 or 20 years. Several contributors examined the social and intellectual processes by which sciences and technologies converge (M. Gorman, J. Batterson and A. Pope, and Y. Bar-Yam); others focused on the special education opportunities offered by integrating sciences from the nanoscale (W. Tolles and A. Cohen); on fully involving human resources (D. Akins); and on enhancing human abilities using biological language (J. Klein-Seetharaman and R. Reddy).

References

Bransford, J.D., A.L. Brown, and R.R. Cocking, eds. 1999. *How people learn: Brain, mind, experience, and school.* Washington, D.C.: National Research Council.

Hilton, M., ed. 2002. *Enhancing undergraduate learning with information technology.* Washington, D.C.: Center for Education, National Research Council.

National Academy of Sciences. 1995. *Reshaping the graduate education of scientists and engineers.* Washington, D.C.: National Academies Press.

National Research Council (NRC). 1996. *The role of scientists in the professional development of science teachers.* Washington, D.C.: National Academies Press.

NRC. 1997. *Developing a digital national library for undergraduate science, mathematics, engineering and technology education.* Washington D.C.: National Academies Press.

NRC. 1999a. *Global perspectives for local action: Using TIMSS to improve U.S. mathematics and science education.* Washington D.C.: National Academies Press.

NRC. 1999b. *Transforming undergraduate education in science, mathematics, engineering, and technology.* Washington, D.C.: National Academies Press.

NRC. 2000. *Strengthening the linkages between the sciences and the mathematical sciences.* Washington, D.C.: National Academies Press.

National Science Foundation (NSF). 2000. *Science and engineering indicators.* Arlington, VA: NSF.

Olson, S., and S. Loucks-Horsley, eds. 2000. *Inquiry and the National Science Education Standards.* Washington, D.C.: National Research Council.

Pellegrino, J.W., N. Chudowsky, and R. Glaser, eds. 2001. *Knowing what students know: The science and design of educational assessment.* Washington, D.C.: Center for Education, National Research Council.

Shavelson, R.J., and L. Towne, eds. 2002. *Scientific research in education.* Washington, D.C.: National Research Council.

Weiss, I.R., M.S. Knapp, K.S. Hollweg, and G. Burrill, eds. 2001. *Investigating the influence of standards: A framework for research in mathematics, science, and technology education.* Washington, D.C.: Center for Education, National Research Council.

STATEMENTS

COMBINING THE SOCIAL AND THE NANOTECHNOLOGY: A MODEL FOR CONVERGING TECHNOLOGIES

Michael E. Gorman, University of Virginia

The National Science Foundation (NSF) is considering societal implications as the new field of nanotechnology emerges, rather than wait for major problems to occur before attempting a fix. This concern for ethics at the earliest stages of discovery and invention needs to be extended to converging technologies as well, a theme to which I will return. But at the outset, I will limit my remarks to nanotechnology, following up on the 2001 NSF meeting on this topic (Roco and Bainbridge 2001).

H. Glimell (2001) has discussed how new fields like nanotechnology create the need for work at the boundaries between fields:

> Consider for example molecular electronics compared with bio-nano
> (or the interface of biological and organic nano materials). The actors,
> nodes and connections to appear in the extension of these NSE
> subareas obviously constitute two very different networks of

innovation. Nanoelectronics is being negotiated and molded in between two camps — the conservative mainstream of the microelectronics industry with its skepticism towards anything popping up as a challenger to the three-decade-old CMOS technology trajectory, and the camp committed to a scenario where that trajectory might come to its end within some five years from now. (Glimell 2001, 199)

Peter Galison (1997) uses the metaphor of a trading zone between different cultures to describe cooperative work at boundaries. One of his examples is the collaboration between physicists and engineers in the Radiation Laboratory at MIT during World War II: "Each of the different subcultures was forced to set aside its longer term and more general symbolic and practical modes of work in order to construct the hybrid of practices that all recognized as 'radar philosophy.' Under the gun, the various subcultures coordinated their actions and representations in ways that had seemed impossible in peacetime; thrown together they began to get on with the job of building radar" (Galison 1997, 827). Despite differences in training and expertise, engineers and physicists of varying backgrounds were able to trade important information.

The current debates about nanotechnology are signs of an expanded trading zone. As Etkowitz has pointed out (2001), the physical sciences need to find a way to emulate the success of the life sciences while avoiding the ethical and social problems that have emerged as genetically modified organisms hit the market. Hence, several extravagant promises have been made about nanotechnology, promises that lead to concerns about what would happen if these promises were fulfilled — if, for example, self-replicating nanobots were ever created. The hardest thing to predict about a new technology is the interaction effect it will have with other evolving social and technical systems.

Thomas Park Hughes, a historian of technology who has spent a lifetime studying the invention of large technological systems, discusses how reverse salients attract inventors: "A salient is a protrusion in a geometric figure, a line of battle, or an expanding weather front. As technological systems expand, reverse salients develop. Reverse salients are components in the system that have fallen behind or are out of phase with the others" (Hughes 1987, 73). In the 1870s, progress in telegraphy was hindered by the fact that only two messages could be sent down a single wire at the same time: the classic problem of bandwidth.

What are the reverse salients that attract researchers and funding to nanotechnology? One is Moore's Law, which reaches asymptote very quickly unless a way can be found to shrink integrated circuits to the nanoscale. This current reverse salient is an instance of a historical one. Earlier, vacuum tubes held up progress in computing. Transistors solved that problem, but then formed their own reverse salient as computing needs expanded to the point where "Production of the first 'second generation' (i.e., completely transistorized) computer — the control data CD 1604, containing 25,000 transistors, 100,000 diodes, and hundreds of thousands of resistors and capacitors — lagged hopelessly behind schedule because of the sheer difficulty of connecting the parts" (Reid 1984, 18). The apparent solution was miniaturization, but there were physical limits. The solution was to transform the problem: instead of building tiny transistors, create an integrated circuit. Nanotechnology offers a similar way to transcend the limits of microchip technology.

Another reverse salient is mentioned by several of contributors to the 2001 Report on the Societal Implications of Nanoscience and Nanotechnology of the Nanoscale Science, Engineering, and Technology (NSET) of the National Science and Technology Council (Roco and Bainbridge 2001). This is the ability to study and emulate fine-grained cellular structures. "Follow the analogy of nature" is a common invention heuristic that depends on an intimate knowledge of nature. Bell used this heuristic to transform the telegraph reverse salient in the 1870s. Instead of an improved device to send multiple messages down a single wire, he created a device to transmit and receive speech, using the human ear as a mental model. Bell's telephone patent formed the basis for one of the great communications start-ups of all time, the Bell Telephone Corporation, which surpassed Western Union, the Microsoft of its day (Carlson 1994). Similarly, detailed understanding of cellular processes at the nanoscale will lead to new devices and technologies that may transform existing reverse salients.

A potential set of reverse salients that came up repeatedly in the 2001 NSET report are environmental problems like ensuring clean water and providing adequate energy.

The terrorist attacks on September 11 will create a new series of reverse salients, as we think about ways of using technology to stop terrorism — and also of protecting against misuses of technology that could contribute to terrorism. Research should be directed towards determining which aspects of these broad reverse salients can be converted into problems whose solutions lie at the nanoscale. One important goal of such research should be separating hype from hope.

Role of Practical Ethics Combined with Social Science

The focus of practical ethics is on collaboration among practitioners to solve problems that have an ethical component. Similarly, social scientists who work in science-technology studies typically establish close links to practice. There are four roles for practical ethics linked to social sciences:

- Prevention of undesirable side effects
- Facilitation of quality research in nanotechnology by social scientists
- Targeting of converging technology areas of social concern
- Incorporation of ethics into science education

Prevention of Undesirable Side Effects

What are the potential negative impacts of nanotechnology, as far as important segments of society are concerned? How can these be prevented? The 2001 NSET report made frequent reference to the negative press received by genetically modified organisms (GMOs) as exactly the kind of problem nanotechnology practitioners wish to avoid. Monsanto, in particular, has developed a variety of genetically modified seeds that improve farmer yields while reducing use of pesticides and herbicides. But Monsanto did not include consumers in its trading zone, particularly in Europe, where potential customers want GMO products labeled so they can decide whether to buy. The best prevention is a broad trading zone that includes potential users as well as interested nongovernmental organizations like

Greenpeace in a dialogue over the future of new nanotechnologies. Social scientists and practical ethicists can assist in creating and monitoring this dialogue.

A related area of concern is the division between the rich and poor, worldwide. If new nanotechnologies are developed that can improve the quality of life, how can they be shared across national boundaries and economic circumstances in ways that also protect intellectual property rights and ensure a sufficient return on investment? Consider, for example, the struggle to make expensive AIDS medications available in Africa. Again, proper dissemination of a new technology will require thinking about a broad trading zone from the beginning. Social scientists can help establish and monitor such a trading zone.

Nanotechnology offers potential national security benefits (Tolles 2001). It might be possible, for example, to greatly enhance the performance of Special Forces by using nano circuitry to provide each individual soldier with more information. However, there are limits to how much information a human being can process, especially in a highly stressful situation. This kind of information might have to be accompanied by intelligent agents to help interpret it, turning human beings into cyborgs (Haraway 1997). Kurzweil (1999) speculates that a computer will approximate human intelligence by about 2020. If so, our cyborg soldiers could be accompanied by machines capable of making their own decisions. It is very important that our capacity for moral decision-making keep pace with technology.

Therefore, practical ethicists and social scientists need to be involved in the development of these military technologies. For example, cognitive scientists can do research on how a cyborg system makes decisions about what constitutes a legitimate target under varying conditions, including amount of information, how the information is presented, processing time, and quality of the connection to higher levels of command. Practical ethicists can then work with cognitive scientists to determine where moral decisions, such as when to kill, should reside in this chain of command.

Military technology faces barriers to sharing that are much higher than intellectual property concerns. The cyborg soldier is much more likely to come from a highly developed country and face a more primitive foe. However, technological superiority does not guarantee victory — nor does it guarantee moral superiority. Practical ethicists and social scientists need to act as stand-ins for other global stakeholders in debates over the future of military nanotechnology.

Facilitation of Quality Research in Nanotechnology by Social Scientists

Improving the quality of research is one area of convergence between the nano and the cogno. Cognitive scientists can study expertise in emerging technological areas and can help expert nanotechnology practitioners monitor and improve their own problem-solving processes. Experts rely heavily on tacit knowledge, especially on the cutting-edge areas (Gorman n.d.). Portions of this knowledge can be shared across teams; other portions are distributed, with individuals becoming experts in particular functions. Cognitive scientists can help teams reflect on this division of labor in ways that facilitate collaboration and collective learning (Hutchins 1995). Cognitive methods can therefore be used to study and improve multidisciplinary convergence, including the development of new trading zones.

Targeting of Converging Technology Areas of Social Concern

Practical ethics and social sciences should not be limited to anticipating and preventing problems. Both can play an important role in facilitating the development of nanotechnology, by encouraging reflective practice (Schon 1987).

An important goal of this reflection is to eliminate the compartmentalization between the technical and the social that is so predominant in science and engineering (Gorman, Hertz et al. 2000). Most of the engineers and applied scientists I work with are solutions seeking problems. They are generally people of personal integrity who, however, do not see that ethics and social responsibility should be factors in their choice of problems. Technology can evolve without improving social conditions, but true technological progress requires social progress. Indeed, focusing on social benefits opens up a range of interesting new technological problems.

Practical ethicists can work with engineers and scientists to identify interesting and worthy social concerns to which the latest developments in nanotechnology could be applied. Philosophers and social scientists cannot simply dictate which problems practitioners should try to solve, because not all social problems will benefit from the application of nanotechnology, and not all future technologies are equally likely.

Directing a technology towards a social problem does not eliminate the possibility of undesirable side effects, and a technology designed to produce harm may have beneficial spin-offs. For example, Lave (2001) does an admirable job of discussing the possibility of unforeseen, undesirable effects when nanotechnology is applied to environmental sustainability. The probability of truly beneficial environmental impacts is increased by taking an earth systems perspective (Allenby 2001). Similar high-level systems perspectives are essential for other nanotechnology applications; in order to achieve this kind of perspective, scientists, engineers, ethicists, and social scientists will have to collaborate.

Incorporation of Ethics into Science Education

How can practical ethicists and social scientists work with science and engineering educators to turn students into reflective nanotechnology researchers? I am Chair of a Division of Technology, Culture, and Communication at the University of Virginia, inside the Engineering School, which gives us a great opportunity to link social responsibility directly to engineering practice. We rely heavily on the case method to accomplish this (Gorman, Mehalik, et al. 2000). We also co-supervise every engineering student's senior thesis; we encourage students to think about the social impact of their work. But we need to go a step further and encourage more students to pursue work linking the social, the ethical, and the technical.

This kind of linkage can attract students into engineering and science, especially if this sort of education is encouraged at the secondary level. Unfortunately, our secondary and elementary educational systems are now focused more on the kind of accountability that can be measured in examinations and less on the kind of creativity and perseverance that produces the best science and engineering. New educational initiatives in nanotechnology can play an important role in changing this climate.

A New Kind of Engineering Research Center

Several years ago, NSF sponsored an Engineering Research Center (ERC) that combined bioengineering and educational technology. Why not also sponsor an ERC that combines research and teaching on the societal implications of nanotechnology? Parts of this center could be distributed, but it should include one or more nanotechnology laboratories that are willing to take their fundamental science and apply it in directions identified as particularly beneficial by collaborating social scientists and practical ethicists. The goal would be "to infuse technological development with deeper, more thoughtful and wide-ranging discussions of the social purposes of nanotechnology...putting socially beneficial technologies at the top of the research list" (Nardi 2001, 318-19). Deliberations and results should be shared openly, creating an atmosphere of transparency (Weil 2001).

This center could combine graduate students in science and engineering with those trained in social sciences and ethics, thus forming a "living bridge" connecting experts from a variety of disciplines. Some graduate students could even receive training that combines engineering, ethics, and social sciences, as we do in a graduate program at the University of Virginia (Gorman, Hertz, et al. 2000).

The center should hold annual workshops bringing other ERCs and other kinds of research centers involved with nanotechnology together with applied ethicists and social scientists. There should be a strong educational outreach program designed to encourage students concerned with making the world a better place to consider careers in nanotechnology. Hopefully, the end-result would be a model for creating trading zones that encourage true technological progress.

This kind of a center need not be limited to nanotechnology. What about a science and technology center on the theme of converging nano, bio, info and cogno (NBIC) technologies directed towards maximum social benefit? One example of a potential NBIC product is of a smart agent able to look up the price and availability of a particular item and identify the store where it can be found while a consumer walks through the mall. This kind of technology has no benefits for the millions all over the world who are dying of AIDS, suffering from malnutrition, and/or being oppressed by dictators.

References

Allenby, B. 2001. Earth systems engineering and management. *IEEE Technology and Society* 19(4): 10-21.

Carlson, W.B. 1994. Entrepreneurship in the early development of the telephone: How did William Orton and Gardiner Hubbard conceptualize this new technology? *Business and Economic History* 23(2):161-192.

Galison, P.L. 1997. *Image and logic: A material culture of microphysics*. Chicago: U. of Chicago Press.

Glimell, H. 2001. Dynamics of the emerging field of nanoscience. In *Societal implications of nanoscience and nanotechnology*, ed. M.C. Roco and W.S. Bainbridge. Dordrecht, Neth.: Kluwer Academic Press.

Gorman, M., M. Hertz, et al. 2000. Integrating ethics and engineering: A graduate option in systems engineering, ethics, and technology studies. *Journal of Engineering Education* 89(4):461-70.

Gorman, M.E. 1998. *Transforming nature: Ethics, invention and design*. Boston: Kluwer Academic Publishers.

Gorman, M.E. N.d. Types of knowledge and their roles in technology transfer. *J. of Technology Transfer* (in press).

Gorman, M.E., M.M. Mehalik, et al. 2000. *Ethical and environmental challenges to engineering*. Englewood Cliffs, NJ: Prentice-Hall.

Haraway, D. 1997. *Modest_Witness@Second_Millennium.FemaleMan_Meets_OncoMouse*. London: Routledge.

Hughes, T.P. 1987. The evolution of large technological systems. In *The Social Construction of Technological Systems*, ed. W.E. Bjiker, T.P. Hughes, and T.J. Pinch. Cambridge, MA: MIT Press.

Hutchins, E. 1995. *Cognition in the wild*. Cambridge, MA: MIT Press.

Kurzweil, R. 1999. *The age of spiritual machines*. New York: Penguin Books.

Lave, L.B. 2001. Lifecycle/sustainability implications of nanotechnology. In *Societal implications of nanoscience and nanotechnology*, ed. M.C. Roco and W.S. Bainbridge. Dordrecht, Neth.: Kluwer Academic Press.

Nardi, B.A. 2001. A cultural ecology of nanotechnology. In *Societal implications of nanoscience and nanotechnology*, ed. M.C. Roco and W.S. Bainbridge. Dordrecht, Neth.: Kluwer Academic Press.

Reid, T.R. 1984. *The chip*. New York: Simon and Schuster.

Roco, M.C., and W.S. Bainbridge. 2001. *Societal implications of nanoscience and nanotechnology*. Dordrecht, Neth.: Kluwer Academic Press.

Schon, D.A. 1987. *Educating the reflective practitioner: Toward a new design for teaching and learning in the professions*. San Francisco: Jossey-Bass.

Tolles, W.M. 2001. National security aspects of nanotechnology. In *Societal implications of nanoscience and nanotechnology*. ed. M.C. Roco and W.S. Bainbridge. Dordrecht, Neth: Kluwer Academic Press.

Weil, V. 2001. Ethical issues in nanotechnology. *Societal implications of nanoscience and nanotechnology*, ed. M.C. Roco and W.S. Bainbridge. Dordrecht, Neth: Kluwer Academic Press.

BREADTH, DEPTH, AND ACADEMIC NANO-NICHES

W.M. Tolles, Consultant

The report to the President titled *Science: The Endless Frontier* (Bush 1945) ushered in a period of rapid growth in research for two to three decades. This was stimulated further by the launch of Sputnik and programs to explore the moon. Over the past 56 years, research has moved from an environment where there was unquestioned acceptance of academic-style research by both academia and industry to an environment in which industry, in its effort to maintain profit margins in the face of global competition, has rejected the academic model of research and now focuses on short-term objectives. The need for industry to hire new blood and to generate new ideas is a major stimulus for cooperation between industry and academia. Academia has mixed reactions to these more recent trends. Universities are concerned about a loss of some independence and freedom to pursue new ideas in conjunction with industry, primarily due to the proprietary nature of maturing research/development. The pressures on academia to "demonstrate relevance" have continued for decades. In the search for "relevance," the concept of nanotechnology has emerged to satisfy a large community of researchers in both academia and industry.

The discovery of a new suite of experimental tools (beginning with scanning tunneling microscopy) with which to explore ever smaller features, to the level of the atom, reopened the doors joining the progress of academia to that of industry. The nanotechnology concept fulfilled the pressures of both the commercial world (pursuing continuation of the fruits of miniaturization) and academia (pursuing opportunities to research the many new pathways opened by these tools). The umbrella term "nanotechnology" covers programs already underway in both communities, thus giving a stamp of approval to many existing efforts. The goals and expectations of nanotechnology have been chosen in such a way that the march of the science and technology will yield new systems in many technological markets. There is little chance of disappointing the public (and Congress), due to the productivity of these endeavors. Yet, there appears to be more to the umbrella term than simply a new label for existing research directions. It has generated a new stimulus for academic pursuits in subtle ways that will have a lasting impact on our educational system.

Depth and Breadth a Bonus for Nanotechnology in Academia

University graduates must have skills in depth within a particular subject, a necessary aspect of pursuing the frontier of new knowledge with sufficient dedication to advance these frontiers. Yet, industry, concerned with satisfying consumers, is responsive to new opportunities that continually change. A university graduate may offer just what a given industrial position desires at a given time, but inevitable change may render those skills obsolete. Choosing new research directions more often, even within academic endeavors, is an inevitable part of a world characterized by rapidly expanding frontiers of new knowledge. *Depth* is an essential ingredient in the university experience, but *breadth* provides for greater flexibility when change occurs. The challenge to academia is to retain its strength in creating new knowledge while offering increasingly important breadth in its educational programs. Pursued separately, adding breadth to a student's experience can be satisfied by extending the time on campus, but this is costly and not particularly productive towards developing new knowledge, one of the primary goals of academia. Both professors and students are reluctant to substitute nondisciplinary courses in a curriculum already heavily laden with disciplinary material. What would be ideal in the academic experience would be to introduce *breadth* while simultaneously pursuing *depth*.

The subject of nanotechnology offers this opportunity, due to the multidisciplinary nature of the field. Researcher #1 in the field of chemistry or physics, for example, may wish to obtain knowledge of the structure of self-assembled particles, which may be of interest to researcher #2 in the field of electronics, who is interested in examining ways to fabricate quantum dots or novel structures for transistors. This is but one small example of the many opportunities that arise for joint objectives bridging disciplines. Such opportunities are labeled "nano-niches." These situations offer the student not only the opportunity to examine a phenomenon in depth, but to exchange results with similar activities in neighboring fields, where new perspectives may be obtained about other disciplines with relatively little additional effort (see schematic in Figure F.1).

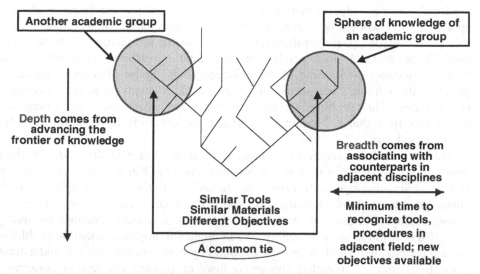

Figure F.1. Nanotechnology offers hope of depth plus breadth.

The value of multidisciplinary research has been extolled for years. However, it is impossible to have multidisciplinary research without having disciplines! Organizational changes that universities are now introducing include structures that encourage multidisciplinary pursuits, with consequent benefits to the student and the educational process. Breadth may be introduced by pursuing research objectives that have common features across disciplinary lines and by associating with more than one discipline. By encouraging "social interaction" (use of common instrumentation, materials, and theory) with peers in neighboring disciplines, the related frontiers in other disciplines may be easily introduced. This provides a graduate with stronger career opportunities, having the combined ability to pursue research in depth, but also having the ability to recognize additional options when the inevitable need for alternate opportunities arises.

Sharing expensive instrumentation in a common facility is one way to stimulate overlap of the academic disciplines, and this has been introduced extensively for nanotechnology. The National Nanofabrication Users Network (http://www.nnun.org/) consists of instrumentation centers at five major universities. A number of centers and institutes (http://www.nano.gov/centers.htm) have been introduced that stimulate the overlap of disciplines pursuing common goals. These organizations focus on objectives such as chemical and biological sensors, electron transport in molecules, nanoelectronics, assembly of nanostructures, and nanoscale devices/systems and their applications. These "academic nano-niches" are already established, and they will generate the benefits of multidisciplinary programs, with the concurrent advantages of depth and breadth. Other means of stimulating overlap involve common courses, seminars, and temporary exchanges of personnel.

Vision in Nanotechnology: How to Achieve it

One virtue of multidisciplinary research is the introduction of more comprehensive goals that may be achieved by several interactive research programs. A statement of these goals, along with the consequences, is frequently referred to as "vision." Occasionally, a research group sets out to conquer the larger goals with approaches that worked well with the previous in-depth methodology alone. That is, they pursue a larger goal with limited knowledge of the full picture. With the urgent need for faculty to obtain research funds, less time is available to examine the full picture associated with some of these larger goals. Some directions chosen by groups with a limited perspective may ignore the wisdom of more experienced communities. This problem is more severe when goals include "legions" of researchers from many disciplines, such as those currently being pursued by the computer industry.

Thus, the call for vision has generated its own unease in the midst of these transformations. Articulating a vision is tricky. As Yogi Berra stated, "It's tough to make predictions, especially about the future" (http://www.workinghumor.com/ quotes/yogi_berra.shtml). This difficulty has been exacerbated by the introduction of virtual reality. Images can be readily drawn that conjure phenomena totally inconsistent with the world of reality. When applied to apparent scientific problems, misperceptions may result in groups expounding concepts they do not understand; perceptions may even violate the usual laws of physics (or related constraints recognized through years of experience).

Nevertheless, vision statements are important for the research world, and Congressional appropriations for research are increasingly tied to (1) a linear extrapolation of past success and (2) visions that portend significant impact for the nation. The concepts associated with nanotechnology support these criteria in many ways. Most notably, enhanced electronics, enhanced medical diagnostics, improved medical procedures, and new materials are major areas that meet these two criteria. Stating a goal, pursuing it, and reaching it generate credibility. This is achieved best by those well versed in scientific principals and methods and the ramifications of potential paths to be pursued. It is not achieved by visionaries who appear to understand the world only through the images of virtual reality, without the sound knowledge of the basic principles drawn from the experimental world and experience with the perversity of Mother Nature. In addition, although serendipity has its place, it is not to be depended upon for productivity in research or for setting goals at the initiation of a program. The plethora of paths to follow in research exceeds by far the number of researchers. Consequently, a judicious choice of directions is essential, and the process of choosing these goals is vitally important to the health of the enterprise.

In light of the controversy surrounding discussion of the hazards of the so-called "self-replicating nanobots" (Tolles 2001, 173), a few words of caution seem in order. The nanotechnology community should show some restraint when releasing articles to the press about any major impact on an already established field. Setting scientific goals that may be achieved within a career (or within a decade) seems preferable to choosing goals that appear incompatible with the behavior of the physical world. The hazards of the so-called "self-replicating nanobots" seem to have already generated far more discussion than they warrant (Tolles 2001). Visions of ultra-fast and powerful computers the size of poppy seeds conjure unrealistic expectations, feeding further the fears that the products of our creation may be smarter than we are and that we may sow the seeds of our own destruction. "The rub in exploring the borderlands is finding that balance between being open-minded *enough to accept radical new ideas but not so open-minded that your brains fall out"* (Shermer 2001, 29). We must recognize that it is difficult to predict the future; in particular, there is no reason to raise hopes for a device or a phenomenon that violates the basic laws of physics and chemistry. Another perspective: *"... the burden of proof is not on those who know how to make chips with 10^7 transistors and connect them together with millions of wires, it is up to those who show something in a laboratory to prove that it is better"* (Keyes 2001b).

The Academic Nano-Niches

Several "nano-niches" that appear most obvious today are outlined below. There are, of course, many other concepts emerging from the fertile frontier of miniaturization that are not easily categorized. Perhaps other significant niches will emerge in this new dimension of material control and behavior.

Nano-Niche #1

Objectives for enhancing electronic devices have been the basis for many nanotechnology programs. The nanotechnology efforts in programs such as molecular electronics have been pursued for decades with little impact on the electronics industry thus far. The more conservative microelectronics industry continues to pursue CMOS and is skeptical of radically new ideas that may deviate from its International Technology Roadmap for Semiconductors (ITRS) (Semiconductor Industry Association 2001) for a number of years in the future (Glimmell 2001). This is one area of nanotechnology that could benefit from a significant overlap with expertise in the electronics and information technology communities. Goals of forming molecular computers have appeared in a number of places. The physical realities one must meet to achieve such goals have been mentioned in a number of papers (e.g., Keyes 2001a; Meindl 1995, 1996; Meindl, Chen, and Davis 2001; Semiconductor Industry Association 2001). Molecular transistors have recently been fabricated (Bachtold et al. 2001; Schön, Meng, and Bao 2001). They have even been incorporated into circuits that can be used for logic operations (Bachtold et al. 2001). The challenges facing this nano-community now are very similar to those facing the semiconductor industry (see the Roadmap). These two communities will begin to work together cooperatively for a common goal. Innovative methods for incorporating new nanostructures into more conventional circuits will probably be the outcome of these interactions. The chemical and biological influences on the nanostructure of semiconductors is just beginning to be recognized (Whaley et al. 2000). Of course, alternative architectures for computational tasks represent a likely path for new breakthroughs. The brain of living species represents proof that such alternative architectures exist. It is through the innovation of these communities that such advances are likely to be introduced.

Nano-Niche #2

Research in nanostructures associated with biomolecular science is well recognized and proves to be a fertile field for a nano-niche. Biomolecules are often large and qualify as "nanostructures." Introduction of the tools and experience of chemists and physicists, even electrical engineers, in pursuing this mainstream of nanotechnology offers many opportunities for the synergism of multidisciplinary research in biology, biotechnology, and medicine. A biology student pursuing research with the tools of nanotechnology enters biomedical frontiers that include ability to fabricate sensors for the rapid, inexpensive detection of environmental hazards and disease organisms and to fabricate biomolecules with an objective to target selective cells (such as cancer cells) for modification of their function (Alivisatos 2001). Miniature chemistry laboratories are being fabricated on chips. These tools are likely to find applications in the task of sequencing genetic codes, of importance for medical purposes. This nano-niche includes the disciplines of chemistry, physics, biomolecular engineering, and even electrical engineering. One caution is worth noting. The ability to create new microbes, viruses, etc., in this field could lead to new biological species that present risks. As stated elsewhere, "The main risks for negative societal implications of nanotechnology will probably continue to be in the area of biotechnology rather than electronics" (Doering 2001, 68).

Nano-Niche #3

The field of materials science has always been a multidisciplinary endeavor. This is no less true for materials composed of nanostructures. One recent article points out the value of porous silicon as a stimulus to educational opportunities in electronics, optoelectronics, microoptics, sensors, solar cells, micromachining, acoustics, medicine, biotechnology, and astrophysics (Parkhutik and Canham 2000). A new material may be prepared using a variety of fabrication techniques from a number of disciplines and find applications in a number of technologies, accounting for the value of such a field for introducing breadth to the student experience. Of course, the depth from such an endeavor comes from advancing the knowledge about a given material using the tools from various scientific disciplines. Since new materials are of interest due to the possible substitution in an existing science or technology, the multidisciplinary aspect of materials will always exist.

Nanotechnology as a Stimulus to Inquiring Minds

As a stimulus for education in the sciences, nanotechnology has led to a wealth of fascinating scientific revelations. Attracting young inquiring minds has been the subject of an NSF-supported consortium project at Arizona State University in conjunction with other universities. This project, Interactive Nano-Visualization in Science and Engineering Education (IN-VSEE), may be viewed at http://invsee.asu.edu/. The goal of this program is to bring the excitement of discovery with electron and scanning tunneling microscopy into the classroom, targeting students in upper-level high school through college. At this level, the attraction of the multidisciplinary aspects is obvious. The subject of nanotechnology as a basis to illustrate scientific principals is likewise clear.

Summary

In summary, nanotechnology provides an impetus for transforming the academic experience, introducing a new stimulus for breadth in the career of a student while minimizing the additional time to assimilate that breadth. The historical functions of creating new knowledge through in-depth study need not be compromised with such programs. Programs in nanotechnology represent excellent areas of research to demonstrate this and will be one basis for a subtle transformation of the academic environment. Philosophers, business schools, psychologists, and many of the "soft sciences" may debate the implications of nanotechnology. However, without a realistic view of what may be expected from this fertile research frontier, there may be unnecessary discussions about unrealistic expectations. Information released to the media and studies of a social nature should follow careful assessments by technically qualified research teams presenting rational projections for the future potential of this fascinating field.

References

Alivisatos, A.P. 2001. Less is more in medicine: Sophisticated forms of nanotechnology will find some of their first real-world applications in biomedical research, disease diagnosis and, possibly, therapy. *Sci. Am.* 285: 66.

Bachtold. A., P. Hadley, T. Nakanishi, and C. Dekker. 2001. Logic circuits with carbon nanotube transistors. *Science* 294:1317.

Bush, V. (Director of OSRD). 1945. *Science, the endless frontier.* Report to the President for Postwar Scientific Research. Washington, D.C.: U.S. Government Printing Office.

Doering, R. 2001. Societal implications of scaling to nanoelectronics. In *Societal implications of nanoscience and nanotechnology*, ed. M.C. Roco and W.S. Bainbridge. Dordrecht, Neth.: Kluwer Academic Press.

Glimmell, H. 2001. Dynamics of the emerging field of nanoscience. In *Societal implications of nanoscience and nanotechnology*, ed. M.C. Roco and W.S. Bainbridge. Dordrecht, Neth.: Kluwer Academic Press.

Keyes, R.W. 2001a. The cloudy crystal ball: Electronic devices for logic. *Phil. Mag.*, 81(9):1315-1330.

_____. 2001b. Private communication.

Meindl, J.D. 1995. Low power microelectronics: Retrospect and prospect. *Proc. IEEE* 83:619-635.

_____. 1996. Physical Limits on gigascale integration. *J. Vac. Sci. Technol.* B 14:192-195.

Meindl, J.D., Q. Chen, and J.A. Davis. 2001. Limits on silicon nanoelectronics for terascale integration. *Science* 293: 2044.

Parkhutik, V.P., and L.T. Canham. 2000. Porous silicon as an educational vehicle for introducing nanotechnology and interdisciplinary materials science. *Phys. Stat. Sol.* 182: 591.

Schön, J.H., H. Meng, and Z. Bao. 2001. Field-effect modulation of the conductance of single molecules. *Science Online* Nov. 8.

Semiconductor Industry Association. 2001. *International technology roadmap for semiconductors.* Online: http://public.itrs.net/Files/2001ITRS/Home.htm.

Shermer, M. 2001. Nano nonsense and cryonics. *Scientific American* September.

Stern, P.C., and L.L. Carstensen, eds. 2000. The aging mind: Opportunities in cognitive research. National Research Council, http://www.nap.edu/catalog/9783.html.

Tolles, W.M. 2001. National security aspects of nanotechnology. In *Societal implications of nanoscience and nanotechnology.* ed. M.C. Roco and W.S. Bainbridge. Dordrecht, Neth: Kluwer Academic Press.

Whaley, S.R., D.S. English, E.L. Hu, P.F. Barbara, and A.M. Belcher. 2000. Selection of peptides with semiconductor binding specificity for directed nanocrystal assembly. *Nature* 405: 665.

UNIFYING PRINCIPLES IN COMPLEX SYSTEMS

Yaneer Bar-Yam, New England Complex Systems Institute

The ability of science and technology to augment human performance depends on an understanding of systems, not just components. The convergence of technologies is an essential aspect of the effort to enable functioning systems that include human beings and technology, serving the human beings to enhance their well-being directly and indirectly through what they do and what they do for other human beings. The recognition today that human beings function in teams rather than as individuals implies that technological efforts are essential that integrate human beings across scales of tools, communication, and biological and cognitive function.

Understanding the role of complex systems concepts in technology integration requires a perspective on how the concept of complexity is affecting science, engineering, and finally, technology integration.

Complex Systems and Science

The structure of scientific inquiry is being challenged by the broad relevance of complexity to the understanding of physical, biological, and social systems (Bar-Yam 2000; Bar-Yam and Minai 2002; Gallagher and Appenzeller 1999). Cross-disciplinary interactions are giving way to transdisciplinary and unified efforts to address the relevance of large amounts of information to describing, understanding, and controlling complex systems. From the study of biomolecular interactions (Service 1999; Normile 1999; Weng, Bhalla, and Iyengar 1999) to the strategy tactics of 21st century Information Age warfare and the war on terrorism, complexity has arisen as a unifying description of challenges to understanding and action. In this arena of complex systems, information, and action, structure and function are entangled. New approaches that recognize the importance of patterns of behavior, the multiscale space of possibilities, and evolutionary or adaptive processes that select systems or behaviors that can be effective in a complex world are central to advancing our understanding and capabilities (Bar-Yam 1997).

Complex Systems and Engineering

The failure of design and implementation of a new air traffic control system, failures of Intel processors, medical errors (IOM 2000), failures of medical drugs, even the failure of the Soviet Union, can be described as failures of large, complex systems. Systematic studies of large-scale engineering projects have revealed a remarkable proportion of failures in major high-investment projects. The precursors of such failures (multisystem integration, high-performance constraints, many functional demands, high rates of response, and large, context-specific protocols), are symptomatic of complex engineering projects. The methods for addressing and executing major engineering challenges must begin from the recognition of the role of complexity and the specific tools that can guide the design, or self-organization, of highly complex systems. Central to effective engineering are evaluation of the complexity of system functions; recognition of fundamental engineering tradeoffs of structure, function, complexity, and scale in system capabilities; and application of indirection to specification, design, and control of system development and the system itself.

Defining Complex Systems and Complex Tasks

One way to define a complex task is as a problem where the number of distinct possibilities that must be considered, anticipated, or dealt with is substantially larger than can be reasonably named or enumerated. We can casually consider in an explicit way tens of possibilities, a professional can readily deal with hundreds of possibilities, and a major project deals with thousands. The largest projects deal with tens of thousands. For larger numbers of possibilities, we must develop new strategies (Bar-Yam 1997). Simplifying a complex task by ignoring the need for different responses is what leads to errors or failures that affect the success of the entire effort, leaving it as a gamble with progressively higher risks.

The source of complex tasks is complex systems. Complex systems are systems with interdependent parts. Interdependence means that we cannot identify the system behavior by just considering each of the parts and combining them. Instead we must consider how the relationships between the parts affect the behavior of the whole. Thus, a complex task is also one for which many factors must be considered to determine the outcome of an action.

Converging Technologies

The rapid development of nanotechnology and its convergence with biological, information, and cognitive sciences is creating a context in which complex systems concepts that enable effective organizations to meet complex challenges can be realized through technological implementation. At the same time, complex systems concepts and methods can describe the framework in which this convergence is taking place. From the fine-scale control of systems based upon nanotechnology to understanding the system properties of the integrated socio-technical system consisting of human beings and computer information networks, the synergy of complex systems theory and converging technologies is apparent as soon as we consider the transition between components and functions.

Looking Forward

Human civilization, its various parts (including its technology), and its environmental context may be described as complex. The most reliable prediction possible is that this complexity will continue to increase. The great opportunity of the convergence of nanotechnology, biomedical, information, and cognitive sciences is an explosive increase in what is possible through combining advances in all areas. This is, by definition, an increase in the complexity of the systems that will be formed out of technology and of the resulting behaviors of people who use them directly or are affected by them. The increasing complexity suggests that there will be a growing need for widespread understanding of complex systems as a counterpoint to the increasing specialization of professions and professional knowledge. The insights of complex systems research and its methodologies may become pervasive in guiding what we build, how we build it, and how we use and live with it. Possibly the most visible outcome of these developments will be an improved ability of human beings, aided by technology, to address global, social, and environmental problems: third world development, poverty in developed countries, war, and natural disasters. At an intermediate scale, the key advances will dramatically change how individuals work together in forming functional teams that are more directly suited to the specific tasks they are performing. In the context of individual human performance, the key to major advances is recognizing that the convergence of technology will lead to the possibility of designing (or, more correctly, adapting) the environment of each individual for his or her individual needs and capabilities in play and work.

The Practical Need

Complex systems studies range from detailed studies of specific systems to studies of the mechanisms by which patterns of collective behaviors arise, to general studies of the principles of description and representation of complex systems. These studies are designed to enable us to understand and modify complex systems, design new ones for new functions, or create contexts in which they self-organize to serve our needs without direct design or specification. The need for applications to biological, cognitive, social, information, and other engineered systems is apparent.

Biology has followed an observational and reductionistic approach of accumulating large bodies of information about the parts of biological systems and looking for interpretations of system behavior in terms of these parts. It has become increasingly clear that biological systems are intricate, spatially structured, biochemically based networks. The role of information in biological action and the relationships of structure and function are only beginning to be probed by mathematicians, physicists, and engineers who are interested in biological systems as systems designed by nature for their functional capabilities. While biologists are increasingly looking to mathematical approaches and perspectives developed in physics and engineering, engineers are increasingly looking to biological systems for inspiration in designing artificial systems. Underlying these systems are a wealth of design principles in areas that include the biochemical networks (Gallagher and Appenzeller 1999; Service 1999; Normile 1999; Weng, Bhalla, and Iyengar 1999); immune systems (Perelson and Wiegel 1999; Noest 2000; Segel and Cohen 2001; Pierre et al. 1997) and neural systems (Anderson and Rosenfeld 1988; Bishop 1995; Kandel, Schwartz, and Jessell 2000); and animal behaviors such as the swimming mechanisms of fish (Triantafyllou and Triantafyllou 1995) and the gaits of animals (Golubitsky et al. 1999). These systems and architectures point to patterns of function that have a much higher robustness to failure and error and a higher adaptability than conventional human engineered systems.

Computers have made a transition from systems with tightly controlled inputs and outputs to systems that are networked and respond on demand as part of interactive information systems (Stein 1999). This has changed radically the nature of the issues facing their design. The collective behaviors of these networked computer systems, including the Internet, limit their effectiveness. Whether these have to do with the dynamics of packet loss in Internet traffic or the effect of computer viruses or worms (Forrest, Hofmeyr, and Somayaji 1997; Kephart et al. 1997; Goldberg et al. 1998), that at times have incapacitated a large fraction of the Internet, these effects are not small. The solution to these problems lies in understanding collective behaviors and in designing computer systems to be effective in environments with complex demands and to have higher defensive robustness.

The human brain is often considered the paradigmatic complex system. The implications of this recognition are that cognitive function is distributed within the brain, and mechanisms may vary from individual to individual. Complete explanations of cognitive function must themselves be highly complex. Major advances in cognitive science are currently slowed by a combination of efforts on the one hand to explain cognitive function directly from the behavior of individual molecular and cellular components, and on the other hand to aggregate or average the cognitive mechanisms of different human beings. Still, diverse advances that are being made are pointing the way to improvements in education (NIMH 2002), man-machine interfaces (Norman and Draper 1986; Nielsen 1993; Hutchins 1995), and retention of capabilities during aging (Stern and Carstensen 2000; Mandell and Schlesinger 1990; Davidson, Teicher, and Bar-Yam 1997).

The recognition of the complexity of conflict in the war on terrorism is another indication that the basic concept of complexity in social systems or problems has begun to be recognized. Unfortunately, this understanding has yet to be transferred to address other diverse major fundamental social system problems, as found in medical system cost containment, education system reform, and alleviation of poverty. In each case, current approaches continue to be dominated by large-scale strategies that are ineffective in addressing complex problems. Even with the appearance of more holistic approaches to, for example, third world development (World Bank 1998), the basic concept of existing strategy remains weakly informed by complex systems insights. This gap is an opportunity for major contributions by the field of complex systems at both the conceptual and technical levels. Further contributions can be made based upon research projects that emphasize the intrinsic complexity of these systems.

Understanding complex global physical and biological systems is also a major challenge. Many key problems today have to do with indirect effects of human activities that may have substantial destructive effects on the human condition. These include global warming and ecological deterioration due to overexploitation of resources. Effective approaches to these problems require understanding both the environmental and socioeconomic implications of our current actions and of actions that are designed to alleviate these problems (NSF n.d.). For example, the problem of global warming includes the effects of large-scale human activity interacting with both the linear and potentially nonlinear climactic response. Despite the grave risks associated with global warming, a key factor impeding actions to alleviate it is fear of major impacts of such efforts on socioeconomic systems. Better understanding of the potential effects of such interventions should enable considered actions to be taken.

Interest

Study of complex systems has become recognized as a basic scientific endeavor whose inquiry has relevance to the management of complex organizations in a complex world (Herz 2001). More specific attention has been gained in information technology (Horn 2001), biotechnology (Strausberg and Austin 1999; NSF n.d.; NIGMC n.d.; NSF 2001), healthcare industries, and the military.

Information technology companies building computer hardware and software have begun to recognize the inherently interactive and distributed nature of the systems they are designing. A significant example is the IBM "Autonomic Computing" initiative (Horn 2001), which is inspired by the biological paradigm of the autonomic nervous system and is conceptually based upon modeling robustness through biologically inspired system design. In a different perspective, Apple Computer has demonstrated the relevance of human factors, ranging from hardware design to ease-of-use and facilitation of creativity, as essential aspects of the role of computers in computer-human systems.

The major advances in biotechnology, including the genome project and other high-throughput data acquisition methods, have led to a dramatic growth in the importance of modeling and representation tools to capture large bodies of information and relate them to system descriptions and properties. Many private companies at the forefront of biotechnology are developing bioinformatics tools that strive to relate information to functional descriptions also described as "functional genomics" (Srausberg and Austin 1999). This is one facet of a broader recognition of the importance of capturing the multiscale properties of biological systems as reflected in NSF's biocomplexity initiative (NSF n.d.) and the complex biological systems programs at NIH (NIGMS 2002), as well as in joint programs.

For several years, the interest in complex systems as a conceptual and quantitative management tool has led consulting companies to work on practical implementations of strategy and more specific modeling efforts (Ernst and Young 2000, Gleick 1987). One of the areas of particular interest has been in the healthcare management community, where rapid organizational change has led to a keen interest in complex systems insights.

In the military and intelligence communities, there has been increasing realization of the relevance of networked distributed control and information systems. All branches of the military and the joint chiefs of staff have adopted vision statements that focus on complex systems concepts and insights as guiding the development of plans for information age warfare. These concepts affect both the engineering of military sensors, effectors, and information networks, and the underlying nature of military force command and control.

More broadly, the public's attention has been widely attracted to the description of complex systems research and insights. Indeed, many popular descriptions of complex systems research existed before the first textbook was written (Gleick 1987; Lewin 1992; Waldrop 1992; Gell-Mann 1994; Casti 1994; Goodwin 1994; Kauffman 1995; Holland 1995; Coveney and Highfield 1995; Bak 1996). The excitement of scientists as well as the public reflects the potential impact on our ability to understand questions that affect everyday life, perspectives on the world around us, fundamental philosophical disputes, and issues of public concern such as major societal challenges, the dynamics of social networks, global computer networks (the WWW), biomedical concerns, psychology, and ecology.

The Goals

The goals of complex systems research are to understand the following:

- Understand the development and mechanisms of patterns of behavior and their use in engineering

- Understand the way to deal with complex problems (engineering, management, economic, sociopolitical) using strategies that relate the complexity of the challenge to the complexity of the system that must respond to them

- Understand the unifying principles of organization, particularly for systems that deal with large amounts of information (physical, biological, social, and engineered)

- Understand the interplay of behaviors at multiple scales and between the system and its environment

- Understand what is universal and what is not, when averaging applies and when it does not, what can be known and what cannot, what are the classes of universal behavior and the boundaries between them, and what are the relevant parameters for describing or affecting system behaviors

- Develop the ability to capture and represent specific systems rather than just accumulate data about them: (in this context) to describe relationships, know key behaviors, recognize relevance of properties to function, and simulate dynamics and response.

- Achieve a major educational shift toward unified understanding of systems and patterns of system behavior.

The traditional approach of science of taking things apart and assigning the properties of the system to its parts has been quite successful, but the limits of this approach have become apparent in recent years. When properties of a system result from dependencies and relationships but we assign them to their parts, major obstacles arise to understanding and control. Once the error of assignment is recognized, some of the obstacles can be overcome quickly, while others become subjects of substantive inquiry. Many scientists think that the parts are universal but the way parts work together is specific to each system. However, it has become increasingly clear that how parts work together can also be studied in general, and by doing so, we gain insight into every kind of system that exists, including physical systems like the weather as well as biological, social, and engineered systems.

Understanding complex systems does not mean that we can predict their behavior exactly; it is not just about massive databases or massive simulations, even though these are important tools of research in complex systems. The main role of research in the study of complex systems is recognizing what we can and cannot say about complex systems given a certain level (or scale) of description and knowing how we can generalize across diverse types of complex systems. It is just as important to know what we can know, as to know. Thus the concept of deterministic chaos appears to be a contradiction in terms: how can a deterministic system also be chaotic? It is possible because there is a rate at which the system behavior becomes dependent on finer and finer details (Cvitanovic 1989; Strogatz 1994; Ott 1993). Thus, how well we know a system at a particular time determines how well we can predict its behavior over time. Understanding complexity is neither about prediction or lack of predictability, but rather a quantitative knowledge of how well we can predict, and only within this constraint, what the prediction is.

Fundamental Research in Complex Systems: Theorems and Principles

Fundamental research in complex systems is designed to obtain characterizations of complex systems and relationships between quantities that characterize them. When there are well-defined relationships, these are formalized as theorems or principles. More general characterizations and classifications of complex systems are described below in major directions of inquiry. These are only a sample of the ongoing research areas.

A theorem or principle of complex systems should apply to physical, biological, social, and engineered systems. Similar to laws in physics, a law in complex systems should relate various quantities that characterize the system and its context. An example is Newton's second law that relates force, mass, and acceleration. Laws in complex systems relate qualities of system, action, environment, function, and information. Three examples follow.

Functional Complexity

Given a system whose function we want to specify, for which the environmental (input) variables have a complexity of $C(e)$, and the actions of the system have a complexity of $C(a)$, then the complexity of specification of the function of the system is

$$C(f)=C(a)\,2^{C(e)}$$

where complexity is defined as the logarithm (base 2) of the number of possibilities or, equivalently, the length of a description in bits.

The proof follows from recognizing that complete specification of the function is given by a table whose rows are the actions ($C(a)$ bits) for each possible input, of which there are $2^{C(e)}$. Since no restriction has been assumed on the actions, all actions are possible, and this is the minimal length description of the function. Note that this theorem applies to the complexity of description as defined by the observer, so that each of the quantities can be defined by the desires of the observer for descriptive accuracy. This theorem is known in the study of Boolean functions (binary functions of binary variables) but is not widely understood as a basic theorem in complex systems (Bar-Yam 1997).

The implications of this theorem are widespread and significant to science and engineering. The exponential relationship between the complexity of function and the complexity of environmental variables implies that systems that have environmental variables (inputs) with more than a few bits (i.e., 100 bits or more of relevant input) have functional complexities that are greater than the number of atoms in a human being and thus cannot be reasonably specified. Since this is true about most systems that we characterize as "complex," the limitation is quite general. The implications are that fully phenomenological approaches to describing complex systems, such as the behaviorist approach to human psychology, cannot be successful. Similarly, the testing of response or behavioral descriptions of complex systems cannot be performed. This is relevant to various contexts, including testing computer chips, and the effects of medical drugs in double-blind population studies. In each case, the number of environmental variables (inputs) is large enough that all cases cannot be tested.

Requisite Variety

The Law of Requisite Variety states that the larger the variety of actions available to a control system, the larger the variety of perturbations it is able to compensate (Ashby 1957). Quantitatively, it specifies that a well-adapted system's probability of success in the context of its environment can be bounded:

$$-\text{Log}_2(P) < C(e)-C(a)$$

Qualitatively, this theorem specifies the conditions in which success is possible: a matching between the environmental complexity and the system complexity, where success implies regulation of the impact of the environment on the system.

The implications of this theorem are widespread in relating the complexity of desired function to the complexity of the system that can succeed in the desired function. This is relevant to discussions of the limitations of specific engineered control system structures, of the limitations of human beings, and of human organizational structures.

Note that this theorem, as formulated, does not take into account the possibility of avoidance (actions that compensate for multiple perturbations because they anticipate and thus avoid the direct impact of the perturbations), or the relative measure of the space of success to that of the space of possibilities. These limitations can be compensated for.

Non-averaging

The Central Limit Theorem specifies that collective or aggregate properties of *independent* components with bounded probability distributions are Gaussian, distributed with a standard deviation that diminishes as the square root of the number of components. This simple solution to the collective behavior of non-interacting systems does not extend to the study of interacting or interdependent systems. The lack of averaging of properties of complex systems is a statement that can be used to guide the study of complex systems more generally. It also is related to a variety of other formal results, including Simpson's paradox (Simpson 1951), which describes the inability of averaged quantities to characterize the behavior of systems, and Arrow's Dictator Theorem, which describes the generic dynamics of voting systems (Arrow 1963; Meyer and Brown 1998).

The lack of validity of the Central Limit Theorem has many implications that affect experimental and theoretical treatments of complex systems. Many studies rely upon unjustified assumptions in averaging observations that lead to misleading, if not false, conclusions. Development of approaches that can identify the domain of validity of averaging and use more sophisticated approaches (like clustering) when they do not apply are essential to progress in the study of complex systems.

Another class of implications of the lack of validity of the Central Limit Theorem is the recognition of the importance of individual variations between different complex systems, even when they appear to be within a single class. An example mentioned above is the importance of individual differences and the lack of validity of averaging in cognitive science studies. While snowflakes are often acknowledged as individual, research on human beings often is based on assuming their homogeneity.

More generally, we see that the study of complex systems is concerned with their universal properties, and one of their universal properties is individual differences.

This apparent paradox, one of many in complex systems (see below), reflects the importance of identifying when universality and common properties apply and when they do not, a key part of the study of complex systems.

Major Directions of Inquiry

How Understanding Self-Organization & Pattern Formation Can be Used to Form Engineered Systems

Self-organization is the process by which elements interact to create spatio-temporal patterns of behavior that are not directly imposed by external forces. To be concrete, consider the patterns of spontaneous traffic jams or heart beats. For engineering applications, the promise of understanding such pattern formation is the opportunity to use the natural dynamics of the system to create structures and impose functions rather than to construct them element by element. The robustness of self-organized systems is also a desired quality in conventional engineered systems — and one that is difficult to obtain. For biomedical applications, the promise is to understand developmental processes like the development of the fertilized egg into a complex physiological organism, like a human being. In the context of the formation of complex systems through development or through evolution, elementary patterns are the building blocks of complex systems. This is diametrically opposed to considering parts as the building blocks of such systems.

Spontaneous (self-organizing) patterns arise through symmetry breaking in a system when there are multiple inequivalent static or dynamic attractors. In general, in such systems, a particular element of a system is affected by forces from more than one other element, and this gives rise to "frustration" as elements respond to aggregate forces that are not the same as each force separately. Frustration contributes to the existence of multiple attractors and therefore of pattern formation.

Pattern formation can be understood using simple rules of local interaction, and there are identifiable classes of rules (universality) that give rise to classes of patterns. These models can be refined for more detailed studies. Useful illustrative examples of pattern forming processes are local-activation, long-range inhibition models that can describe patterns on animal skins, magnets, dynamics of air flows in clouds, wind-driven ocean waves, and swarm behaviors of insects and animals. Studies of spontaneous and persistent spatial pattern formation were initiated a half century ago by Turing (1952), and the wide applicability of patterns has gained increasing interest in recent years (Bar-Yam 1997; Meinhardt 1994; Murray 1989; Nijhout 1992; Segel 1984; Ball 1999).

The universality of patterns has been studied in statistical physics, where dynamic patterns arise in quenching to a first-order phase transition both in cases of conserved (spinodal decomposition, e.g., oil-water separation) and nonconserved (coarsening, e.g., freezing water) order parameters (Bray 1994) and also in growing systems (self-organized criticality, e.g., roughening). Generic types of patterns are relevant for such contexts and are distinguished by their spatio-temporal behaviors. Classic models have characteristic spatial scales (Turing patterns, coarsening, spinodal decomposition); others are scale invariant (self-organized criticality, roughening). Additional classes of complex patterns arise in networks with long-range interactions (rather than just spatially localized interactions) and are used for modeling spin glasses, neural networks (Anderson and Rosenfeld 1988; Bishop 1995; Kandel, Schwartz, and Jessell 2000), or genetic networks (Kauffman 1969).

Understanding Description and Representation

The study of how we describe complex systems is itself an essential part of the study of such systems. Since science is concerned with describing reproducible phenomena and engineering is concerned with the physical realization of described functions, description is essential to both. A description is some form of identified map of the actual system onto a mathematical or linguistic object. Shannon's information theory (Shannon 1963) has taught us that the notion of description is linked to the space of possibilities. Thus, while description appears to be very concrete, any description must reflect not only what is observed but also an understanding of what might be possible to see. An important practical objective is to capture information and create representations that allow human or computer-based inquiry into the properties of the system.

Among the essential concepts relevant to the study of description is the role of universality and non-universality (Wilson 1983) as a key to the classification of systems and of their possible representations. In this context, effective studies are those that identify the class of models that can capture properties of a system, rather than those of a single model of a system. Related to this issue is the problem of testability of representations through validating the mapping of the system to the representation. Finally, the practical objective of achieving human-usable representations must contend with the finite complexity of a human being, as well as other human factors due to both "intrinsic" properties of complex human function and "extrinsic" properties that are due to the specific environment in which human beings have developed their sensory and information processing systems.

The issue of human factors can be understood more generally as part of the problem of identifying the observer's role in description. A key issue is identifying the scale of observation: the level of detail that can be seen by an observer, or the degree of distinction between possibilities (NIGMS 2002; Bar-Yam 1997). Effective descriptions have a consistent precision so that all necessary but not a lot of unnecessary information is used, irrelevant details are eliminated, and all relevant details are included. A multiscale approach (Bar-Yam 1997) relates the notion of scale to the properties of the system and relates descriptions at different scales.

The key engineering challenge is to relate the characteristics of a description to function. This involves relating the space of possibilities of the system to the space of possibilities of the environment (variety, adaptive function). Complexity is a logarithmic measure of the number of possibilities of the system, equivalently the length of the description of a state. The Law of Requisite Variety (Ashby 1957) limits the possible functions of a system of a particular complexity.

Understanding Evolutionary Dynamics

The formation of complex systems and the structural/functional change of such systems is the process of adaptation. Evolution (Darwin 1859) is the adaptation of populations through intergenerational changes in the composition of the population (the individuals of which it is formed), and learning is a similar process of adaptation of a system through changes in its internal patterns, including (but not exclusively) the changes in its component parts.

Characterizing the mechanism and process of adaptation, both evolution and learning, is a central part of complex systems research (Holland 1992; Kauffman 1993; Goodwin 1994; Kauffman 1995; Holland 1995). This research generalizes the problem of biological evolution by recognizing the relevance of processes of incremental change to the formation of all complex systems. It is diametrically opposed to the notion of creation in engineering that typically assumes new systems are invented without precursor. The reality of incremental changes in processes of creativity and design reflect the general applicability of evolutionary concepts to all complex systems.

The conventional notion of evolution of a population based upon replication with variation and selection with competition continues to be central. However, additional concepts have become recognized as important and are the subject of ongoing research, including the concepts of co-evolution (Kauffman 1993), ecosystems (Kauffman 1993), multiple niches, hierarchical or multilevel selection (Brandon and Burian 1984; Bar-Yam 2000), and spatial populations (Sayama, Kauffman, and Bar-Yam 2000). Ongoing areas of research include the traditional philosophical paradoxes involving selfishness and altruism (Sober and Wilson 1999), competition and cooperation (Axelrod 1984), and nature and nurture (Lewontin 20001). Another key area of ongoing inquiry is the origin of organization, including the origins of life (Day 1984), which investigate the initial processes that give rise to the evolutionary process of complex systems.

The engineering applications of evolutionary process are often mostly associated with the concept of evolutionary programming or genetic algorithms (Holland 1992; Fogel, Owens, and Walsh 1966). In this context, evolution is embodied in a computer. Among the other examples of the incorporation of evolution into engineering are the use of artificial selection and replication in molecular drug design (Herschlag and Cech 1990; Beaudry and Joyce 1992; Szostak 1999), and the human-induced variation with electronic replication of computer viruses, worms, and Trojan horses in Internet attacks (Goldberg et al. 1998). The importance of a wider application of evolution in management and engineering is becoming apparent. The essential concept is that evolutionary processes may enable us to form systems that are more complex than we can understand but that will still serve the functions we need. When high complexity is necessary for desired function, the system should be designed for evolvability: e.g., smaller components (subdivided modular systems) evolve faster (Simon 1998). We note, however, that in addition to the usual concept of modularity, evolution should be understood to use patterns, not elements, as building blocks. The reason for this is that patterns are more directly related to collective system function and are therefore testable in a system context.

Understanding Choices and Anticipated Effects: Games and Agents

Game theory (von Neumann and Morgenstern 1944; Smith 1982; Fudenberg and Tirole 1991; Aumann and Hart 1992) explores the relationship between individual and collective action using models where there is a clear statement of consequences (individual payoffs), that depend on the actions of more than one individual. A paradigmatic game is the "prisoner's dilemma." Traditionally, game theory is based upon logical agents that make optimal decisions with full knowledge of the possible outcomes, though these assumptions can be usefully relaxed. Underlying game theory is the study of the role of anticipated effects on actions and the paradoxes that arise because of contingent anticipation by multiple anticipating agents, leading to choices that are undetermined within the narrow definition of the game and thus are sensitive to additional properties of the system. Game theory is relevant to fundamental studies of various aspects of collective behavior: altruism and selfishness, for example, and cooperation and competition. It is relevant to our understanding of biological evolution, socio-economic systems, and societies of electronic agents. At some point in the increasing complexity of games and agents, the models become agent-based models directed at understanding specific systems.

Understanding Generic Architectures

The concept of a network as capturing aspects of the connectivity, accessibility, or relatedness of components in a complex system is widely recognized as important to understanding aspects of these systems — so much so that many names of complex systems include the term "network." Among the systems that have been identified thus are artificial and natural transportation networks (roads, railroads, waterways, airways) (Maritan et al. 1996; Banavar, Maritan, and Rinaldo 1999; Dodds and Rothman 2000), social networks (Wasserman and Faust 1994), military forces (INSS 1997), the Internet (Cheswick and Burch n.d.; Zegura, Calvert, and Donahoo 1997), the World Wide Web (Lawrence and Giles 1999; Huberman et al. 1998; Huberman and Lukose 1997), biochemical networks (Service 1999; Normile 1999; Weng, Bhalla, and Iyengar 1999), neural networks (Anderson and Rosenfeld 1988; Bishop 1995; Kandel, Schwartz, and Jessell 2000), and food webs (Williams and Martinez 2000). Networks are anchored by topological information about nodes and links, with additional information that can include nodal locations and state variables, link distances, capacities, and state variables, and possibly detailed local functional relationships involved in network behaviors.

In recent years, there has been significant interest in understanding the role played by the abstract topological structure of networks represented solely by nodes and links (Milgram 1967; Milgram 1992; Watts and Strogatz 1998; Barthélémy and Amaral 1999; Watts 1999; Latora and Marchiori 2001; Barabási and Albert 1999; Albert, Jeong, and Barabási 1999; Huberman and Adamic 1999; Albert, Jeong, and Barabási 2000; Jeong et al. 2001). This work has focused on understanding the possible relationships between classes of topological networks and their functional capacities. Among the classes of networks contrasted recently are locally connected, random, small-world (Milgram 1967, 1992; Watts and Strogatz 1998; Barthélémy and Amaral 1999; Watts 1999), and scale-free networks (Latora and Marchiori 2001; Barabási and Albert 1999; Albert, Jeong, and Barabási 1999; Huberman and Adamic 1999; Albert, Jeong, and Barabási 2000; Jeong et al. 2001). Other network architectures include regular lattices, trees, and hierarchically decomposable networks (Simon 1998). Among the issues of functional capacity are which networks are optimal by some measure, e.g., their efficiency in inducing connectivity, and the robustness or sensitivity of their properties to local or random failure or directed attack. The significance of these studies from an engineering perspective is in answering such questions as, What kind of organizational structure is needed to perform what function with what level of reliability? and What are the tradeoffs that are made in different network architectures? Determining the organizational structures and their tradeoffs is relevant to all scales and areas of the converging technologies: nanotechnology, biomedical, information, cognition, and social networks.

Understanding (Recognizing) the Paradoxes of Complex Systems

The study of complex systems often reveals difficulties with concepts that are used in the study of simpler systems. Among these are conceptual paradoxes. Many of the paradoxes take the form of the coexistence of properties that, in simpler contexts, appear to be incompatible. In some cases it has been argued that there is a specific balance of properties; for example, the "edge-of-chaos" concept suggests a specific balance of order and chaos. However, in complex systems, order and chaos often coexist, and this is only one example of the wealth of paradoxes that are present. A more complete list would include paired properties such as the following:

- Stable and adaptable
- Reliable and controllable
- Persistent and dynamic
- Deterministic and chaotic
- Random and predictable
- Ordered and disordered

- Cooperative and competitive
- Selfish and altruistic
- Logical and paradoxical
- Averaging and non-averaging
- Universal and unique

While these pairs describe paradoxes of properties, the most direct paradox in complex systems is a recognition that more than one "cause" can exist, so that A causes B, and C causes B are not mutually incompatible statements. The key to understanding paradox in complex systems is to broaden our ability to conceive of the diversity of possibilities, both for our understanding of science and for our ability to design engineered systems that serve specific functions and have distinct design tradeoffs that do not fit within conventional perspectives.

Developing Systematic Methodologies for the Study of Complex Systems

While there exists a conventional "scientific method," the study of complex systems suggests that many more detailed aspects of scientific inquiry can be formalized. The existence of a unified understanding of patterns, description, and evolution as relevant to the study of complex systems suggests that we adopt a more systematic approach to scientific inquiry. Components of such a systematic approach would include experimental, theoretical, modeling, simulation, and analysis strategies. Among the aspects of a systematic strategy are the capture of quantitative descriptions of structure and dynamics, network analysis, dynamic response, information flow, multiscale decomposition, identification of modeling universality class, and refinement of modeling and simulations.

Major Application Areas of Complex Systems Research

The following should provide a sense of the integral nature of complex systems to advances in nanotechnology, biomedicine, information technology, cognitive science, and social and global systems. A level of complexity is found in their convergence.

Nanotechnology

Development of functional systems based on nanotechnological control is a major challenge beyond the creation of single elements. Indeed, the success of nanotechnology in controlling small elements can synergize well with the study of complex systems. To understand the significance of complex systems for nanotechnology, it is helpful to consider the smallest class of biological machines, also considered the smallest complex systems — proteins (Fersht 1999). Proteins are a marvel of engineering for design and manufacture. They also have many useful qualities that are not common in artificial systems, including robustness and adaptability through selection. The process of manufacturing a protein is divided into two parts, the creation of the molecular chain and the collapse of this chain to the functional form of the protein. The first step is ideal from a manufacturing point of view, since it enables direct manufacture from the template (RNA), which is derived from the information archive (DNA), which contains encoded descriptions of the protein chain. However, the chain that is formed in manufacture is not the functional form. The protein chain "self-organizes" (sometimes with assistance from other proteins) into its functional (folded) form. By manufacturing proteins in a form that is not the functional form, key aspects of the manufacturing process can be simplified, standardized, and made efficient while allowing a large variety of functional machines to be described in a simple language. The replication of DNA provides a mechanism for creating many equivalent information archives (by exponential growth) that can be transcribed to create templates to manufacture proteins in a massively parallel way when mass production is necessary. All of these processes rely upon rapid molecular dynamics. While proteins are functionally robust in any particular function, their functions can also be changed or adapted by changing the archive, which "describes" their function, but in an indirect and non-obvious way. The rapid parallel process of creation of proteins allows adaptation of new machines through large-scale variation and selection.

A good example of this process is found in the immune system response (Perelson and Wiegel 1999; Noest 2000; Segel and Cohen 2001; Pierre et al. 1997). The immune system maintains a large number of different proteins that serve as antibodies that can attach themselves to harmful antigens. When there is an infection, the antigens that attach most effectively are replicated in large numbers, and they are also subjected to a process of accelerated evolution through mutation and selection that generates even better-suited antibodies. Since this is not the evolutionary process of organisms, it is, in a sense, an artificial evolutionary process optimized (engineered) for the purpose of creating well-adapted proteins (machines). Antibodies are released into the blood as free molecules, but they are also used as tools by cells that hold them attached to their membranes so that the cells can attach to, or "grab hold of," antigens. Finally, proteins also form complexes and are part of membranes and biochemical networks, showing how larger functional structures can be built out of simple machines. An artificial analog of the immune system's use of evolutionary dynamics is the development of ribozymes by *in vitro* selection, now being used for drug design (Herschlag and Cech 1990; Beaudry and Joyce 1992; Szostak 1999).

Proteins and ribozymes illustrate the crossover of biology and nanotechnology. They also illustrate how complex systems concepts of self-organization, description, and evolution are important to nanotechnology. Nanotechnological design and manufacturing may take advantage of the system of manufacture of proteins or other approaches may be used. Either way, the key insights into how proteins work show the importance of understanding various forms of description (DNA); self-reproduction of the manufacturing equipment (DNA replication by polymerase chain reaction or cell replication); rapid template-based manufacture (RNA transcription to an amino-acid chain); self-organization into functional form (protein folding); and evolutionary adaptation through replication (mutation of DNA and selection of protein function) and modular construction (protein complexes). Understanding complex systems concepts thus will enable the development of practical approaches to nanotechnological design and manufacture and to adaptation to functional requirements of nanotechnological constructs.

Biomedical Systems

At the current time, the most direct large-scale application of complex systems methods is to the study of biochemical networks (gene regulatory networks, metabolic networks) that reveal the functioning of cells and the possibilities of medical intervention (Service 1999; Normile 1999; Weng, Bhalla and Iyengar 1999). The general studies of network structure described above are complementary to detailed studies of the mechanisms and function of specific biochemical systems (von Dassow et al. 2001). High-throughput data acquisition in genomics and proteomics is providing the impetus for constructing functional descriptions of biological systems (Strausberg and Austin 1999). This, however, is only the surface of the necessary applications of complex systems approaches that are intrinsic to the modern effort to understand biological organisms, their relationships to each other, and their relationship to evolutionary history. The key to a wider perspective is recognizing that the large quantities of data currently being collected are being organized into databases that reflect the data acquisition process rather than the potential use of this information. Opportunities for progress will grow dramatically when the information is organized into a form that provides a description of systems and system functions. Since cellular and multicellular organisms, including the human being, are not simply biochemical soups, this description must capture the spatiotemporal dynamics of the system as well as the biochemical network and its dynamics. In the context of describing human physiology from the molecular scale, researchers at the Oak Ridge National Laboratory working towards this goal call it the Virtual Human Project (Appleton 2000). This term has also been used to describe static images of a particular person at a particular time (NLM 2002).

The program of study of complex systems in biology requires not only the study of a particular organism (the human being) or a limited set of model organisms, as has been done in the context of genomics until now. The problem is to develop comparative studies of systems, understanding the variety that exists within a particular type of organism (e.g., among human beings) and the variety that exists across types of organisms. Ultimately, the purpose is to develop an understanding or description of the patterns of biological systems today as well as throughout the evolutionary process. The objective of understanding variety and evolution requires us to understand not just any particular biochemical system, but the space of possible biochemical systems filtered to the space of those that are found today, their general properties, their specific mechanisms, how these general properties carry across organisms, and how they are modified for different contexts. Moreover, new approaches that consider biological organisms through the relationship of structure and function and through information flow are necessary to this understanding.

Increasing knowledge about biological systems is providing us with engineering opportunities and hazards. The great promise of our biotechnology is unrealizable without a better understanding of the systematic implications of interventions that we can do today. The frequent appearance of biotechnology in the popular press through objections to genetic engineering and cloning reveals the great specific knowledge and the limited systemic knowledge of these systems. The example of corn genetically modified for feed and its subsequent appearance in corn eaten by human beings (Quist and Chapela 2001) reveals the limited knowledge we have of indirect effects in biological systems. This is not a call to limit our efforts, simply to focus on approaches that emphasize the roles of indirect effects and explore their implications scientifically. Without such studies, not only are we shooting in the dark, but in addition we will be at the mercy of popular viewpoints.

Completion of the virtual human project would be a major advance toward creating models for medical intervention. Such models are necessary when it is impossible to test multidrug therapies or specialized therapies based upon individual genetic differences. Intervention in complex biological systems is an intricate problem. The narrow bridge that currently exists between medical double blind experiments and the large space of possible medical interventions can be greatly broadened through systemic models that reveal the functioning of cellular systems and their relationship to cellular function. While today individual medical drugs are tested statistically, the main fruit of models will be as follows:

- to reveal the relationship between the function of different chemicals and the possibility of multiple different types of interventions that can achieve similar outcomes
- the possibility of discovering small variations in treatment that can affect the system differently
- possibly most importantly, to reveal the role of variations among human beings in the difference of response to medical treatment

A key aspect of all of these is the development of complex systems representations of biological function that reveal the interdependence of biological system and function.

Indeed, the rapid development of medical technologies and the expectation of even more dramatic changes should provide an opportunity for, even require, a change in the culture of medical practice. Key to these changes should be understanding the dynamic state of health. Conventional *homeostatic* perspectives on health are being modified to *homeodynamic* perspectives (Goldberger, Rigney, and West 1990; Lipsitz and Goldberger 1992). What is needed is a better understanding of the functional capabilities of a healthy individual to respond to changes in the external and internal environment for self-repair or -regulation. This is essential to enhance the individual's ability to maintain his or her own health. For example, while physical decline is a problem associated with old age, it is known that repair and regulatory mechanisms begin to slow down earlier, e.g., in the upper 30s, when professional athletes typically end their careers. By studying the dynamic response of an individual and changes over his/her life cycle, it should be possible to understand these early aspects of aging and to develop interventions that maintain a higher standard of health. More generally, understanding the network of regulatory and repair mechanisms should provide a better mechanism for dynamic monitoring — with biomedical sensors and imaging — health and disease and the impact of medical interventions. This would provide key information about the effectiveness of interventions for each individual, enabling feedback into the treatment process that can greatly enhance its reliability.

Information Systems

Various concepts have been advanced over the years for the importance of computers in performing large-scale computations or in replacing human beings through artificial intelligence. Today, the most apparent role of computers is as personal assistants and as communication devices and information archives for the socioeconomic network of human beings. The system of human beings and the Internet has become an integrated whole leading to a more intimately linked system. Less visibly, embedded computer systems are performing various specific functions in information processing for industrial age devices like cars. The functioning of the Internet and the possibility of future networking of embedded systems reflects the properties of the network as well as the properties of the complex demands upon it. While the Internet has some features that are designed, others are self-organizing, and the dynamic behaviors of the Internet reflect problems that may be better solved by using more concepts from complex systems that relate to interacting systems adapting in complex environments rather than conventional engineering design approaches.

Information systems that are being planned for business, government, military, medical, and other functions are currently in a schizophrenic state where it is not clear whether distributed intranets or integrated centralized databases will best suit function. While complex systems approaches generally suggest that creating centralized databases is often a poor choice in the context of complex function, the specific contexts and degree to which centralization is useful must be understood more carefully in terms of their functions and capabilities, both now and in the future (Bar-Yam 2001).

A major current priority is enabling computers to automatically configure themselves and carry out maintenance without human intervention (Horn 2001). Currently, computer networks are manually configured, and often the role of various choices in configuring them are not clear, especially for the performance of networks. Indeed, evidence indicates that network system performance can be changed dramatically using settings that are not recognized by the users or system administrators until chance brings these settings to their attention. The idea of developing more automatic processes is a small part of the more general perspective of developing adaptive information systems. This extends the concept of self-configuring and self-maintenance to endowing computer-based information systems with the ability to function effectively in diverse and variable environments. In order for this functioning to take place, information systems must, themselves be able to recognize patterns of behavior in the demands upon them and in their own activity. This is a clear direction for development of both computer networks and embedded systems.

Development of adaptive information systems in networks involves the appearance of software agents. Such agents range from computer viruses to search engines and may have communication and functional capabilities that allow social interactions among them. In the virtual world, complex systems perspectives are imperative in considering such societies of agents. As only one example, the analogy of software agents to viruses and worms has also led to an immune system perspective in the design of adaptive responses (Forrest, Hofmeyr, and Somayaji 1997; Kephart et al. 1997).

While the information system as a system is an important application of complex systems concepts, complex systems concepts also are relevant to considering the problem of developing information systems as effective repositories of information for human use. This involves two aspects, the first of which is the development of repositories that contain descriptions of complex systems that human beings would like to understand. The example of biological databases in the previous section is only one example. Other examples are socio-economic systems, global systems, and astrophysical systems. In each case, the key issue is to gain an understanding of how such complex systems can be effectively represented. The second aspect of designing such information repositories is the recognition of human factors in the development of human-computer interfaces (Norman and Draper 1986; Nielsen 1993; Hutchins 1995). This is important in developing all aspects of computer-based information systems, which are used by human beings and designed explicitly or implicitly to serve human beings.

More broadly, the networked information system that is being developed serves as part of the human socio-economic-technological system. Various parts of this system, which includes human beings and information systems, as well as the system as a whole, are functional systems. The development and design of such a self-organizing system and the role of science and technology is a clear area of application of complex systems understanding and methods. Since this is a functional system based upon a large amount of information, among the key questions is how should the system be organized when action and information are entangled.

Cognitive Systems

The decade of the 1990s was declared by President George Bush, senior (1990), the "decade of the brain," based, in part, on optimism that new experimental techniques such as Positron Emission Tomography (PET) imaging would provide a wealth of insights into the mechanisms of brain function. However, a comparison of the current experimental observations of cognitive processes with those of biochemical processes of gene expression patterns reveals the limitations that are still present in these observational techniques in studying the complex function of the brain. Indeed, it is reasonable to argue that the activity of neurons of a human being and their functional assignment is no less complex than the expression of genes of a single human cell.

Current experiments on gene expression patterns allow the possibility of knocking out individual genes to investigate the effect of each gene on the expression pattern of all other genes measured individually. The analogous capability in the context of cognitive function would be to incapacitate an individual neuron and investigate the effect on the firing patterns of all other neurons individually. Instead, neural studies are based upon sensory stimulation and measures of the average activity of large regions of cells. In gene expression studies, many cells are used with the same genome and a controlled history through replication, and averages are taken of the behavior of these cells. In contrast, in neural studies averages are often taken of the activity patterns of many individuals with distinct genetic and environmental backgrounds. The analogous biochemical experiment would be to average behavior of many cells of different types from a human body (muscle, bone, nerve, red blood cell, etc.) and different individuals to obtain a single conclusion about the functional role of the genes.

The more precise and larger quantities of genome data have revealed the difficulties in understanding genomic function and the realization that gene function must be understood through models of genetic networks (Fuhrman et al. 1998). This is to be contrasted with the conclusions of cognitive studies that investigate the aggregate response of many individuals to large-scale sensory stimuli and infer functional assignments. Moreover, these functional assignments often have limited independently verifiable or falsifiable implications. More generally, a complex systems perspective suggests that it is necessary to recognize the limitations of the assignment of function to individual components ranging from molecules to subdivisions of the brain; the limitations of narrow perspectives on the role of environmental and contextual effects that consider functioning to be independent of effects other than the experimental stimulus; and the limitations of expectations that human differences are small and therefore that averaged observations have meaning in describing human function.

The problem of understanding brain and mind can be understood quite generally through the role of relationships between patterns in the world and patterns of neuronal activity and synaptic change. While the physical and biological structure of the system is the brain, the properties of the patterns identify the psychofunctioning of the mind. The relationship of external and internal patterns are further augmented by relationships between patterns within the brain. The functional role of patterns is achieved through the ability of internal patterns to represent both concrete and abstract entities and processes, ranging from the process of sensory-motor response to internal dialog. This complex nonlinear dynamic system has a great richness of valid statements that can be made about it, but identifying an integrated understanding of the brain/mind system cannot be captured by perspectives that limit their approach through the particular methodologies of the researchers involved. Indeed, the potential contributions of the diverse approaches to studies of brain and mind have been limited by the internal dynamics of the many-factioned scientific and engineering approaches.

The study of complex systems aspects of cognitive systems, including the description of patterns in the world and patterns in mind, the construction of descriptions of complex systems, and the limitations on information processing that are possible for complex systems are relevant to the application of cognitive studies to the understanding of human factors in man-machine systems (Norman and Draper 1986; Nielsen 1993; Hutchins 1995) and more generally to the design of systems that include both human beings and computer-based information systems as functional systems. Such hybrid systems, mentioned previously in the section on information technology, reflect the importance of the converging technology approach.

The opportunity for progress in understanding the function of the networked, distributed neuro-physiological system also opens the possibility of greater understanding of development, learning, and aging (NIMH n.d.; Stern and Carstensen 2000; Mandell and Schlesinger 1990; Davidson, Teicher, and Bar-Yam 1997). While the current policy of education reform is using a uniform measure of accomplishment and development through standardized testing, it is clear that more effective measures must be based on a better understanding of cognitive development and individual differences. The importance of gaining such knowledge is high because evaluation of the effectiveness of new approaches to education typically requires a generation to see the impact of large-scale educational changes on society. The positive or negative effects of finer-scale changes appear to be largely inaccessible to current research. Thus, we see the direct connection between complex systems approaches to cognitive science and societal policy in addressing the key challenge of the education system. This in turn is linked to the solution of many other complex societal problems, including poverty, drugs, and crime, and also to effective functioning of our complex economic system requiring individuals with diverse and highly specialized capabilities.

Studies of the process of aging are also revealing the key role of environment in the retention of effective cognitive function (Stern and Carstensen 2000; Mandell and Schlesinger 1990; Davidson, Teicher, and Bar-Yam 1997). The notion of "use it or lose it," similar to the role of muscular exercise, suggests that unused capabilities are lost more rapidly than used ones. While this is clearly a simplification, since losses are not uniform across all types of capabilities and overuse can also cause deterioration, it is a helpful guideline that must be expanded upon in future research. This suggests that research should focus on the effects of the physical and social environments for the elderly and the challenges that they are presented with.

We can unify and summarize the complex systems discussion of the cognitive role of the environment for children, adults, and the elderly by noting that the complexity of the environment and the individual must be matched for effective functioning. If the environment is too complex, confusion and failure result; if the environment is too simple, deterioration of functional capability results. One approach to visualizing this process is to consider that the internal physical parts and patterns of activity are undergoing evolutionary selection dictated by the patterns of activity that result from environmental stimulation. This evolutionary approach also is relevant to the recognition that individual differences are analogous to different ecological niches. A more detailed research effort would not only consider the role of complexity but also the effect of specific patterns of environment and patterns of internal functioning, individual differences in child development, aging, adult functioning in teams, and hybrid human-computer systems.

Social Systems and Societal Challenges

While social systems are highly complex, there are still relatively simple collective behaviors that are not well understood. These include commercial fads, market cycles and panics, bubbles and busts. Understanding the fluctuating dynamics and predictability of markets continues to be a major challenge. It is important to emphasize that complex systems studies are not necessarily about predicting the market, but about understanding its predictability or lack thereof.

More generally, there are many complex social challenges associated with complex social systems ranging from military challenges to school and education system failures, healthcare errors, and problems with quality of service. Moreover, other major challenges remain in our inability to address fundamental social ills such as poverty (in both developed and undeveloped countries), drug use, and crime. To clarify some aspects of social systems from a complex systems perspective, it is helpful to focus on one of these, and the current military context is a convenient focal point.

Wars are major challenges to our national abilities. The current war on terrorism is no exception. In dealing with this challenge, our leadership, including the president and the military, has recognized that this conflict is highly complex. Instead of just sending in tens to hundreds of thousands of troops, as was done in the Gulf War, there is a strategy of using small teams of special forces to gain intelligence and lay the groundwork for carefully targeted, limited, and necessary force.

A large-scale challenge can be met by many individuals doing the same thing at the same time or by repeating the same action, similar to a large military force. In contrast, a complex challenge must be met by many individuals doing many different things at different times. Each action has to directly match the local task that must be done. The jungles of Vietnam and the mountains of Afghanistan, reported to have high mountains and deep narrow valleys, are case studies in complex terrains. War is complex when targets are hidden, not only in the terrain but also among people — bystanders or friends. It is also complex when the enemy can itself do many different things, when the targets are diverse, the actions that must be taken are specific, and the difference between right and wrong action is subtle.

While we are still focused on the war on terrorism, it seems worthwhile to transfer the lessons learned from different kinds of military conflicts to other areas where we are trying to solve major problems. Over the past 20 years, the notion of war has been used to describe the War on Poverty, the War on Drugs, and other national challenges. These were called wars because they were believed to be challenges requiring the large force of old-style wars. They are not. They are complex challenges that require detailed intelligence and the application of the necessary forces in the right places. Allocating large budgets for the War on Poverty did not eliminate the problem; neither does neglect. The War on Drugs has taken a few turns, but even the recent social campaign "Just say no!" is a large-scale approach. Despite positive intentions, we have not won these wars because we are using the wrong strategy.

There are other complex challenges that we have dealt with using large forces. Third World development is the international version of the War on Poverty to which the World Bank and other organizations have applied large forces. Recently, more thoughtful approaches are being taken, but they have not gone far enough. There is a tendency to fall into the "central planning trap." When challenges become complex enough, even the very notion of central planning and control fails. Building functioning socioeconomic systems around the world is such a complex problem that it will require many people taking small and targeted steps — like the special forces in Afghanistan.

There are other challenges that we have not yet labeled wars, which are also suffering from the same large-force approach. Among these are cost containment in the medical system and improving the education system. In the medical system, the practice of cost controls through managed care is a large-force approach that started in the early 1980s. Today, the medical system quality of care is disintegrating under the stresses and turbulence generated by this strategy. Medical treatment is clearly one of the most complex tasks we are regularly engaged in. Across-the-board cost control should not be expected to work. We are just beginning to apply the same kind of large-scale strategy to the education system through standardized testing. Here again, a complex systems perspective suggests that the outcomes will not be as positive as the intentions.

The wide applicability of lessons learned from fighting complex wars and the effective strategies that resulted should be further understood through research projects that can better articulate the relevant lessons and how they pertain to solving the many and diverse complex social problems we face.

Global and Larger Systems

Global systems — physical, biological, and social — are potentially the most complex systems studied by science today. Complex systems methods can provide tools for analyzing their large-scale behavior. Geophysical and geobiological systems, including meteorology, plate tectonics and earthquakes, river and drainage networks, the biosphere and ecology, have been the motivation for and the application of complex systems methods and approaches (Dodds and Rothman 2000; Lorenz 1963; Bak and Tang 1989; Rundle, Turcotte, and Klein 1996; NOAA 2002). Such applications also extend to other planetary, solar, and astrophysical systems. Converging technologies to improve human performance may benefit from these previous case studies.

Among the key problems in studies of global systems is understanding the indirect effects of global human activity, which in many ways has reached the scale of the entire earth and biosphere. The possibility of human impact on global systems through overexploitation or other by-products of industrial activity has become a growing socio-political concern. Of particular concern is the impact of human activity on the global climate (climate change and global warming) and on the self-sustaining properties of the biosphere through exploitation and depletion of key resources (e.g., food resources like fish, energy resources like petroleum, deforestation, loss of biodiversity). Other global systems include global societal problems that can include the possibility of global economic fluctuations, societal collapse, and terrorism. Our effectiveness in addressing these questions will require greater levels of understanding and representations of indirect effects, as well as knowledge of effective mechanisms for intervention, if necessary. In this context, the objective is to determine which aspects of a system can be understood or predicted based upon available information, along with the level of uncertainty in such predictions. In some cases, the determination of risk or uncertainty is as important as the prediction of the expected outcome. Indeed, knowing "what is the worst that can happen" is often an important starting point for effective decision-making.

In general, the ability of humanity to address global problems depends on the collective behavior of people around the world. Global action is now typical in response to local natural disasters (earthquakes, floods, volcanoes, droughts); man-made problems from wars (Gulf War, Bosnia, Rwanda, the war on terrorism); and environmental concerns (international agreements on environment and development). In addition, there is a different sense in which addressing global concerns requires the participation of many individuals: The high complexity of these problems implies that many individuals must be involved in addressing these problems, and they must be highly diverse and yet coordinated. Thus, the development of complex systems using convergent technologies that facilitate human productivity and cooperative human functioning will be necessary to meet these challenges.

What is to be Done?

The outline above of major areas of complex systems research and applications provides a broad view in which many specific projects should be pursued. We can, however, single out three tasks that, because of their importance or scope, are worth identifying as priorities for the upcoming years: (1) transform education; (2) develop sets of key system descriptions; and (3) design highly complex engineering projects as evolutionary systems.

Transform Education

The importance of education in complex systems concepts for all areas of science, technology, and society at large has been mentioned above but should be reemphasized. There is need for educational materials and programs that convey complex systems concepts and methods and are accessible to a wide range of individuals, as well as more specific materials and courses that explain their application in particular contexts. A major existing project on fractals can be used as an example (Buldyrev et al. n.d.). There are two compelling reasons for the importance of such projects. The first is the wide applicability of complex systems concepts in science, engineering, medicine, and management. The second is the great opportunity for engaging the public in exciting science with a natural relevance to daily life and enhancing their support for ongoing and future research. Ultimately, the objective is to integrate complex systems concepts throughout the educational system.

Develop Sets of Key System Descriptions

There are various projects for describing specific complex systems (NOAA 2002; Kalra et al. 1988; Goto, Kshirsagar, and Magnenat-Thalmann 2001; Heudin 1999; Schaff et al. 1997; Tomita et al. 1999), ranging from the earth to a single cell, which have been making substantial progress. Some of these focus more on generative simulation, others on representation of observational data. The greatest challenge is to merge these approaches and develop system descriptions that identify both the limits of observational and modeling strategies and the opportunities they provide jointly for the description of complex systems. From this perspective, some of the most exciting advances are in representation of human forms in computer-based animation (Kalra et al. 1988; Goto, Kshirsagar, and Magnenat-Thalmann 2001; Heudin 1999), and particularly, in projecting human beings electronically. Pattern recognition is performed on realtime video to obtain key information about dynamic facial expression and speech, which is transmitted electronically to enable animation of a realistic computer-generated image that represents, in real time, the facial expression and speech of the person at a remote location (Goto, Kshirsagar, and Magnenat-Thalmann 2001). Improvement in such systems is measured by the growing bandwidth necessary for the transmission, which reflects our inability to anticipate system behavior from prior information.

To advance this objective more broadly, developments in systematic approaches (including quantitative languages, multiscale representations, information capture, and visual interfaces) are necessary, in conjunction with a set of related complex systems models. For example, current computer-based tools are largely limited to separated procedural languages (broadly defined) and databases. A more effective approach may be to develop quantitative descriptive languages based on lexical databases that merge the strength of human language for description with computer capabilities for manipulating and visually representing quantitative attributes (Smith, Bar-Yam, and Gelbart 2001). Such extensible quantitative languages are a natural bridge between quantitative mathematics, physics, and engineering languages and qualitative lexicons that dominate description in biology, psychology, and social sciences. They would facilitate describing structure, dynamics, relationships, and functions better than, for example, graphical extensions of procedural languages. This and other core complex systems approaches should be used in the description of a set of key complex systems under a coordinating umbrella.

For each system, an intensive collection of information would feed a system representation whose development would be the subject and outcome of the project. For example, in order to develop a representation of a human being, there must be intensive collection of bio-psycho-social information about the person. This could include multisensor monitoring of the person's physical (motion), psycho-social (speech, eye-motion), physiological (heart rate), and biochemical (food and waste composition, blood chemistry) activity over a long period of time, with additional periodic biological imaging and psychological testing. Virtual world animation would be used to represent both the person and his/her environment. Models of biological and psychological function representing behavioral patterns would be incorporated and evaluated. Detailed studies of a particular individual along with comparative studies of several individuals would be made to determine both what is common and what is different. As novel relevant convergent technologies become available that would affect human performance or affect our ability to model human behavior, they can be incorporated into this study and evaluated. Similar coordinating projects would animate representations of the earth, life on earth, human civilization, a city, an animal's developing embryo, a cell, and an engineered system, as suggested above. Each such project is both a practical application and a direct test of the limits of our insight, knowledge, and capabilities. Success of the projects is guaranteed because their ultimate objective is to inform us about these limits.

Design Highly Complex Engineering Projects as Evolutionary Systems

The dramatic failures in large-scale engineering projects such as the Advanced Automation System (AAS), which was originally planned to modernize air traffic control, should be addressed by complex systems research. The AAS is possibly the largest engineering project to be abandoned. It is estimated that several billion dollars were spent on this project. Moreover, cost overruns and delays in modernization continue in sequel projects. One approach to solving this problem, simplifying the task definition, cannot serve when the task is truly complex, as it appears to be in this context. Instead, a major experiment should be carried out to evaluate implementation of an evolutionary strategy for large-scale engineering. In this approach, the actual air traffic control system would become an evolving system, including all elements of the system, hardware, software, the air traffic controllers, and the designers and manufacturers of the software and hardware. The system context would be changed to enable incremental changes in various parts of the system and an evolutionary perspective on population change.

The major obstacle to any change in the air traffic control system is the concern for safety of airplanes and passengers, since the existing system, while not ideally functioning, is well tested. The key to enabling change in this system is to introduce redundancy that enables security while allowing change. For example, in the central case of changes in the air traffic control stations, the evolutionary process would use "trainers" that consist of doubled air traffic control stations, where one has override capability over the other. In this case, rather than an experienced and inexperienced controller, the two stations are formed of a conventional and a modified station. The modified station can incorporate changes in software or hardware. Testing can go on as part of operations, without creating undue risks. With a large number of trainers, various tests can be performed simultaneously and for a large number of conditions. As a particular system modification becomes more extensively tested and is found to be both effective and reliable, it can be propagated to other trainers, even though testing would continue for extended periods of time. While the cost of populating multiple trainers would appear to be high, the alternatives have already been demonstrated to be both expensive and unsuccessful. The analogy with paired chromosomes in DNA can be seen to reflect the same design principle of redundancy and robustness. These brief paragraphs are not sufficient to explain the full evolutionary context, but they do resolve the key issue of safety and point out the opening that this provides for change. Such evolutionary processes are also being considered for guiding other large-scale engineering modernization programs (Bar-Yam 2001).

Conclusions

The excitement that is currently felt in the study of complex systems arises not from a complete set of answers but rather from the appearance of a new set of questions, which are relevant to NBIC. These questions differ from the conventional approaches to science and technology and provide an opportunity to make major advances in our understanding and in applications.

The importance of complex systems ideas in technology begins through recognition that novel technologies promise to enable us to create ever more complex systems. Even graphics-oriented languages like OpenGL are based on a procedural approach to drawing objects rather than representing them. Moreover, the conventional boundary between technology and the human beings that use them is not a useful approach to thinking about complex systems of human beings and technology. For example, computers as computational tools have given way to information technology as an active interface between human beings that are working in collaboration. This is now changing again to the recognition that human beings and information technology are working together as an integrated system.

More generally, a complex systems framework provides a way in which we can understand how the planning, design, engineering, and control over simple systems gives way to new approaches that enable such systems to arise and be understood with limited or indirect planning or control. Moreover, it provides a way to better understand and intervene (using technology) in complex biological and social systems.

References

Albert, R., H. Jeong, and A-L. Barabási. 2000. Error and attack tolerance of complex networks. *Nature* 406: 378-382.

Albert, R., H. Jeong, and A.-L. Barabási. 1999. Diameter of the World-Wide Web. *Nature* 401:130–131.

Anderson, J.A., and E. Rosenfeld, eds. 1988. *Neurocomputing*. Cambridge: MIT Press.

Arrow, K.J. 1963. *Social choice and individual values*. New York: Wiley.

Ashby, W.R. 1957. *An introduction to cybernetics*. London: Chapman and Hall.

Aumann, R.J., and S. Hart, eds. 1992. *Handbook of game theory with economic applications*, Vols. 1, 2. Amsterdam: North-Holland.

Axelrod, R.M. 1984. *The evolution of cooperation*. New York: Basic Books.

Bak, P. 1996. *How Nature works: The science of self-organized criticality*. New York: Copernicus, Springer-Verlag.

Bak, P., and C. Tang. 1989. Earthquakes as a self-organized critical phenomenon, *J. Geophys. Res.*, 94(15):635-37.

Ball, P. 1999. *The self-made tapestry: Pattern formation in Nature*. Oxford: Oxford Univ. Press.

Banavar, J.R., A. Maritan, and A. Rinaldo. 1999. Size and form in efficient transportation networks. *Nature* 399: 130-132.

Barabási, A.-L., and R. Albert. 1999. Emergence of scaling in random networks. *Science* 286: 509–511.

Barthélémy, M., and L.A.N. Amaral. 1999. Small-world networks: Evidence for a crossover picture. *Phys. Rev. Lett.* 82: 3180–3183.

Bar-Yam, Y. 1997. *Dynamics of complex systems*. Reading, MA: Addison-Wesley.

Bar-Yam, Y. 2000. Formalizing the gene-centered view of evolution. *Advances in Complex Systems* 2: 277-281.

Bar-Yam, Y., and A. Minai, eds. 2002. *Unifying themes in complex systems II: Proceedings of the 2nd International Conference on Complex Systems*. Perseus Press.

Bar-Yam, Y., ed. 2000. *Unifying themes in complex systems: Proceedings of the International Conference on Complex Systems*. Perseus Press.

Beaudry, A., and G.F. Joyce. 1992. Directed evolution of an RNA enzyme. *Science* 257: 635-641.

Bishop, M. 1995. *Neural networks for pattern recognition*. New York: Oxford University Press.

Brandon, R.N., and R.M. Burian, eds. 1984. *Genes, organisms, populations: Controversies over the units of selection*. Cambridge: MIT Press.

Bray, J. 1994. *Advances in Physics* 43: 357.

Buldyrev, S.V., M.J. Erickson, P. Garik, P. Hickman, L.S. Shore, H.E. Stanley, E.F. Taylor, and P.A. Trunfio. N.d. "Doing science" by learning about fractals. Working Paper, Boston Univ. Center for Polymer Science.

Bush, G.H.W. 1990. By the President of the United States of America A Proclamation, Presidential Proclamation 6158.

Casti, J.L. 1994. *Complexification: Explaining a paradoxical world through the science of surprise.* New York: Harper Collins.

Cheswick, W., and H. Burch. N.d. Internet mapping project. Online: http://www.cs.bell-labs.com/who/ches/map/.

Coveney, P., and R. Highfield. 1995. *Frontiers of complexity: The search for order in a chaotic world.* New York: Fawcett Columbine.

Cvitanovic, P., ed. 1989. *Universality in chaos: A reprint selection.* 2d ed. Bristol: Adam Hilger.

Darwin, C. 1964. *On the origin of species (by means of natural selection).* A facsimile of the first edition, 1859. Cambridge: Harvard University Press.

Davidson, A., M.H. Teicher, and Y. Bar-Yam. 1997. The role of environmental complexity in the well-being of the elderly. *Complexity and Chaos in Nursing* 3: 5.

Day, W. 1984. *Genesis on Planet Earth: The search for life's beginning.* 2nd ed. New Haven: Yale Univ. Press.

Devaney, R.L. 1989. *Introduction to chaotic dynamical systems,* 2d ed. Reading, MA: Addison-Wesley.

Dodds, P.S., and D.H. Rothman. 2000. Scaling, universality, and geomorphology. *Annu. Rev. Earth Planet. Sci.* 28:571-610.

Ernst and Young. 2000. *Embracing complexity.* Vols. 1-5, 1996-2000. Ernst and Young, Ctr. for Business Innovation. http://www.cbi.cgey.com/research/current-work/biology-and-business/complex-adaptive-systems-research.html.

Fersht, A.. 1999. *Structure and mechanism in protein science: A guide to enzyme catalysis and protein folding.* New York: W.H. Freeman.

Fogel, L.J., A.J. Owens, and M.J. Walsh. 1966. *Artificial intelligence through simulated evolution.* New York: Wiley.

Forrest, S.S., A. Hofmeyr, and A. Somayaji. 1997. Computer immunology. *Communications of the ACM* 40:88-96.

Fudenberg, D., and J. Tirole. 1991. *Game theory.* Cambridge: MIT Press.

Fuhrman, S., X. Wen, G. Michaels, and R. Somogyi. 1998. Genetic network inference. *InterJournal* 104.

Gallagher, R., and T. Appenzeller. 1999. Beyond reductionism. *Science* 284:79.

Gell-Mann, M. 1994. *The quark and the jaguar.* New York: W.H. Freeman.

Gleick, J. 1987. *Chaos: Making a new science.* New York: Penguin.

Goldberg, L.A., P.W. Goldberg, C.A. Phillips, and G.B. Sorkin. 1998. Constructing computer virus phylogenies. *Journal of Algorithms* 26 (1): 188-208.

Goldberger, L., D.R. Rigney, and B.J. West. 1990. Chaos and fractals in human physiology. *Sci. Amer.* 262:40-49.

Golubitsky, M., I. Stewart, P.L. Buono, and J.J. Collins. 1999. Symmetry in locomotor central pattern generators and animal gaits. *Nature* 401:675), 693-695 (Oct 14).

Goodwin, B.C. 1994. *How the leopard changed its spots: The evolution of complexity.* New York: C. Scribner's Sons.

Goto, T., S. Kshirsagar, and N. Magnenat-Thalmann. 2001. Automatic face cloning and animation. *IEEE Signal Processing Magazine* 18 (3) (May):17-25.

Herschlag, D., and T.R. Cech. 1990. DNA cleavage catalysed by the ribozyme from Tetrahymena. *Nature* 344:405-410.

Herz, J.C. 2001. The allure of chaos. *The Industry Standard* (Jun 25). Online: http://www.thestandard.com/article/0,1902,27309,00.html

Heudin, J.C., ed. 1998. *Virtual worlds: Synthetic universes, digital life, and complexity.* Reading, MA: Perseus.

Holland, J.H. 1992. *Adaptation in natural and artificial systems.* 2d ed. Cambridge: MIT Press.

_____. 1995. *Hidden order: How adaptation builds complexity.* Reading, MA: Addison-Wesley.

Horn, P. 2001. *Autonomic computing*. IBM.

Huberman, A., and L.A. Adamic. 1999. Growth dynamics of the World-Wide Web. *Nature* 401:131.

Huberman, A., and R.M. Lukose. 1997. *Science* 277:535-538.

Huberman, A., P. Pirolli, J. Pitkow, and R.M. Lukose. 1998. *Science* 280:95-97.

Hutchins, E. 1995. *Cognition in the wild*. Cambridge: MIT Press.

INSS. 1997. *1997 Strategic assessment.* http://www.ndu.edu/inss/sa97/sa97exe.html. Institute for National Strategic Studies. Washington, D.C.: U.S. Government Printing Office (National Defense University Press).

IOM. 2000. *To err is human: Building a safer health system*. Washington, D.C.: Institute of Medicine.

Jeong, H., B. Tombor, R. Albert, Z. Oltvai, and A.-L. Barabási. 2001. The large-scale organization of metabolic networks. *Nature* 407, 651 - 654 (05 Oct 2000).

Kalra, P., N. Magnenat-Thalmann, L. Moccozet, G. Sannier, A. Aubel, and D. Thalmann. 1988. RealTime animation of realistic virtual humans. *Computer Graphics and Applications* 18(5):42-56.

Kandel, E.R., J.H. Schwartz ,and T.M. Jessell, eds. 2000. *Principles of neural science*. 4th ed. NY: McGraw-Hill.

Kauffman, S. 1969. Metabolic stability and epigenesis in randomly constructed genetic nets. *J. Theor. Biol.* 22:437.

Kauffman, S.A. 1993. *The origins of order: Self organization and selection in evolution*. NY: Oxford Univ. Press.

_____. 1995. *At home in the universe*. NY: Oxford Univ. Press.

Kephart, J.O., G.B. Sorkin, D.M. Chess, and S.R. White. 1997. Fighting computer viruses: Biological metaphors offer insight into many aspects of computer viruses and can inspire defenses against them. *Scientific American.* November.

Latora, V., and M. Marchiori. 2001. Efficient behavior of small-world networks. *Phys. Rev. Lett.* 87:198701.

Lawrence, S., and C.L. Giles. 1999. Accessibility of information on the web. *Nature* 400:107-109.

Lewin, R. 1992. *Complexity: Life at the edge of chaos*. New York: Macmillan.

Lewontin, R. 2000. *The triple helix : Gene, organism, and environment*. Cambridge: Harvard Univ. Press.

Lipsitz, L.A., and A.L. Goldberger. 1992. Loss of "complexity" and aging. *JAMA* 267:1806-1809.

Lorenz, E.N. 1963. Deterministic nonperiodic flow. *J. Atmosph. Sci.* 20:130-141.

Mandell, J., and M.F. Schlesinger. 1990. Lost choices: Parallelism and topo entropy decrements in neuro-biological aging. In *The ubiquity of chaos*, ed. S. Krasner. Washington, D.C.: Amer. Assoc. Adv. of Science.

Maritan, A., F. Colaiori, A. Flammini, M. Cieplak, and J. Banavar. 1996. Universality classes of optimal channel networks. *Science* 272: 984-986.

Meinhardt, H. 1994. *The algorithmic beauty of sea shell patterns*. New York: Springer-Verlag.

Meyer, A., and T.A. Brown. 1998. Statistical mechanics of voting. *Phys. Rev. Lett.* 81:1718-1721.

Milgram, S. 1967. The small-world problem. *Psychol. Today* 2:60–67.

_____. 1992. The small world problem. In *The individual in a social world: Essays and experiments*. 2nd ed., ed. S. Milgram, J. Sabini, and M. Silver. New York: McGraw Hill.

Murray, J.D. 1989. *Mathematical biology*. New York: Springer-Verlag.

Nielsen, J. 1993. *Usability engineering*. Boston: Academic Press.

NIGMS. 2002. Complex Biological Systems Initiative, National Institute of General Medical Science, NIH, http://www.nigms.nih.gov/funding/complex_systems.html.

Nijhout, H.F. 1992. *The development and evolution of butterfly wing patterns*. Washington, D.C.: Smithsonian Institution Press.

NIMH. 2002. *Learning and the brain*, National Institute of Mental Health, NIH. Online: http://www.edupr.com/brain4.html.

NLM. 2002. The Visible Human Project, National Library of Medicine. http://www.nlm.nih.gov/research/visible/visible_human.html.

NOAA. 2002. Ecosystems and Global Change, NOAA National Data Centers, NGDC http://www.ngdc.noaa.gov/seg/eco/eco_sci.shtml.

Noest, J. 2000. Designing lymphocyte functional structure for optimal signal detection: Voilà, T cells. *Journal of Theoretical Biology*, 207(2):195-216.

Norman, D.A., and S. Draper, eds. 1986. *User centered system design: New perspectives in human-computer interaction*. Hillsdale, NJ: Erlbaum.

Normile, D. 1999. Complex systems: Building working cells "in silico." *Science* 284: 80.

NSF. N.d. Biocomplexity initiative: http://www.nsf.gov/pubs/1999/nsf9960/nsf9960.htm; http://www.nsf.gov/pubs/2001/nsf0134/nsf0134.htm; http://www.nsf.gov/pubs/2002/nsf02010/nsf02010.html.

NSF. 2001. Joint DMS/NIGMS Initiative to Support Research Grants in Mathematical Biology. National Science Foundation Program Announcement NSF 01-128, online at http://www.nsf.gov/cgi-bin/getpub?nsf01128.

Appleton, W. 2000. Science at the interface. *Oak Ridge National Laboratory Review* (Virtual human) 33: 8-11.

Ott, E. 1993. *Chaos in dynamical systems*. Cambridge: Cambridge University Press.

Perelson, W., and F.W. Wiegel. 1999. Some design principles for immune system recognition. *Complexity* 4: 29-37.

Pierre, D.M., D. Goldman, Y. Bar-Yam, and A.S. Perelson. 1997. Somatic evolution in the immune system: The need for germinal centers for efficient affinity maturation. *J. Theor. Biol.* 186:159-171.

Quist, D., I.H. Chapela. 2001. Transgenic DNA introgressed into traditional maize landraces in Oaxaca, Mexico. *Nature* 414:541-543 (29 Nov).

Rundle, B., D.L. Turcotte, and W. Klein, eds. 1996. *Reduction and predictability of natural disasters*. Reading, MA: Perseus Press.

Sayama, H., L. Kaufman, and Y. Bar-Yam. 2000. Symmetry breaking and coarsening in spatially distributed evolutionary processes including sexual reproduction and disruptive selection. *Phys. Rev. E* 62:7065.

Schaff, J., C. Fink, B. Slepchenko, J. Carson and L. Loew. 1997. A general computational framework for modeling cellular structure and function. *Biophys. J.* 73:1135-1146.

Segel, L.A. 1984. *Modeling dynamic phenomena in molecular and cellular biology*. Cambridge: Cambridge Univ. Press.

Segel, L.A., and I.R. Cohen, eds. 2001. *Design principles for the immune system and other distributed autonomous systems*. New York: Oxford University Press.

Service, R.F. 1999. Complex systems: Exploring the systems of life. *Science* 284: 80.

Shannon, E. 1963. A mathematical theory of communication. In *Bell Systems Technical Journal*, July and October 1948; reprinted in C.E. Shannon and W. Weaver, *The mathematical theory of communication*. Urbana: University of Illinois Press.

Simon, H.A. 1998. *The sciences of the artificial*. 3rd ed. Cambridge: MIT Press.

Simpson, H. 1951. The interpretation of interaction in contingency tables. *Journal of the Royal Statistical Society, Ser. B* 13:238-241.

Smith, J.M. 1982. *Evolution and the theory of games*. Cambridge: Cambridge University Press.

Smith, M.A., Y. Bar-Yam, and W. Gelbart. 2001. Quantitative languages for complex systems applied to biological structure. In *Nonlinear dynamics in the life and social sciences*, ed. W. Sulis and I. Trofimova, NATO Science Series A/320. Amsterdam: IOS Press.

Sober, E., and D.S. Wilson. 1999. *Unto others*. Cambridge: Harvard Univ. Press.

Stacey, R.D. 1996. *Complexity and creativity in organizations*. San Francisco: Berrett-Koehler.

_____. 2001. *Complex responsive processes in organizations*. New York: Routledge.

Stein, L.A. 1999. Challenging the computational metaphor: Implications for how we think. *Cybernetics and Systems* 30 (6):473-507.

Sterman, J.D. 2000. *Business dynamics: Systems thinking and modeling for a complex world*. Irwin Professional.

Stern, P.C., and L.L. Carstensen, eds. 2000. *The aging mind: Opportunities in cognitive research*. Washington, D.C.: National Academy Press.

Strausberg, R.L., and M.J.F. Austin. 1999. Functional genomics: Technological challenges and opportunities. *Physiological Genomics* 1:25-32.

Strogatz, S.H. 1994. *Nonlinear dynamics and chaos with applications to physics, biology, chemistry, and engineering*. Reading, MA: Addison-Wesley.

Szostak, J.W. 1999. In vitro selection and directed evolution, *Harvey Lectures* 93:95-118. John Wiley & Sons.

Tomita, M., K. Hashimoto, K. Takahashi, T. Shimizu, Y. Matsuzaki, F. Miyoshi, K. Saito, S. Tanida, K. Yugi, J.C. Venter, and C. Hutchison. 1999. E-CELL: Software environment for whole cell simulation. *Bioinformatics* 15:316-317.

Triantafyllou, G.S., and M.S. Triantafyllou. 1995. An efficient swimming machine. *Scientific American* 272: 64-70.

Turing, A.M. 1952. The chemical basis of morphogenesis, *Phil. Trans. R. Soc. Lond. B* 237(641):37-72.

von Dassow, G., E. Meir, E.M Munro, and G.M Odell. 2001. The segment polarity is a robust developmental module. *Nature* 406:188-192.

von Neumann, J., and O. Morgenstern. 1944. *Theory of games and economic behavior*. Princeton Univ. Press.

Waldrop, M.M. 1992. *Complexity: The emerging science at the edge of order and chaos*. NY: Simon & Schuster.

Wasserman, S., and K. Faust. 1994. *Social network analysis*. Cambridge: Cambridge University Press.

Watts, J. 1999. *Small worlds*. Princeton: Princeton Univ. Press.

Watts, J., and S.H. Strogatz. 1998. Collective dynamics of 'small-world' networks. *Nature* 393:440–442.

Weng, G., U.S. Bhalla, and R. Iyengar. 1999. Complexity in biological signaling systems. *Science* 284:92.

Williams, R.J., and N.D. Martinez. 2000. Simple rules yield complex food webs. *Nature* 404:180–183.

Wilson, K.G. 1983. The renormalization-group and critical phenomena. *Reviews of Modern Physics* 55(3):583-600.

World Bank. 1998. *Partnership for development: Proposed actions for the World Bank* (May).

Zegura, E.W., K.L. Calvert, and M.J. Donahoo. 1997. A quantitative comparison of graph-based models for internet topology. *IEEE/ACM Trans. Network.* 5: 770–787.

MIND OVER MATTER IN AN ERA OF CONVERGENT TECHNOLOGIES

Daniel L. Akins, City University of New York

Within the next 10 to 15 years, economically viable activities connected with nanoscience, bioscience, information technology, and cognitive science (NBIC) will have interlaced themselves within ongoing successful technologies, resulting in new and improved commercial endeavors. The impact of such eventualities would be enormous even if the emerging activities were developing independently, but with a range of synergies, their overlapping emergence and transitioning into the applied engineering arena promises to result in industrial products and technologies that stretch our imaginations to the point that they appear fanciful. Indeed, it is becoming more widely acknowledged that the potential of the new convergent NBIC technologies for influencing and defining the future is unlimited and likely unimaginable.

Nevertheless, leading personalities and recognized experts have attempted to gaze into the future to look at the character of the emerging technologies. What they herald are enterprises that dramatically impact mankind's physical environment, commerce, and, indeed, the performance of the human species itself. Intellectual leaders have divined some of the very likely near-term outcomes that will help determine the technologies that flourish beyond the 10-15 year timeframe. Examples of products of such technologies have ranged over the full panoply of futuristic outcomes, from unbelievably fast nanoprocessors to the creation of nanobots. Even more resolution to what we can anticipate is being provided in various forums associated with the present workshop focusing on NBIC technologies.

However, the emerging NBIC technologies — figuratively speaking, our starships into our future — will only take us as far as the skills of those who captain and chart the various courses. But acquisition of skills depends on many things, including most assuredly the existence of a positive social environment that allows creative juices to flow. As a result, educational issues, both *pedagogy* and *people*, surface as ingredients fundamental to the realization of successful technologies.

Pedagogy

It seems clear that progress in the NBIC arena will necessitate contributions from several fields whose practitioners have tended to address problems in a sequential manner. The operative approach has been as follows: first something useful is found; then, if providence allows it, someone else gets involved with new insights or new capabilities; ultimately, commercial products are realized. In this era of convergent technologies, such a recipe can no longer be accepted, and practitioners must be taught in a new way.

The new pedagogy involves multidisciplinary training at the intersection of traditional fields, and it involves scientists, engineers, and social scientists. Although we still will need the ivory tower thinker, we will especially need to engage the intellects of students and established researchers in multidisciplinary, multi-investigator pursuits that lead to different ways of looking at research findings as well as to use different research tools. In acknowledgement of the necessity for multidiscipline skills and the participation in cross-discipline collaborations, nearly all of the funding agencies and private foundations provide substantial funding for research as well as for education of students in projects that are multidisciplinary and cross-disciplinary in character. A case in point is the Integrative Graduate Education and Research Training (IGERT) project (established by NSF in 1999), housed at The City University of New York, which involves three colleges from CUNY (the City College, Hunter College, and the College of Staten Island); Columbia University; and the University of Rochester.

IGERT participants are dedicated to the creation of research initiatives that span disciplinary and institutional boundaries and to the objective that such initiatives be reflected in the education and training of all its students. The overall goal is to educate and train the next generation of scientists in an interdisciplinary environment whereby a graduate student may participate in all the phases of a research project: synthesis, materials fabrication, and characterization. Our students, though trained as described, will be rigorously educated in a field of chemistry, engineering, or materials science. It is expected that such students will develop imaginative problem-solving skills and acquire a broad range of expertise and fresh, interdisciplinary outlooks to use in their subsequent positions. Our students will not just be sources of samples or instrument technicians but full partners with multidisciplinary training.

Without dealing with the specific science focus, the value-added elements of the CUNY-IGERT are described below:

- Multidisciplinary training (with choice of home institution after initial matriculation period at CUNY)
- IGERT-focused seminar program (via video-teleconferencing)
- Reciprocal attendance at annual symposia
- Expanded training opportunities (rotations and extended visits to appropriate collaborating laboratories)
- Formalized special courses (utilizing distance learning technology)
- Credit-bearing enrichment activities and courses
- Collaborative involvement with industry and national laboratories
- International partnerships that provide a global perspective in the research and educational exposures of students

Such a model for coupling research and education will produce individuals capable of creatively participating in the NBIC arena.

The People

The second key educational issue concerns the people who make the science and engineering advances that will form the bedrock of new technologies. If these individuals are not equitably drawn from the populace at large, then one can predict with certitude that social equity and displacement issues will gain momentum with every advance and can, in fact, dissipate or forestall the anticipated benefits of any endeavor.

It is thus clearly in America's best interest to ensure equitable participation of all elements in the front-line decision-making circles, in particular, to include groups that are historically underrepresented in leading-edge science and engineering during this era of anticipated, unbridled growth of NBIC technologies. The rich opportunities to make contributions will help members of underrepresented groups, especially, to reassert and revalidate their forgotten and sometimes ignored historical science and technological prowess. Success here would go a long way to avoid an enormous challenge to a bright future. What we stand to gain is the inclusion of the psychology and intellectual talents of an important segment of our society in solutions of ongoing and future world-shaping events. Two important activities immediately come to mind that make the point. One represents an opportunity lost; the second, a challenge we dare not ignore.

The first was NASA's space-venturing time capsule to other worlds several decades ago. Among many good things associated with this undertaking was one I consider unfortunate, a single-race representation of the inhabitants of the Earth. Clearly, a different psychological view, one more inclusive, should have prevailed and probably would have if minorities had had a say.

The second is the mapping of the human genome. The resultant data bank, I should think, will reflect the proclivities and prejudices of its creators, and its exploitation in the battle against genetic diseases. Clearly we should all have a hand in what it looks like and how it is to be used.

Summary

Only by utilizing new educational approaches for providing NBIC practitioners with the skills and insights requisite for success and also by making sure that historically underrepresented citizens are not left behind can the full promise of this era of convergence be realized.

References

Karim, M.A. 2001. Engineering: Diversity of disciplines and of students. *The Interface* (Newsletter of the IEEE Education Society and ASEE ECE Division) November:12-15.
Roy, R. 1977. Interdisciplinary science on campus — the elusive dream. *Chemical and Engineering News* August 29:29.

CONVERGING TECHNOLOGY AND EDUCATION FOR IMPROVING HUMAN PERFORMANCE

Avis H. Cohen, University of Maryland

This statement will address two general issues. One relates to potential uses for nanotechnology in neuroscience and biomedical engineering. The other addresses suggested issues in the education of potential scientists who will be most effective in the development of the new technologies.

Potential Uses for Nanotechnology in Neuroscience Research and Biomedical Engineering

The following areas have the highest potential for application:

a) Basic Neuroscience

- Exploration of single neurons (see Zygmond et al. 1999, a graduate-level reference for the concepts presented below):
    - Develop nanoscale delivery systems for compounds relevant to the nervous system such as neurotransmitters or receptor blockers, etc.

These would be used for distributed application to single cells in culture and *in situ*.

— Develop nanoscale sensors, conductive fibers for stimulating and recording the electrical activity from the surface of single neurons.

— Combine delivery and sensing nanofibers with exploration of single neurons in culture, both soma and dendrites, both spread over surface of neuron

b) Observation and Study of Growing Cells

• Use sensors and delivery systems to study neuronal development or regenerating fibers *in situ*. This requires that nanosensors and nano-optical devices be placed in a developing or injured nervous system, either alone or in combination with MEMS or aVLSI devices

c) Development

• Monitor growth cones with nano-optical devices

• Provide growth factors with nanoscale delivery systems

d) Regeneration

• Study processes as neurons are attempting or failing to regenerate. How do neurons behave as they try to grow? What happens as they encounter obstacles or receptors?

e) Applications in Biomedical Engineering

The following applications assume that nanofibers can be grown or extruded from the tips of microwires *in situ:*

• Monitor spinal cord injury or brain injury

— use nanofibers to assess the local levels of calcium in injury sites

— use nano delivery systems to provide local steroids to prevent further damage

• Neuroprosthetic devices

— Use nanofibers in conjunction with MEMS or aVLSI devices as delivery systems and stimulating devices for neuroprosthetic devices — make them more efficient.

— Use CPG prosthetic device in conjunction with microwires to stimulate locomotion

— Develop artificial cochlea with more outputs

— Develop artificial retina with more complex sensors – in combination with aVLSI retinas

Figure F.2 illustrates the positioning of a cochlear implant in the human cochlea (Zygmond et al. 1999). These devices are in current use. The electrode array is inserted through the round window of the cochlea into the fluid-filled space called *scala tympani*. It likely stimulates the peripheral axons of the primary auditory neurons, which carry messages via the auditory nerve into the brain. It is presently known that the information encoded by the sparsely distributed electrodes is

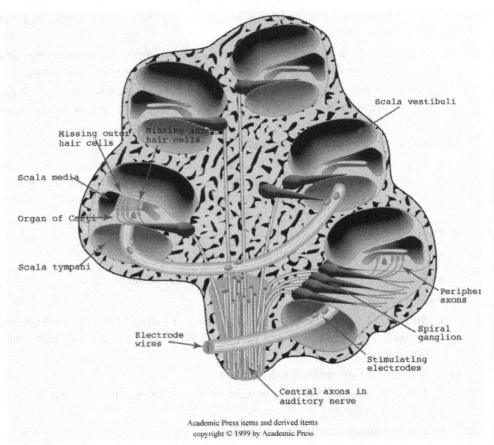

Scala vestibuli

Missing outer
hair cells

Missing inner
hair cells

Scala media

Organ of Corti

Scala tympani

Peripher
axons

Spiral
ganglion

Electrode
wires

Stimulating
electrodes

Central axons in
auditory nerve

Figure F.2. The positioning of a cochlear implant in the human cochlea.

nowhere near that carried by the human cochlea. The device, therefore, is of limited value for hearing-impaired individuals with long-term auditory nerve damage that predates their normal speech learning (Moller 2001). If nanofibers could be deployed from each electrode to better distribute the information, it would likely improve the quality of the device considerably. This would be a relative easy use of the new technology, with easy testing to affirm its usefulness.

Training the Future Developers of Nanotechnology

In the new era of converging technologies, one can become either a generalist and be superficially capable in many fields, or one can become a specialist and master a single field. If one chooses the former route, one is unlikely to produce deep, insightful work. If one chooses the latter route, then it is only possible to take full advantage of the convergence of the technologies by working in collaboration with others who are expert in the other relevant fields. Unfortunately, our present educational system does not foster the type of individual who works well in collaborations.

To achieve the training of good scientists who have the capacity to work well in multidisciplinary groups, there are several new kinds of traits necessary. The first and perhaps most difficult is to learn to communicate across the disciplines. We learn the technical language of our respective disciplines and use it to convey our thoughts as clearly and precisely as possible. However, researchers in other disciplines are unfamiliar with the most technical language we prefer to use. When talking across the bridges we seek to build, we must learn to translate accurately but clearly to intelligent listeners who will not know our respective languages. We must begin to train our students to learn the skill of communicating across the disciplinary divides. We must develop programs in which students are systematically called upon to explain their work or the work of others to their peers in other areas. Thus, the best programs will be those that throw the students from diverse disciplines together. Narrowly focused programs may turn out neuroscientists superbly trained for some functions, but they will not be good at collaborative efforts with scientists in other fields without considerable additional work. They will not easily produce the next generation of researcher who successfully forms collaborative efforts to use the new converging technologies.

We should also begin to systematically pose challenges to our students such that they must work in teams of mixed skills, teams of engineers, mathematicians, biologists, chemists, and cognitive scientists. This will provide the flavor of the span that will be required. We cannot train our students to be expert in this broad a range of fields; therefore, we must train and encourage them to communicate across the range and to seek out and work with experts who offer the expertise that will allow the best science to be done. Funding agencies must continue to enlarge the mechanisms that support this type of work if they want to have a unique position in fostering the development and optimal utilization of the new technologies as applied to neuroscience, among other fields.

My experience with the Telluride Workshop on Neuromorphic Engineering has given me some important insights into the optimal methods for educating for the future. It has shown me that it will be easier to train engineers to understand biology, than to train biologists to comprehend engineering. There are some notable exceptions, fortunately, like Miguel Nicolelis and Rodolfo Llinás. Among biologists, there is beginning to be curiosity and enthusiasm for engineering, robotics, and the new emerging technologies. This must be fostered through showcasing technological accomplishments such as successful robotic efforts and the analog VLSI retinas and cochleas developed using neuromorphic engineering. We must also try harder to get biologists to attend the Telluride Workshop and to stay long enough to gain some insights into the power of the approach. The field of nanobiotechnology is growing much faster among engineers than among biologists. We must work harder to improve our outreach to biologists.

The formation of workshops such as Telluride is a good way to begin to put together the necessary groups for the exploitation of the new methods being developed in nanotechnology. It is likely that the full potential for nanodevices will only be reached by uniting engineers with biologists. Biologists presently have little exposure to information about nanotechnology. Comparatively, the engineers know relatively little about the real neuronal substrate with which they seek to interface. It will not be a trivial task to actually understand what will emerge when nanotubes are directly contacting neurons, stimulating them, and recording from them. It will require considerable expertise and imagination. Exposing biologists to the potential power and usefulness of the technology, and exposing engineers to the complexity of the biological substrate, can only come about through intense interactions; it cannot come about through groups operating alone. The journal *Science* has done a great deal to bring nanotechnology to the attention of the general scientist. However, no true understanding can come without hard work.

Development of novel bioengineering programs will be another approach to development of nanotechnology. Training biologists and engineers in the same educational program will go a long way to overcoming some of the present ignorance. Nanotechnology is difficult. The underlying chemistry and physics will not come easily to everyone. It is most likely that the best method of developing it is through explicit programmatic efforts to build collaborative teams of engineers and biologists. Summer workshops can provide incentives by exposing individuals to the potentials of the union, but only through full-fledged educational programs can the efforts move forward effectively.

References
Moller, A.R., 2001. Neurophysiologic basis for cochlear and auditory brainstem implants. *Am. J. Audiol* 10(2):68-77.
Zygmond, M.J., F.E. Bloom, S.C. Landis, J.L. Roberts, and L.R. Squire, eds. 1999. *Fundamental Neuroscience*, New York: Academic Press.

VISIONARY PROJECTS

CONVERGING TECHNOLOGIES: A K-12 EDUCATION VISION

James G. Batterson and Alan T. Pope, NASA Langley Research Center

Over the next 15 years, converging technologies (CT), the synergistic interplay of nano-, bio-, information, and cognitive technologies (NBIC) will enable significant improvements in how, where, and what is taught in grades K-12 and will also support the lifelong learning required by a rapidly developing technological economy. Through national and state standards, half the schools in the United States will be teaching science based on the unifying principles of science and technology (NRC 1995) rather than the isolated subjects taught since before the industrial revolution. New tools for learning such as neuroscience sensors, increased quality of Internet service via guaranteed bandwidth, and a new understanding of biological feedback for self-improvement will provide new, highly efficient learning methods for all, in particular guaranteeing that all children can read by age five. Students will no longer be dependent on rigid regimentation of the classroom or schoolhouse and class schedules, as they will have courses and supplemental information available to them from numerous venues around the clock. Consider the following scenario.

The year is 2015. You enter a public school. From the outside, it appears to be much the same physical structure as schools were for 50 years. But inside is a totally different world. Teachers are busily meeting with one another and engaged in e-learning to stay current on the latest developments in education and their disciplines. They are contributing their experiences to a databank that parses the data into information and places it on an information website for other teachers and researchers to use. Science teachers are working in a cross-disciplinary program that has been particularly fruitful — NBIC — a wonderful stew of nanotechnology, biotechnology, information technology, and cognitive technologies. NBIC has allowed these teachers to productively access and continually learn new information through advances in small biological and neurological sensors and the biofeedback they produce. A number of special needs students are working in rooms, receiving cues from a wireless network that are appropriate for their individual cognitive and physical needs as developed through NBIC. Advances in NBIC research allow for better meeting the requirements of more and more special needs students each year with fewer human resources. Each student in the community can interact with other students worldwide to share information, language, and culture. While the student population of more than 50 million students has been joined by millions of parents as lifelong learning requirements are realized, no new buildings have been required, as many students take advantage of 24/7 availability of coursework at their homes, in work areas, and at the school. The capital investment savings have been redirected into increased pay to attract and retain the highest quality teachers and curriculum developers. The line between education and recreation has blurred as all citizens visit the school building throughout the day to better their lives.

The Critical Roles of Converging Technologies

Converging technologies hold true promise to revolutionize the teaching in grades K-12 and beyond. The interplay of these technologies, each with the other, provides the opportunity for extraordinary advances in K-12 education on three fronts: content, process, and tools

Content

The recent extraordinary and rapid results of the Human Genome Project (HGP) provide for a revolution in the content of biology curriculum for K-12. The rapid completion of this project was due in a large part to the availability of IT-supported and -inspired experimental, analytical, and observational capability. While known as a "biology" project, the revolutionary advances are truly due to cross-disciplinary fertilization. CT offers K-12 education a focus that builds on the HGP accomplishments and provides content that folds in nanotechnology to understand the interactions of and to physically manipulate particles and entities at the fundamental sizes of the building blocks of life. New course content must be created that is sensitive to these developments and can be updated on an annual basis to be relevant to students' needs and the rapidly growing state of knowledge in the research fields. New courses that delve into the aspects of intelligent, sentient life and cognitive processes must also be developed. These courses must be created in the context of state-of-the-art and state-of-the-practice biotechnology, information technology, and nanotechnology. The state of Texas has already altered its formerly strictly discipline-structured curriculum with the insertion of an Integrated Physics and Chemistry Course. The content advances called for in this essay are in the same vein as the Texas advance but a quantum jump into the future – a jump necessary to serve students of the United States in a globally competitive economy (NAP 1995).

Process

A fundamental understanding of the physical or biological basis for cognition developed in CT will allow for a revolution in the individualization of the K-12 educational process. Psychologists currently study people's responses to stimuli and their ability to control their responses given certain physical data from their bodies (popularly known as biofeedback). However, to map the various learning modalities of children, physical and biological characteristics must be associated with a child's cognitive behaviors in such a way that genotypic or phenotypic mitigations can be identified and applied. The analysis of such data will require nano-, cogno-, bio-, and information technologies that are years beyond today's capabilities, as will the presentation of educational media once the appropriate intervention or course of treatment is identified.

Technologies for measuring brain activity and assessing cognitive function, representing advances in usability and sensitivity over the current electro-, magneto-, and hemo-encephalographic technologies, will be developed that have the ability to go beyond diagnosing disorders to assessing students' learning strengths and weaknesses. This enhanced sensitivity will be enabled by advanced biotechnologies that are tuned to monitor cognitive function and will support the selection of appropriate remediation. Neurologically-based technologies will be available to assist in the remediation of learning impairments as well as to enhance the cognitive abilities of children. These technologies will extend a student's ability to concentrate and focus, to remember and retain, and to deal with stress.

Attention and memory enhancement technologies will be built upon computer-based cognitive rehabilitation technologies that are already available, as indicated in an NIH Consensus Statement (1998): "Cognitive exercises, including computer-assisted strategies, have been used to improve specific neuropsychological processes, predominantly attention, memory, and executive skills. Both randomized controlled studies and case reports have documented the success of these interventions using intermediate outcome measures. Certain studies using global outcome measures also support the use of computer-assisted exercises in cognitive rehabilitation."

Other education-related technologies include improvement of a student's attention and stress management abilities using brainwave and autonomic nervous system (ANS) biofeedback technologies. The Association for Applied Psychophysiology and Biofeedback (AAPB) has initiated a program "to assist educational and health professionals to teach children and youth to regulate their own bodies, emotions, relationships, and lives" (AAPB 2001).

Foreshadowing and early beginnings of this trend can already be seen, and it will gather momentum rapidly in the next few years. Computer software that simultaneously trains cognitive abilities directly relevant to academic performance and delivers brainwave biofeedback is used in school settings and is commercially available (Freer 2001). Biofeedback enrichment of popular video games (Palsson et al. 2001) has already been demonstrated to work as well as traditional clinical neurofeedback for attention deficit disorder. This same technology is also designed to deliver autonomic self-regulation training for stress management. Instrument functionality feedback, developed at NASA Langley Research Center, is a novel training concept for reducing pilot error during demanding or unexpected events in the cockpit by teaching pilots self-regulation of excessive autonomic nervous system reactivity during simulated flight tasks (Palsson and Pope 1999). This training method can also teach stressed youngsters to practice autonomic physiological self-regulation while playing video games without the need for conscious attention to such practice.

Embedding physiological feedback training into people's primary daily activities, whether work or play, is a largely untapped and rich opportunity to foster health and growth. It may soon be regarded to be as natural and expected as is the addition of vitamins to popular breakfast cereals. Toymakers of the future might get unfavorable reviews if they offer computer games that only provide "empty entertainment."

Twenty years from now, physiological feedback will be embedded in most common work tasks of adults and will be integral to the school learning and play of children. Interactions with computers or computer-controlled objects will be the predominant daily activity of both adults and children, and physiological feedback will be embedded in these activities to optimize functioning and to maintain well-being and health.

Tools

CT brings distance learning of today to a true 24/7 educational resource. Telepresence and intelligent agents will allow students to investigate fundamental biological questions through online laboratories and high-fidelity simulations. The simulations will be extensions of today's state-of-the-art distance surgery and robotic surgery. Actual data and its expected variations in physical attributes such as color, density, location, and tactile tension will be available in real time. Students in cities, suburbs, and remote rural areas will all have access to the same state-of-the-art content and delivery. These tools will first be available at central locations such as schools or libraries. As hardware cost and guaranteed available bandwidth allows, each home will become a school unto itself — providing lifelong learning for children and adult family members.

Delivery of learning experiences will be designed to enhance student attention and mental engagement. This goal will be supported in the classroom and at home by digital game-based learning (DGBL) experiences that provide (1) meaningful game context, (2) effective interactive learning processes including feedback from failure, and (3) the seamless integration of context and learning (Prensky 2001). Entertaining interactive lessons are available (Lightspan Adventures™) that run on a PC or a PlayStation® game console so that they can be used both in school and after school and in students' homes.

Patented technologies are also available that "use the latest brain research to develop a wide range of early learning, language, and reading skills: from letter identification and rhyming to vocabulary and story analysis" for "children who struggle with basic language skills or attention problems" (Scientific Learning 2001).

Another set of educational tools enabled by CT, physiological monitoring, will be used to guide complex cognitive tasks. The recent proposal for NASA's Intelligent Synthesis Environment (ISE) project included an animation of a computer-aided design system responding to a user's satisfaction about a design iteration, measured via remote sensing of brainwaves. Similarly, a student's engagement in and grasp of educational material will be monitored by brain activity measurement technology, and the presentation can be adjusted to provide challenge without frustration.

Virtual reality technologies, another tool set, will provide the opportunity for immersive, experiential learning in subjects such as history and geography. Coupled with interactive simulations, VR environments will expand the opportunities for experiences such as tending of ecosystems and exploring careers. A NASA invention called "VISCEREAL" uses skin-surface pulse and temperature measurements to create a computer-generated VR image of what is actually happening to blood vessels under the skin (Severance and Pope 1999). Just as pilots use artificial vision to "see" into bad weather, students can use virtual reality to see beneath their skin. Health education experiences will incorporate realtime physiological monitoring integrated with VR to enable students to observe the functioning of their own bodies.

Transforming Strategy

The major technical barrier for instituting CT into the K-12 curriculum is the political complexity of the curriculum development process. Curriculum is the result of the influence of a number communities, both internal and external to the school district, as shown in Figure F.3.

The CT Initiative must identify and work with all the appropriate K-12 communities to successfully create and integrate new curriculum — perhaps addressing a K-16 continuum. While teacher institutes occasionally can be useful, participatory partnering in real curriculum development promises to leave a lasting mark on more students and faculty. It is key to successful curriculum development to put together a coalition of teachers, administrators, students, parents, local citizens, universities, and industry for curriculum development. The virtual lack of any interdepartmental or cross-discipline courses in K-12 curricula is indicative of

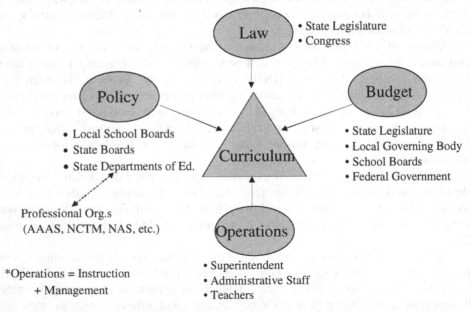

Figure F.3. The curriculum communities.

the gap that must be bridged to teach CT.

From the CT Initiative, courses can be created, but for *curriculum development*, the courses must be institutionalized or put into the context of the other courses in the school district. This institutionalization requires the involvement and support of the entire range of communities shown in Figure F.3.

There are approximately 50 million K-12 students in 15,000 school districts in the United States, its territories, and the District of Columbia. Reaching these districts or students individually would be virtually impossible. Rather, a major strategy should be to take advantage of the leverage available through impacting the national science education standards and emerging state standards (Figure F.4). At the national level, development and inclusion of CT curriculum involves development of national CT standards as a part of the national science education standards developed by the National Resource Council (NRC 1995). CT scientists should work for a regular review of the current standards and be prepared to provide CT standards as members of the review and standards committees.

Because there is no national U.S. curriculum, having national CT standards serves only an advisory function. For these standards to be used in curriculum development, they need to be accepted by state boards of education in development of their separate state standards (Figure F.3 and Figure F.4). Each state must then have courses available that meet the standards it adopts. Many states have developed statewide assessments or tests for various subjects. A major step toward implementation of CT curricula would be positioning CT questions on statewide science assessment tests.

Complementary to the development of a K-12 curriculum *per se* is the development of a CT mentality in the general population and in the next generation of teachers and parents. Thus, development of CT courses at colleges in general, and in their teacher preparation departments in particular, is desirable.

Thus the transforming strategy for educational content has the following components:

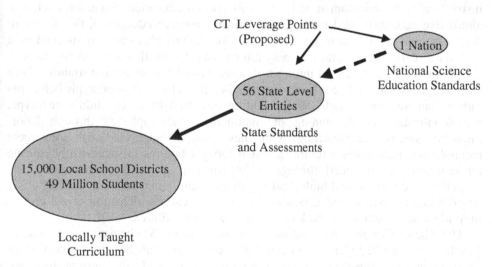

Figure F.4. Relationships between national and state standards and local school districts.

- Influence over the National Science Education Standards (NRC)
- Development of CT science content standards
- Development of CT courses for K-12 to support the CT standards
- Influence on each state's science standards and assessment instruments
- Development of CT courses for schools of education and in the general education of the next generation of university students
- Development, in cooperation with a writer of children's books, of "early reader" (ages 1-5) books containing CT concepts

Ethics

Ethical issues regarding the ability to analyze each child's capacity to learn and develop will arise. Categorization of humans relating to their abilities, and perhaps to their inferred potential in any area, may challenge many of our Western traditions and ethical values.

Implications

The implications of CT content, process, and tools for education of all children are dramatic. A specific focus would be the population of students today classified as "special education" students under IDEA (the Individuals with Disabilities Education Act – PL94-142). This includes approximately 10 percent of the entire age 3-17 cohort in the United States, or almost five and a half million children in the 6-21 year age bracket. More than one million of these children are diagnosed with speech or language impairment; 2.8 million with specific learning disabilities such as dyslexia; 600,000 with mental retardation; 50,000 with autism; and 450,000 with emotional disturbance.

In K-12 education, school district visions commonly aspire to educate all children to their full potential. The reality has been that many children are not educated to a level that allows them to be productive members of adult society, let alone to reach their own full potential. While there is some differentiation of instruction and curriculum strands (such as special education, governor's schools, alternative education, and reading and hearing resource education), the ability to diagnose individual student needs is based on failure of a child to succeed in a "standard" early curriculum. It is only after such a failure that analysis begins with the possibility of a placement into one of several available alternative strands. These strands again treat a bulk condition identified empirically from phenotypic behaviors rather than treating an individual condition analyzed from the child's genotype. Individualization or fine-tuning of treatment is accomplished through labor-intensive one-on-one teaching. Our new vision, supported by convergent technologies, anticipates a future in which today's failures to successfully educate all children are mitigated through a fundamental physical understanding and modeling of cognitive and biological capabilities and processes in the young child. Appropriate mitigation and direction are based on early anticipation of the child's individual needs rather than bulk treatment after early failures.

The Glenn Commission (National Commission on Mathematics and Science Teaching for the 21st Century, Glenn 2000) estimated that the cost of meeting its three goals of improving science teaching quality with the current teachers, developing more science and math teachers, and improving the science and math teaching environment would cost approximately $5 billion in the first year. Roughly, this money would be used to provide teacher summer institutes, leadership training, incentives, scholarships, assessments, and coordination. Since this is aimed at all science and math teachers over a five-year program (there are 1.5 million science and math teachers for grades K-12 in the United States), CT could take early advantage of any implementation of a plan such as that proposed by the Glenn Commission.

Revisions in curriculum standards seem to take about five to ten years to develop, absent a major sea change in what is being taught. CT is a major change, and it further moves curriculum to stay current with scientific and technological advances. This will require regularly occurring curriculum reviews at the state level and the ability to adjust content and assessment with a factor of ten more efficiency than is done today. As a guide to the states, a national curriculum must also be reviewed and updated in a similarly regular way.

References

American Association for the Advancement of Science (AAAS). 1993. *Benchmarks for science literacy*. New York: Oxford University Press.

Association for Applied Psychophysiology and Biofeedback (AAPB). 2001. http://www.aapb.org/.

Freer, P. 2001. Scientific research: Case study #1. Retrieved October 5, 2001, from http://www.playattention.com/studies.htm.

Glenn, J. et al. 2000. *Before it's too late: A report to the nation from the National Commission on Mathematics and Science Teaching for the 21st Century*. (Colloquially known as the "Glenn Commission") (September). http://www.ed.gov/americacounts/glenn/.

NAP. 1995. *National science education standards: An overview*. Washington, D.C.: National Academy Press.

NCTM (National Council of Teachers of Mathematics). 1995. *Assessment standards for school mathematics*. Reston, VA.: National Council of Teachers of Mathematics.

_____. 1991. *Professional standards for teaching mathematics*. Reston, VA.: National Council of Teachers of Mathematics.

_____. 1989. *Curriculum and evaluation standards for school mathematics*. Reston, VA.: National Council of Teachers of Mathematics.

NIH. 1998. *Consensus statement: Rehabilitation of persons with traumatic brain injury*. 16(1) (October 26-28). Washington, D.C.: National Institutes of Health, p.17.

National Research Council (NRC). 1995. *National science education standards*. Washington, D.C.: National Academy Press. www.nap.edu/readingroom/books/nses/html.

_____. 2000. *Educating teachers of science, mathematics, and technology: New practices for a new millennium*. Washington, D.C.: National Academy Press. http://books.nap.edu/html/educating_teachers/.

Palsson, O.S., A.T. Pope, J.D. Ball, M.J. Turner, S. Nevin, and R. DeBeus. 2001. Neurofeedback videogame ADHD technology: Results of the first concept study. Abstract, Proceedings of the 2001 Association for Applied Psychophysiology and Biofeedback Meeting, March 31, 2001, Raleigh-Durham, NC.

Palsson, O.S., and A.T. Pope. 1999. Stress counterresponse training of pilots via instrument functionality feedback. Abstract, Proceedings of the 1999 Association for Applied Psychophysiology and Biofeedback Meeting. April 10, 1999, Vancouver, Canada.

Pope, A.T., and O.S. Palsson. 2001. *Helping video games "rewire our minds."* Retrieved November 10, 2001, from http://culturalpolicy.uchicago.edu/conf2001/agenda2.html.

Prensky, M. 2001. *Digital game-based learning*. New York: McGraw-Hill.

Prinzel, L.J., and F.G. Freeman. 1999. Physiological self-regulation of hazardous states of awareness during adaptive task allocation. In Proceedings of the Human Factors and Ergonomics Society, 43rd Annual Meeting.

Scientific Learning, Inc. 2001. http://www.scientificlearning.com.

Severance, K., and A.T. Pope. 1999. VISCEREAL: A Virtual Reality Bloodflow Biofeedback System. Abstract, Proceedings of the 1999 Association for Applied Psychophysiology and Biofeedback (AAPB) Meeting. April 10, 1999, Vancouver, Canada.

EXPANDING THE TRADING ZONES FOR CONVERGENT TECHNOLOGIES

Michael E. Gorman, University of Virginia

Stimulating convergence among nano, bio, info, and cognitive science obviously will require that different disciplines, organizations, and even cultures work together. To make certain this convergence is actually beneficial to society, still other stakeholders will have to be involved, including ethicists, social scientists, and groups affected by potential technologies. To promote this kind of interaction, we first need a vision — supplied, in this case, by a metaphor.

Vision: Developing "Trading Zones," a Metaphor for Working Together

A useful metaphor from the literature on science and technology studies is the trading zone. Peter Galison used it to describe how different communities in physics and engineering worked together to build complex particle detectors (Galison 1997). They had to develop a creole, or reduced common language, that allowed them to reach consensus on design changes:

> Two groups can agree on rules of exchange even if they ascribe
> utterly different significance to the objects being exchanged; they may
> even disagree on the meaning of the exchange process itself.
> Nonetheless, the trading partners can hammer out a *local*
> coordination, despite vast *global* differences. In an even more
> sophisticated way, cultures in interaction frequently establish contact
> languages, systems of discourse that can vary from the most function-
> specific jargons, through semispecific pidgins, to full-fledged creoles
> rich enough to support activities as complex as poetry and
> metalinguistic reflection (Galison 1997, 783).

My colleague Matt Mehalik and I have classified trading zones into three broad categories, on a continuum:

1. *A hierarchical trading zone governed by top-down mandates.* An extreme example is Stalinist agricultural and manufacturing schemes used in the Soviet Union (Graham 1993; Scott 1998) where the government told farmers and engineers exactly what to do. These schemes were both unethical and inefficient, stifling any kind of creativity. There are, of course, top-down mandates where the consequences for disobedience are less severe, but I would argue that as we look to the future of NBIC, we do not want research direction set by any agency or group, nor do we want a hierarchy of disciplines in which one dominates the others.

2. *An equitable trading zone state in which no one group is dominant,* and each has its own distinct perspective on a common problem. This kind of trading zone was represented by the NBIC conference where different people with expertise and backgrounds exchanged ideas and participated jointly in drafting plans for the future.

3. *A shared mental model trading zone based on mutual understanding of what must be accomplished.* Horizontal or lattice styles of business management are designed to promote this kind of state. An example is the group that created the Arpanet (Hughes 1998).

Another example is the multidisciplinary global group that invented a new kind of environmentally intelligent textile. Susan Lyons, a fashion designer in New York, wanted to make an environmental statement with a new line of furniture fabric. Albin Kaelin's textile mill in Switzerland was in an "innovate or die" situation. They started a trading zone around this environmental idea and invited the architect William McDonough, who supplied a mental model based on an analogy to nature, "waste equals food," meaning that the fabric had to fit smoothly back into the natural cycle in the same way as organic waste products. The architect brought in Michael Braungart, a chemical engineer who created and monitored detailed design protocols for producing the fabric. The actual manufacturing process involved bringing still others into the trading zone (Mehalik 2000).

Note that the shared mental model did not mean that the architect understood chemical engineering, or vice-versa. All members arrived at a common, high-level understanding of waste equals food and translated that into their own disciplinary practices, while staying in constant touch with each other. The creoles that arise among Galison's communities are typically devoted to local coordination of practices. In this fabric case, we see a Creole-like phrase, "waste equals food," evolve into a shared understanding that kept different expertises converging on a new technology.

Role of Converging Technologies

Converging technologies designed to benefit society will involve trading zones with a shared mental model at the point of convergence. "Waste equals food" created a clear image of an environmental goal for the fabric network. Similar shared mental models will have to evolve among the NBIC areas.

The process of technological convergence will not only benefit from trading zones, it can play a major role in facilitating them. Consider how much easier it is to maintain a transglobal trading zone with the Internet, cell phones, and air transport. Imagine a future in which convergent technologies make it possible for people to co-locate in virtual space for knowledge exchange, with the full range of nonverbal cues and sensations available. Prototypes of new technological systems could be created rapidly in this virtual space and tested by representatives of stakeholders, who could actually make changes on the fly, creating new possibilities. The danger, of course, is that these virtual prototypes would simply become an advanced form of vaporware, creating an inequitable trading zone where technology is pushed on users who never have full information. But in that case, new trading zones for information would emerge, as they have now — witness the success of *Consumer*

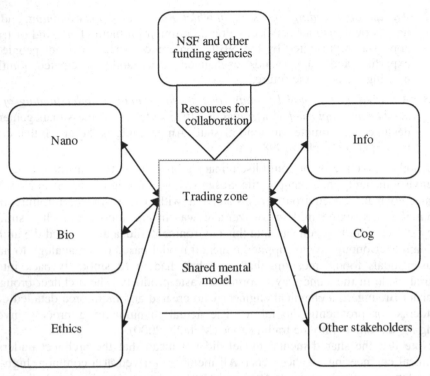

Figure F.5.   Technologies converging on a trading zone seeded by resources that encourage collaboration.

Reports. It is essential that powerful new technologies for disseminating and creating knowledge be widely accessible, not limited to an elite.

Transforming Strategies

Effective trading zones around convergent technologies cannot be created simply by bringing various groups together, although that is a first step. Here, federal agencies and foundations can form a trading zone around resources (see Figure F.5) — like the role of the National Science Foundation in the National Nanotechnology Initiative. This kind of program must not micromanage the sort of research that must be done; instead, it has to provide incentives for real engagement among different cultures of expertise.

Technologies designed to improve human health, increase cognitive performance, and improve security will have to fit into global social systems. We need to create active technological and scientific trading zones built around social problems. These trading zones will require experts with depth in relevant domains. The trading zones will need to provide incentives for them to come together, including opportunities to obtain funding and to work on "sweet" technological problems (Pacey 1989). In addition, each zone will require a core group of practitioners from different disciplines to share a mental model of what ought to be accomplished.

Here, it is worth recalling that mental models are flexible and adaptable (Gorman 1992; Gorman 1998). One good heuristic for creating a flexible shared mental model came up repeatedly during the conference: "follow the analogy of nature." Alexander Graham Bell employed this heuristic in inventing the telephone (Gorman 1997). Similarly, McDonough's "waste equals food" mental model is based on the analogy to living systems, in which all organic waste is used as food by forms of life.

Similarly, as we look at beneficial ways in which human performance can be enhanced, it makes sense to study the processes and results of millions of years of evolution, which have affected not only biological systems, but also the climate cycles of the entire planet (Allenby 2001). The pace of technological evolution is now so fast that it exceeds the human capacity to reason about the consequences. Hence, we have to anticipate the consequences — to attempt to guide new discoveries and inventions in a beneficial direction. Nature's great inventions and failures can be a powerful source of lessons and goals. As Alan Kay said, "The best way to predict the future is to create it."

We see NASA adopting this analogy to nature when it proposes aircraft that function like high-technology birds, with shifting wing-shapes. The human ear served as Alexander Graham Bell's mental model for a telephone; in the same way, a bird might serve as a mental model for this new kind of aircraft. Creating this kind of air transport system will require an active trading zone among all of the NBIC areas, built around a shared mental model of what needs to be accomplished.

Good intellectual trading zones depend on mutual respect. Hard scientists and engineers will have to learn to respect the expertise of ethicists and social scientists, and vice-versa. The ethicist, for example, cannot dictate moral behavior to the scientists and engineers. Instead, s/he has to be ready to trade expertise, learning about the science and engineering while those practitioners get a better understanding of ethical issues.

Consider, for example, a trading zone between the medical system and its users around bioinformatics. Patients will be willing to trade personal information in exchange for more reliable diagnoses. But the patients will also have to feel they are being treated with respect — like human beings, not data points — or else the trading zone will break down.

In terms of education, what this means is that we want to encourage students to go deeply into problems, not necessarily into disciplines. Elementary students do not see the world divided into academic categories; instead, they see interesting questions. As they pursue these questions, they should be encouraged to engage deeply. But the result will be new kinds of expertise, not necessarily easily labeled as "physics," "chemistry," or "biology." The best trading zones are built around exciting problems by practitioners eager to create the knowledge necessary for solutions. And every such trading zone ought to include practitioners concerned about the social dimensions of technology.

Communication is the key to a successful trading zone. Students need to be given opportunities to work together in multidisciplinary teams, sharing, arguing, and solving difficult, open-ended problems together. Teachers need to scaffold communication in such teams, helping students learn how to present, write, and argue constructively. We have a long tradition of doing this in our Division of Technology, Culture, and Communication in the Engineering School at the University of Virginia (http://www.tcc.virginia.edu).

Estimated Implications

The great thing about trading zones is that successful ones expand, creating more opportunities for all of us to learn from each other. Hopefully, the first NBIC meeting has provided a foundation for such a trading zone between nano, bio, info, and cogno practitioners — and those in other communities like ethics, politics, and social relations.

We must remember to accompany the creation of convergent trading zones with detailed studies of their development. One of the most valuable outcomes would be a better understanding of how to encourage the formation of convergent, multidisciplinary trading zones.

References

Allenby, B. 2001. Earth systems engineering and management. *IEEE Technology and Society Magazine* 194:10-21.

Galison, P. 1997. *Image and logic*. Chicago: The University of Chicago Press.

Gorman, M.E. 1992. *Simulating science: Heuristics, mental models and technoscientific thinking*. Bloomington, Indiana University Press.

_____. 1997. Mind in the world: Cognition and practice in the invention of the telephone. *Social Studies of Science* 27 (4): 583-624.

_____. 1998. *Transforming nature: Ethics, invention and design*. Boston: Kluwer Academic Publishers.

Graham, L.R. 1993. *The ghost of the executed engineer: Technology and the fall of the Soviet Union*. Cambridge: Harvard Univ. Press.

Hughes, T.P. 1998. *Rescuing prometheus*. New York: Pantheon books.

Mehalik, M.M. 2000. Sustainable network design: A commercial fabric case study. *Interfaces: Special edition on ecologically sustainable practices* 333.

Pacey, A. 1989. *The Culture of Technology*. Cambridge, MA, MIT Press.

Scott, J.C. 1998. *Seeing like a state: How certain schemes to improve the human condition have failed*. New Haven, Yale University Press.

BIOLOGICAL LANGUAGE MODELING: CONVERGENCE OF COMPUTATIONAL LINGUISTICS AND BIOLOGICAL CHEMISTRY

Judith Klein-Seetharaman and Raj Reddy, Carnegie Mellon University

How can we improve the nation's productivity and quality of life in the next 10 to 20 years? The nation's performance is dependent on functions of the human body, since they directly or indirectly determine human ability to perform various tasks. There are two types of human ability: (1) "inherent abilities," tasks that humans are able to perform, and (2) "external abilities," tasks that we cannot perform *per se*, but for which we can design machines to perform them. Both categories have individually experienced groundbreaking advances during the last decade. Our inherent abilities in terms of fighting diseases, repair of malfunctioning organs through artificial implants, and increased longevity have greatly improved, thanks to advances in the medical and life sciences. Similarly, technology has provided us with remarkable tools such as smaller and more efficient computers; the Internet; and safer, cleaner, and cheaper means of transport.

Integration of Inherent and External Human Abilities

Advancing our society further necessitates a better integration between the inherent and external abilities. For example, interfacing computers with humans need not require keyboard and mouse: ongoing efforts advance utilization of speech interfaces. But ultimately, it would be desirable to directly interface with the human brain and other organs. This will require further advances in elucidating the fundamental biological mechanisms through which humans think, memorize, sense, communicate, and act. Understanding these mechanisms will allow us to (a) modify our inherent abilities where natural evolution does not feel any pressure for improvement and (b) design interfaces that connect our inherent abilities with external abilities.

Grand Challenge: Mapping Genome Sequence Instructions to Inherent Abilities

How can we aim to understand complex biological systems at a level of detail sufficient to improve upon them and build interfaces to external machines? In principle, all the information to build complex biological systems is stored in an "instruction manual," an organism's (e.g., a human's) genome. While we have recently witnessed the elucidation of the entire human genome sequence, the next logical grand challenge for the coming decade is to map the genome sequence information to biological functions. Interfacing between biological functions and artificially manufactured devices will require improved structure-property understanding as well as manufacturability at a multiscale level ranging from Ångstrom-sized individual components of biological molecules to macroscopic responses. This will be possible through existing and future advances in nanotechnology, biological sciences, information technology, and cognitive sciences (NBIC).

Outline

The sequence ➡ function mapping question is conceptually similar to the mapping of words to meaning in linguistics (Figure F.6). This suggests an

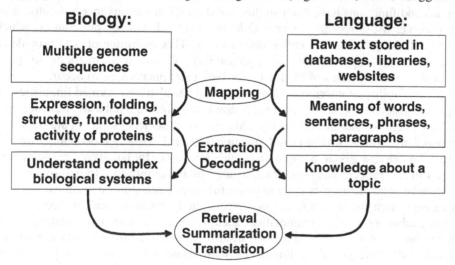

Figure F.6. Analogy between language and biology, which forms the basis for the convergence of computational linguistics and biological chemistry.

opportunity to converge two technologies to address this challenge: computational linguistics and biological chemistry, via "biological language modeling." The term "biological chemistry" is used here to stand for interdisciplinary studies of biological systems, including biochemistry, molecular biology, structural biology, biophysics, genetics, pharmacology, biomedicine, biotechnology, genomics, and proteomics. The specific convergence of linguistics and biological chemistry is described below under the heading, "The Role of Converging Technologies: Computational Linguistics and Biological Chemistry." Its relation to the more general convergence with NBIC is described in section, "The Role of Converging Technologies: NBIC and Biological Language Modeling." Two specific applications of linguistic analysis to biological sequences are given in "The Transforming Strategy," to demonstrate the transforming strategy by example. If we can solve the sequence   function mapping question, the implications for human performance and productivity are essentially unlimited. We have chosen a few practical examples to illustrate the scope of possibilities ("The Estimated Implications"). Implications for society are sketched in "Implications for Society," followed by a brief summary.

The Role of Converging Technologies: Computational Linguistics and Biological Chemistry

Complex biological systems are built from cells that have differentiated to perform specialized functions. This differentiation is achieved through a complicated network of interacting biological molecules. The main action is carried out by proteins, which are essentially nano-sized biological machines that are composed of strings of characteristic sequences of the 20 amino acid building blocks. The sequences of the strings are encoded in their entirety in the genome. The linear strings of amino acids contain in principle all the information needed to fold a protein into a 3-D shape capable of exerting its designated function. With the advent of whole-genome sequencing projects, we now have complete lists of all the protein sequences that define the complex function carried out by the sequenced organisms — hundreds to thousands in bacteria and tens of thousands in humans. Individual proteins and functions have been studied for decades at various levels — atomic to macroscopic. Most recently, a new field has evolved, that of proteomics, which looks at all the proteins in a cell simultaneously. This multitude of data provides a tremendous new opportunity: the applicability of statistical methods to yield practical answers in terms of likelihood for biological phenomena to occur.

The availability of enormous amounts of data has also transformed linguistics. In language, instead of genome sequences, raw text stored in databases, websites, and libraries maps to the meaning of words, phrases, sentences, and paragraphs as compared to protein structure and function Figure F.6). After decoding, we can extract knowledge about a topic from the raw text. In language, extraordinary success in this process has been demonstrated by the ability to retrieve, summarize, and translate text. Examples include powerful speech recognition systems, fast web document search engines, and computer-generated sentences that are preferred by human evaluators in their grammatical accuracy and elegance over sentences that humans build naturally. The transformation of linguistics through data availability has allowed convergence of linguistics with computer science and information technology. Thus, even though a deep fundamental understanding of language is still missing, e.g., a gene for speech has only been discovered a few months ago (Lai et al. 2002), data availability has allowed us to obtain practical answers that fundamentally affect our lives. In direct analogy, transformation of biological chemistry by data availability opens the door to convergence with computer science and information technology. Furthermore, the deeper analogy between biology and language suggests that successful sequence   function mapping is fundamentally similar to the ability to retrieve, summarize, and translate in computational linguistics. Examples for biological equivalents of these abilities are described below under "The Estimated Implications."

The Role of Converging Technologies: NBIC and Biological Language Modeling

The strength of the analogy between biology and language lies in its ability to bridge across scales — atomic, nanostructural, microstructural, and macroscopic — enabling profit from the convergence of other disciplines (Figure F.7). Ideally, we would like to correlate complex biological systems, including their most complex abilities — the cognitive abilities of the brain, such as memory — with the individual atoms that create them. Rapid advances currently occur at all scales because of the convergence of technologies, allowing us to collect more data on natural systems than we were ever able to collect before. The data can be analyzed using information technology at all levels of the hierarchy in scale. Furthermore, mapping can involve any levels of the hierarchy, e.g., atomic   macroscopic or nanostructural   microstructural. The language analogy is useful here because of the hierarchical organization of language itself, as manifested by words, phrases, sentences, and paragraphs.

The Transforming Strategy

One test for convergence of technologies is that their methods are

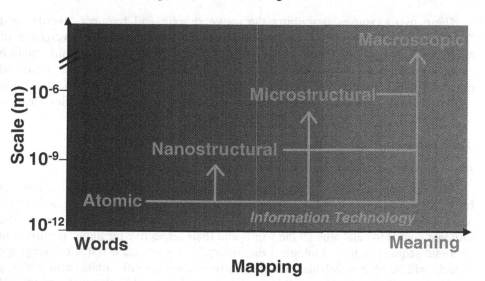

Figure F.7. Biological language modeling allows bridging across scales via the mapping of words to meaning using information technology methods, in particular computational linguistics.

interchangeable, i.e., language technologies should be directly applicable to biological sequences. To date, many computational methods that are used extensively in language modeling have proven successful as applied to biological sequences, including hidden Markov modeling, neural network, and other machine learning algorithms, demonstrating the utility of the methodology. The next step is to fully explore linguistically inspired analysis of biological sequences. Thus, the Carnegie Mellon and Cambridge Statistical Language Modeling (SLM) Toolkit, utilized for natural language modeling and speech recognition in more than 40 laboratories worldwide, was applied to protein sequences, in which the 20 amino acids were treated as words and each protein sequence in an organism as a sentence of a book. Two exemplary results are described here.

1. In human languages, frequent words usually do not reveal the content of a text (e.g., "I", "and", "the"). However, abnormalities in usage of frequent words in a particular text as compared to others can be a signature of that text. For example, in Mark Twain's *Tom Sawyer*, the word "Tom" is amongst the top 10 most frequently used words. When the SLM toolkit was applied to protein sequences of 44 different organisms (bacterial, archaeal, human), specific n-grams were found to be very frequent in one organism, while the same n-gram was rare or absent in all the other organisms. This suggests that there are organism-specific phrases that can serve as "genome signatures."

2. In human languages, rare events reveal the content of a text. Analysis of the distribution of rare and frequent n-grams over a particular protein sequence, that of lysozyme, a model system for protein folding studies, showed that the location of rare n-grams correlates with nucleation sites for protein folding that have been identified experimentally (Klein-Seetharaman 2002). This striking observation suggests that rare events in biological sequences have similar status for the folding of proteins, as have rare words for the topic of a text.

These two examples describing the usage of rare and frequent "words" and "phrases" in biology and in language clearly demonstrates that convergence of computational linguistics and biological chemistry yields important information about the mapping between sequence and biological function. This was observed even when the simplest of computational methods was used, statistical n-gram analysis. In the following, examples for the potential benefits of such information for improving human health and performance will be described.

The Estimated Implications

Implications for Fundamental Understanding of Properties of Proteins

The convergence of linguistics and biology provides a framework to connect biological information gathered in massive numbers of studies, including both large-scale genome-wide experiments and more traditional small-scale experiments. The ultimate goal is to catalogue all the words and their respective meanings occurring in genomic sequences in a "biological dictionary." Sophisticated statistical language models will be able to calculate the probabilities for a specific amino acid within a protein context. It will be possible to examine what combinations of amino acid sequences give a meaningful sentence, and we will be able to predict where spelling mistakes are inconsequential for function and where they will cause dysfunction.

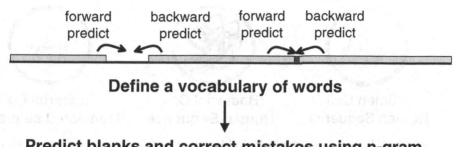

forward   backward   forward   backward
predict   predict   predict   predict

Define a vocabulary of words

Predict blanks and correct mistakes using n-gram statistics

Figure F.8. Opportunity for biological language modeling in genome sequences.

Cataloguing Biological Languages at Hierarchical Levels: Individual Proteins, Cell Types, Organs, and Related and Divergent Species

The language modeling approach is applicable to distinguishing biological systems at various levels, just as language varies among individuals, groups of individuals, and nations. At the most fundamental level, we aim at deciphering the rules for a general biological language, i.e., discovering what aspects are common to all sequences. This will enhance our fundamental understanding of biological molecules, in particular how proteins fold and function. At the second level, we ask how differences in concentrations, interactions, and activities of proteins result in formation and function of different cell-types and ultimately of organs within the same individual. This will allow us to understand the principles underlying cell differentiation. The third level will be to analyze the variations among individuals of the same species, the single nucleotide polymorphisms. We can then understand how differences in characteristics, such as intelligence or predisposition for diseases, are encoded in the genome sequence. Finally, the most general level will be to analyze differences in the biological languages of different organisms, with varying degree of relatedness.

Ideally, all life on earth will be catalogued. The impact on understanding complexity and evolution of species would be profound. Currently, it is estimated that there are 2-100 million species on earth. While it is not feasible to sequence the genomes of all the species, language modeling may significantly speed up obtaining "practical" sequences (Figure F.8). One of the bottlenecks in genome sequencing is the step from draft to finished sequence because of error correction and filling of gaps. However, if we define a vocabulary of the words for an organism from a partial or draft sequence, we should be able to predict blanks and correct mistakes in forward and backward directions using language modeling.

Retrieval, Summarization, and Translation of Biological Sequences

As in human language modeling, success in biological language modeling will be measured by the capacity for efficient (1) retrieval, (2) summarization, and (3) translation:

1. When we desire to enhance the performance of a specific human ability, we can *retrieve* all the relevant biological information required from the vast and complex data available.

Human Cell   **Bacterial Cell**   **Bacterial Cell**
Human Sequence   **Human Sequence**   **Translated sequence**

Figure F.9. Opportunity for biological language modeling to overcome biotechnological challenges. Misfolding of human proteins in bacterial expression systems is often a bottleneck.

2. We can *summarize* which proteins of a pathway are important for the particular task, or which particular part of a key protein is important for its folding to functional 3-D structure. This will allow modifications of the sequences with the purpose of enhancing the original or adding a new function to it. Successful existing examples for this strategy include tagging proteins for purification or identification purposes.

3. Finally, we can *translate* protein sequences from the "language" of one organism into that of another organism. This has very important implications, both for basic sciences and for the biotechnology industry. Both extensively utilize other organisms, i.e., the bacterium *E. coli*, to produce human proteins. However, often proteins cannot be successfully produced in *E. coli* (especially the most interesting ones): they misfold, because the environment in bacterial cells is different from that in human cells (Figure F.9). Statistical analysis of the genomes of human and *E. coli* can demonstrate the differences in rules to be observed if productive folding is to occur. Thus, it should be possible to alter a human protein sequence in such a way that it can fold to its correct functional 3-D shape in *E. coli*. The validity of this hypothesis has been shown for some examples where single point mutations have allowed expression and purification of proteins from *E. coli*. In addition to the traditional use of *E. coli* (or other organisms) as protein production factories, this translation approach could also be used to add functionality to particular organisms.

Implications for Communication Interfaces

The ability to translate highlights one of the most fundamental aspects of language: a means for communication. Knowing the rules for the languages of different organisms at the cellular and molecular levels would also allow us to communicate at this level. This will fundamentally alter (1) human-human, (2) human-other organism, and (3) human-machine interfaces.

1. Human-human communication can be enhanced because the molecular biological language level is much more fundamental than speech, which may in the future be omitted in some cases as intermediary between humans. For example, pictures of memory events could be transmitted directly, without verbal description, through their underlying molecular mechanisms.

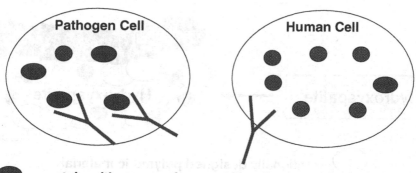

= protein with n-gram, frequent in pathogen and rare in human
= other proteins

Figure F.10. Opportunity for biological language modeling to dramatically speed vaccine or drug development by simultaneously targeting multiple proteins via organism-specific phrases.

2. The differences in language between humans and other organisms can be exploited to "speak" to a pathogen in the presence of its human host (Figure F.10). That this may be possible is indicated by the observation of organism-specific phrases described in "The Transforming Strategy." This has important implications for the fight against bioterrorism and against pathogens in general to preserve and restore human health. The genome signatures should dramatically accelerate vaccine development by targeting pathogen-specific phrases. The advantage over traditional methods is that multiple proteins, unrelated in function, can be targeted simultaneously.

3. Finally, there are entirely novel opportunities to communicate between inherent and external abilities, i.e., human (or other living organisms) and machines. Using nanoscale principles, new materials and interfaces can be designed that are modeled after biological machines or that can interact with biological machines. Of particular importance are molecular receptors and signal transduction systems.

Implications to Rationalize Empirical Approaches

The greatest exploitation of the sequence   structure/function mapping by computational linguistics approaches will be to rationalize empirical observations. Here are two examples.

1. The first example concerns the effect of misfolding of proteins on human health. The correlation between the distribution of rare amino acid sequences in proteins and the location of nucleation sites for protein folding described above is important because misfolding is the cause of many diseases, including Alzheimer's, BSE, and others, either because of changes in the protein sequence or because of alternative structures taken by the same sequences. This can lead to amorphous aggregates or highly organized amyloid fibrils, both interfering with normal cell function. There are databases of mutations that list changes in amyloid formation propensity. Studying the linguistic properties of the sequences of amyloidogenic wild-type and mutant proteins may help rationalizing the mechanisms for

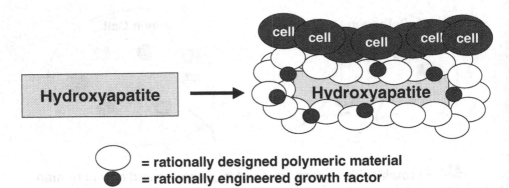

Figure F.11. Opportunity for biological language modeling to rationalize tissue engineering via engineering of growth factors and artificial materials.

misfolding diseases, the first step towards the design of strategies to treat them.

2. The second example is in tissue engineering applications (Figure F.11). The sequence   structure/ function mapping also provides the opportunity to engineer functionality by rationalized directed sequence evolution. Diseased or aged body parts, or organs whose performance we might like to enhance, all need integration of external materials into the human body. One typical application is bone tissue engineering. The current method to improve growth of cells around artificial materials such as hydroxyapatite is by trial and error to change the function of co-polymers and of added growth factors. Mapping sequence to function will allow us to rationally design growth factor sequences that code for altered function in regulating tissue growth.

Implications for Society

The above scenario has important implications for economic benefits, including cheaper and faster drug development, overcoming bottlenecks in biotechnology applications, cheaper and better materials and machines that perform old and new tasks, and environmental benefits. A key challenge will be to maintain reversibility in all changes that are made to living organisms to prevent unwanted environmental catastrophes, such as predominance of new organisms with enhanced capabilities in the environment. These new technologies require drastic changes in education. Human learning, memory, and creativity — which are likely to increase as a result of the revolutions in biology — have to be steered towards attaining literacy in health and biology for all citizens. Close collaboration between academic and industrial partners will allow universities to focus on fundamental advances, keeping in mind the implications and potential applications that will be evaluated and realized by industry.

Summary

Human performance and the nation's productivity will increase drastically if existing and new biological knowledge is exploited by statistical methods to obtain practical answers to the fundamental questions:

- How can we enhance human inherent abilities?
- How can inherent and external abilities be better integrated?

The analogy between language and biology will provide a framework for addressing these questions through convergence of computational linguistics with biological chemistry within the broader context of NBIC. The challenge is to achieve successful mapping of genome sequence to structure and function of biological molecules. It would then be possible to integrate man-made machines into the human body with interfaces at the cellular and molecular level, for example, sensors for biological, chemical, or physical changes in the environment. Artificial organs will perform traditional functions better than youthful, healthy natural organs, or be able to perform new functions. By exploiting differences in languages between different organisms, novel strategies to fight pathogenic infections will emerge. New functions will be built into organisms that lack them. The maximum benefit will be possible if all knowledge is catalogued in a way that it can be accessed efficiently via computers today and in the future by nanomachines of all kinds. The biology-language analogy provides the means to do so if an encyclopedia for vocabulary and rules of biological language can be developed.

References

Klein-Seetharaman, J., M. Oikawa, S.B. Grimshaw, J. Wirmer, E. Duchardt, T. Ueda, T. Imoto, L.J. Smith, C.M. Dobson, and H. Schwalbe. 2002. Long-range interactions within a non-native protein. *Science*. 295: 1719-1722.

Lai, C.S., S.E. Fisher, J.A. Hurst, F. Vargha-Khadem, and A.P. Monaco. 2001. A forkhead-domain gene is mutated in a severe speech and language disorder. *Nature*. 413(6855):519-23.

APPENDICES

APPENDIX A. LIST OF PARTICIPANTS AND CONTRIBUTORS

Government and National Laboratories

James S. Albus
National Institute of Standards
and Technology
Intelligent Systems Division
100 Bureau Drive, Stop 8230
Gaithersburg, MD 20899-8230

Robert Asher
Sandia National Laboratories
Advanced Concepts Group,
Special Projects
P.O. Box 5800
Albuquerque, NM 87185

William Sims Bainbridge
National Science Foundation
4201 Wilson Blvd.
Arlington, VA 22230

James G. Batterson
National Aeronautics and
Space Administration
Langley Research Center
MS 132
Hampton, VA 23681

Ujagar S. Bhachu
Nuclear Regulatory Commission
Mail Stop T-8F5
11545 Rockville Pike
Rockville, MD 20852

Phillip J. Bond
Undersecretary of Commerce
for Technology
1401 Constitution Avenue, NW
Washington, DC 20230

Joseph Bordogna
Deputy Director
National Science Foundation
4201 Wilson Blvd.
Arlington, VA 22230

Tony Fainberg
Defense Threat Reduction Agency

Michael Goldblatt
Defense Advanced Research Projects
Agency
3701 N. Fairfax Dr.
Arlington, VA 22203

Esin Gulari
Assistant Director for Engineering
National Science Foundation
4201 Wilson Blvd., Rm. 505
Arlington, VA 22230

Murray Hirschbein
NASA Headquarters
Rm 6D70
Washington DC 20546-0001

Charles H. Huettner
Senior Policy Advisor for Aviation
National Science and Technology
Council
Executive Office of the President
Old Executive Office Building,
Room 423
Washington, D.C. 20502
(currently Chair of the Presidential
Commission on the Future of the
Aerospace Industry)

Jean M. Johnson
National Science Foundation
4201 Wilson Blvd., Room 965
Arlington, VA 22230

Ken Lasala
National Oceanographic and
Atmospheric Administration
(NOAA)
(currently at DOT)

Cliff Lau
Office of Naval Research
800 N. Quincy St.
Arlington, VA 22217

Tina M. Masciangioli
U.S. Environmental Protection
Agency
Ariel Rios Bldg.
1200 Pennsylvania Ave., N.W.
MC 8722R
Washington, DC 20460

James Murday
Superintendant, Chemistry Division
Naval Research Laboratory
Code 6100
Washington, DC 20375-5342

Robert L. Norwood
Director, Commercial Technology
Division
National Aeronautics and
Space Administration
Washington, DC 20546-0001

Scott N. Pace
Assistant Director for Space and
Aeronautics
Office of Science and
Technology Policy
Room 423 Eisenhower Executive
Office Building
Washington, DC 20502

Lawrence M. Parsons
National Science Foundation
SBE/BCS, Room 995 N
4201 Wilson Boulevard
Arlington, VA 22230

Alan T. Pope
National Aeronautics and
Space Administration
Langley Research Center
MS 152
Hampton, VA 23681

Robert Price
U.S. Department of Energy
Office of Science
1000 Independence Ave.
Code SC-15, Rm. E-438A
Washington, DC 20585

Dave Radzanowski
Office of Management and Budget
New Executive Office Building
725 17th Street, NW
Room 8225
Washington, DC 20503

Mihail C. Roco
Senior Advisor
National Science Foundation
4201 Wilson Blvd.
Arlington, VA 22230

Philip Rubin
Director, Division of Behavioral
and Cognitive Sciences
National Science Foundation
4201 Wilson Blvd., Room 995
Arlington, VA 22230

John Sargent
U.S. Department of Commerce
Office of Technology Policy
1401 Constitution Avenue, NW
Washington, DC 20230

Gary Strong
National Science Foundation
4201 Wilson Blvd.
Arlington, VA 22230

Dave Trinkle
Office of Management and Budget
New Executive Office Building
725 17th Street, NW
Room 8225
Washington, DC 20503

Samuel Venneri
NASA Chief Technologist
NASA Headquarters, Building: HQ,
Room: 9S13
Washington DC 20546-0001

John Watson
National Institutes of Health
National Heart, Lung,
and Blood Institute
Bldg. 31, Room 5A49
31 Center Drive MSC 2490
Bethesda, MD 20892

Barbara Wilson
Chief Technologist
U.S. Air Force Research Laboratory
Arlington, VA

Gerold (Gerry) Yonas
Principal Scientist
Sandia National Laboratory
P. O. Box 5800
Albuquerque, NM 87185-0839

Academic Contributors

Daniel Akins
Director, IGERT on Nanostructures
The City University of New York
Department of Chemistry
Convent Ave at 138th St.
New York, NY 10031

Jill Banfield
University of California, Berkeley
Department of Earth and Planetary
Sciences and Department of
Environmental Science, Policy,
and Management
369 McCone Hall
Berkeley CA 94720-4767

Jeffrey Bonadio
University of Washington
Bioengineering Dept.
466a, Bagley Hall, Box 351720
Seattle, WA 98195-7962

Rudy Burger
MIT
Media Lab Europe
Sugar House Lane
Bellevue, Dublin 8
Ireland

Kathleen Carley
Carnegie Mellon University
Department of Social
& Decision Sciences
Office: PH 219A
208 Porter Hall
Pittsburgh, PA 15213

Lawrence J. Cauller
University of Texas at Dallas
Neuroscience Program, GR41
Richardson, TX 75083-0688

Britton Chance
University of Pennsylvania
250 Anatomy-Chemistry Building
3700 Hamilton Walk
Philadelphia, PA 19104-6059

Avis H. Cohen
University of Maryland
Dept. of Zoology
Bldg. 144, Rm. 1210
College Park, MD 20742-4415

Patricia Connolly
University of Strathclyde
Bioengineering Unit
Wolfson Centre
Glasgow G4 ONW, Scotland
United Kingdom

Delores M. Etter
U.S. Naval Academy
121 Blake Road
Annapolis, MD 21402-5000

Edgar Garcia-Rill
Arkansas Center for Neuroscience
University of Arkansas
for Medical Sciences
4301 West Markham Ave., Slot 510
Little Rock, AR 72205

Reginald G. Golledge
University of California, Santa
Barbara
Dept. of Geography
3616 Ellison Hall
Santa Barbara, CA 93106

Michael E. Gorman
University of Virginia
P.O. Box 400231, Thornton Hall
Charlottesville, VA 22904-4231

Michael Heller
Nanogen
UCSD Dept. of Bioengineering
9500 Gilman Drive
La Jolla, CA 92093-0412

Robert E. Horn
Stanford University
Program on People, Computers
and Design
Center for the Study of Language
and Information
2819 Jackson St. # 101
San Francisco, CA 94115

Kyung A. Kang
University of Louisville
Chemical Engineering Department
106 Ernst Hall
Louisville, KY 40292

Judith Klein-Seetharaman
Carnegie-Mellon University
School of Computer Science
Wean Hall 5317
5000 Forbes Avenue
Pittsburgh, PA 15213

Josef Kokini
Rutgers University
Department of Food Science
and CAFT
Cook College
65 Dudley Road
New Brunswick, NJ 08901-8520

Martha Krebs
University of California, Los Angeles
California Nanosystems Institute
4060 N. Farmonth Dr.
Los Angeles, CA 90027

Abraham Lee
University of California at Irvine
Biomedical Engineering Dept.
204 Rockwell Engineering Center
Irvine, CA 92697-2715

Rodolfo R. Llinás
New York University
School of Medicine
Dept. of Physiology & Neuroscience
550 First Ave
New York, NY 10016

Jack M. Loomis
University of California,
Santa Barbara
Department of Psychology
Santa Barbara, CA 93160-9660

Miguel A. L. Nicolelis
Dept. of Neurobiology and
Biomedical Engineering
Duke University
Durham, NC 27710

Perry Andrew Penz
University of Texas at Dallas
Richardson, TX 75083-0688

Jordan B. Pollack
Brandeis University
Volen Center for Complex Systems
Waltham, MA 02254

Sherry R. Turkle
MIT
Building E51-296C
77 Massachusetts Avenue
Cambridge, MA 02139

William A. Wallace
Rensselaer Polytechnic Institute
DSES
187 Mountain View Avenue
Averill Park, NY 12018

Gregor Wolbring
University of Calgary
Dept. of Medical Biochemistry
3330 Hospital Dr. NW
T2N 4N1, Calgary, Alberta
Canada

Private Sector Contributors

Allen Atkins
The Boeing Company
P.O. Box 516
St. Louis, MO 63166-0516

Yaneer Bar-Yam
New England Complex Systems
Institute
24 Mt. Auburn St.
Cambridge, MA 02138

John H. Belk
Associate Technical Fellow,
Phantom Works
Manager, Technology Planning &
Acquisition
The Boeing Company
S276-1240
P.O. Box 516
St. Louis, MO 63166-0516

James Canton
President
Institute for Global Futures
2084 Union St.
San Francisco, CA 94123

Michael Davey
National Research Council
2101 Constitution Ave., N.W.
Washington, DC 20418

Newt Gingrich
American Entreprise Institute
attn: Anne Beighey, Project Director
1150 17th Street, N.W.
Washington, D.C. 20036

Peter C. Johnson
President and Chief Executive
Officer
Tissue-Informatics, Inc.
711 Bingham St., Suite 202
Pittsburgh, PA 15203

Gary Klein
Klein Associates Inc.
Dayton, OH (Main office)
1750 Commerce Center Blvd. North
Fairborn, Ohio 45324-3987

Philip J. Kuekes
HP Laboratories
3500 Deer Creek Rd.
MS 26U-12
Palo Alto, CA 94304-1126

Thomas Miller
Klein Associates Inc.
1750 Commerce Center Blvd. North
Fairborn, OH 45324-3987

Cherry A. Murray
Senior Vice President
Lucent Technologies
Physical Sciences Research
700 Mountain Avenue
P.O. Box 636
Room 1C-224
Murray Hill, NJ 07974-0636

Brian M. Pierce
Manager, Advanced RF
Technologies Raytheon
2000 East Imperial Hwy.
P.O. Box 902
RE/R01/B533
El Segundo, CA 90245-0902

Warren Robinett
Consultant
719 E. Rosemary St.
Chapel Hill, NC 27514

Ottilia Saxl, CEO
Institute of Nanotechnology
UK

James C. Spohrer
CTO, IBM Venture Capital Relations
Group
Almaden Research Center, E2-302
650 Harry Road
San Jose, CA 95120

William M. Tolles
Consultant
(Naval Research Laboratory, Retired)
8801 Edward Gibbs Place
Alexandria, VA 22309

R. Stanley Williams
HPL Fellow and Director
Quantum Science Research
Hewlett-Packard Laboratories
1501 Page Mill Rd., 1L - 14
Palo Alto, CA 94304

Larry T. Wilson
c/o IEEE
3429 Avalon Road
Birmingham, AL 35209

Dorothy Zolandz
National Research Council
2101 Constitution Ave., N.W.
Washington, DC 20418

APPENDIX B. INDEX OF AUTHORS

APPENDIX C. INDEX OF TOPICS